A Level Biology for OCR

OCR
Oxford Cambridge and RSA
This is an OCR endorsed resource.

A

Year 1 and AS

Series Editor: Ann Fullick

Ann Fullick • Jo Locke • Paul Bircher

OXFORD
UNIVERSITY PRESS

OXFORD
UNIVERSITY PRESS

Great Clarendon Street, Oxford, OX2 6DP, United Kingdom

Oxford University Press is a department of the University of Oxford. It furthers the University's objective of excellence in research, scholarship, and education by publishing worldwide. Oxford is a registered trade mark of Oxford University Press in the UK and in certain other countries

British Library Cataloguing in Publication Data
Data available

978-0-19-835191-7

10 9 8

Paper used in the production of this book is a natural, recyclable product made from wood grown in sustainable forests. The manufacturing process conforms to the environmental regulations of the country of origin.

Printed in China by Golden Cup

This resource is endorsed by OCR for use with specification H020 AS Level GCE Biology A and year 1 of H420 A Level GCE Biology A. In order to gain endorsement this resource has undergone an independent quality check. OCR has not paid for the production of this resource, nor does OCR receive any royalties from its sale. For more information about the endorsement process please visit the OCR website www.ocr.org.uk/

Acknowledgements

The authors would like to thank John Beazley for his reviewing, as well as Amy Johnson, Amie Hewish, Les Hopper, and Sharon Thorn for their tireless work and encouragement. In addition they would like to thank the teams at science and plants for schools (SAPS) and at the Wellcome Trust Sanger Institute for their valuable input to the project, and finally the help received from Dr Jeremy Pritchard and Jennifer Collins.

Ann Fullick would like to thank her partner Tony for his support and amazing photographs, and all of the boys William, Thomas, James, Edward, and Chris for their expert advice and for making her take time off.

Paul Bircher would like to thank his wife, Julie, and the rest of his family for their patience and support throughout the writing of this book. A special mention goes to his irrepressible grandchildren, Leo and Toby, who provided a welcome distraction. Their insanity has kept him sane.

Jo Locke would like to thank her husband Dave for all his support, encouragement, and endless cups of tea, as well as her girls Emily and Hermione who had to wait patiently for Mummy 'to just finish this paragraph'.

AS/A Level course structure

This book has been written to support students studying for OCR AS Biology A and for students in their first year of studying for OCR A Level Biology A. It covers the AS modules from the specification, the content of which will also be examined at A Level. The modules covered are shown in the contents list, which also shows you the page numbers for the main topics within each module. There is also an index at the back to help you find what you are looking for. If you are studying for OCR AS Biology A, you will only need to know the content in the blue box.

AS exam

A level exam

Year 1 content

1 Development of practical skills in biology
2 Foundations in biology
3 Exchange and transport
4 Biodiversity, evolution, and disease

Year 2 content

5 Communication, homeostasis, and energy
6 Genetics, evolution, and ecosystems

A Level exams will cover content from Year 1 and Year 2 and will be at a higher demand. You will also carry out practical activities throughout your course.

Contents

How to use this book vi
Kerboodle ix

Module 1 development of practical skills in biology 2

Module 2 Foundations in Biology 6

Chapter 2 Basic components of living systems 8
2.1 Microscopy 8
2.2 Magnification and calibration 15
2.3 More microscopy 19
2.4 Eukaryotic cell structure 26
2.5 The ultrastructure of plant cells 33
2.6 Prokaryotic and eukaryotic cells 35
Practice questions 38

Chapter 3 Biological Molecules 40
3.1 Biological elements 40
3.2 Water 44
3.3 Carbohydrates 46
3.4 Testing for carbohydrates 51
3.5 Lipids 54
3.6 Structure of proteins 59
3.7 Types of proteins 64
3.8 Nucleic acids 68
3.9 DNA replication and the genetic code 72
3.10 Protein synthesis 76
3.11 ATP 80
Practice questions 82

Chapter 4 Enzymes 84
4.1 Enzyme action 84
4.2 Factors affecting enzyme activity 88
4.3 Enzyme inhibitors 94
4.4 Cofactors, coenzymes, and prosthetic groups 97
Practice questions 99

Chapter 5 Plasma membranes 102
5.1 The structure and function of membranes 102
5.2 Factors affecting membrane structure 106
5.3 Diffusion 108
5.4 Active transport 112
5.5 Osmosis 114
Practice questions 118

Chapter 6 Cell divisions 120
6.1 Cell cycle 120
6.2 Mitosis 124
6.3 Meiosis 128
6.4 The organisation and specialisation of cells 133
6.5 Stem cells 138
Practice questions 143
Module 2 summary 146

Module 3 Exchange and transport 148

Chapter 7 Exchange surfaces and breathing 150
7.1 Specialised exchange surfaces 150
7.2 Mammalian gaseous exchange system 153
7.3 Measuring the process 159
7.4 Ventilation and gas exchange in other organisms 161
Practice questions 168

Chapter 8 Transport in animals 170
8.1 Transport systems in multicellular animals 170
8.2 The blood vessels 174
8.3 Blood, tissue fluid, and lymph 178
8.4 Transport of oxygen and carbon dioxide in the blood 181
8.5 The heart 185
Practice questions 192

Chapter 9 Transport in plants 194
9.1 Transport systems in dicotyledonous plants 194
9.2 Water transport in multicellular plants 198
9.3 Transpiration 201
9.4 Translocation 207

9.5	Plant adaptations to water availability	210
	Practice questions	215
Module 3 summary		218

Module 4 Biodiversity, evolution and disease · 220

Chapter 10 Classification and evolution		222
10.1	Classification	222
10.2	The five kingdoms	227
10.3	Phylogeny	232
10.4	Evidence for evolution	234
10.5	Types of variation	239
10.6	Representing variation graphically	243
10.7	Adaptations	251
10.8	Changing population characteristics	256
	Practice questions	260

Chapter 11 Biodiversity		262
11.1	Biodiversity	262
11.2	Sampling	264
11.3	Sampling techniques	266
11.4	Calculating biodiversity	271
11.5	Calculating genetic biodiversity	274
11.6	Factors affecting biodiversity	278
11.7	Reasons for maintaining biodiversity	283
11.8	Methods for maintaining biodiversity	287
	Practice questions	292

Chapter 12 Communicable diseases		294
12.1	Animal and plant pathogens	294
12.2	Animal and plant diseases	297
12.3	The transmission of communicable diseases	303
12.4	Plant defences against pathogens	306
12.5	Non-specific animal defences against pathogens	308
12.6	The specific immune system	312
12.7	Preventing and treating disease	317
	Practice questions	325
Module 4 summary		326

Paper 1 practice questions	328
Paper 2 practice questions	331
Glossary	336
Answers	349
Index	377
Appendix (Statistics data tables)	383
Acknowledgements	384

How to use this book

This book contains many different features. Each feature is designed to support and develop the skills you will need for your examinations, as well as foster and stimulate your interest in biology.

Terms that you will need to be able to define and understand are highlighted by **bold text**.

Application features

These features contain important and interesting applications of biology in order to emphasise how scientists and engineers have used their scientific knowledge and understanding to develop new applications and technologies. There are also practical application features, with the icon , to support further development of your practical skills. There are also application features with the icon which help support development of your understanding of scientific issues and their impact in society.

1 All application features have a question to link to material covered with the concept from the specification.

Extension features

These features contain material that is beyond the specification. They are designed to stretch and provide you with a broader knowledge and understanding and lead the way into the types of thinking and areas you might study in further education. As such, neither the detail nor the depth of questioning will be required for the examinations. But this book is about more than getting through the examinations.

1 Extension features also contain questions that link the off-specification material back to your course.

Summary Questions

1 These are short questions at the end of each topic.

2 They test your understanding of the topic and allow you to apply the knowledge and skills you have acquired.

3 The questions are ramped in order of difficulty. The icon indicates where a question relates to scientific issues in society.

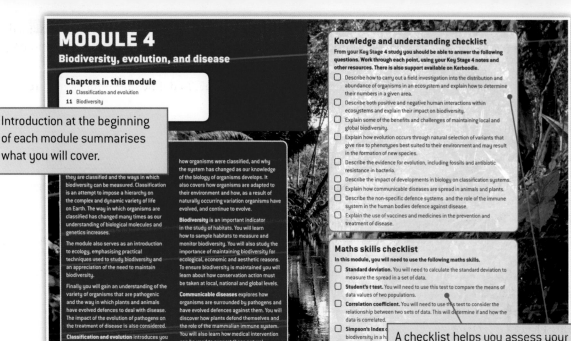

MODULE 4
Biodiversity, evolution, and disease

Chapters in this module
10 Classification and evolution
11 Biodiversity

how organisms were classified, and why the system has changed as our knowledge of the biology of organisms develops. It also covers how organisms are adapted to their environment and how, as a result of naturally occurring variation organisms have evolved, and continue to evolve.

Biodiversity is an important indicator in the study of habitats. You will learn how to sample habitats to measure and monitor biodiversity. You will also study the importance of maintaining biodiversity for ecological, economic and aesthetic reasons. To ensure biodiversity is maintained you will learn about how conservation action must be taken at local, national and global levels.

Communicable diseases explores how organisms are surrounded by pathogens and have evolved defences against them. You will discover how plants defend themselves and the role of the mammalian immune system. You will also learn how medical intervention can be used to support these natural defences such as the role of vaccinations and antibiotics.

they are classified and the ways in which biodiversity can be measured. Classification is an attempt to impose a hierarchy on the complex and dynamic variety of life on Earth. The way in which organisms are classified has changed many times as our understanding of biological molecules and genetics increases.

The module also serves as an introduction to ecology, emphasising practical techniques used to study biodiversity and an appreciation of the need to maintain biodiversity.

Finally you will gain an understanding of the variety of organisms that are pathogenic and the way in which plants and animals have evolved defences to deal with disease. The impact of the evolution of pathogens on the treatment of disease is also considered.

Classification and evolution introduces you to the current system of classification used by scientists. It also explains historically

Knowledge and understanding checklist
From your Key Stage 4 study you should be able to answer the following questions. Work through each point, using your Key Stage 4 notes and other resources. There is also support available on Kerboodle.

☐ Describe how to carry out a field investigation into the distribution and abundance of organisms in an ecosystem and explain how to determine their numbers in a given area.

☐ Describe both positive and negative human interactions within ecosystems and explain their impact on biodiversity.

☐ Explain some of the benefits and challenges of maintaining local and global biodiversity.

☐ Explain how evolution occurs through natural selection of variants that give rise to phenotypes best suited to their environment and may result in the formation of new species.

☐ Describe the evidence for evolution, including fossils and antibiotic resistance in bacteria.

☐ Describe the impact of developments in biology on classification systems.

☐ Explain how communicable diseases are spread in animals and plants.

☐ Describe the non-specific defence systems and the role of the immune system in the human bodies defence against disease.

☐ Explain the use of vaccines and medicines in the prevention and treatment of disease.

Maths skills checklist
In this module, you will need to use the following maths skills.

☐ **Standard deviation.** You will need to calculate the standard deviation to measure the spread in a set of data.

☐ **Student's t test.** You will need to use this test to compare the means of data values of two populations.

☐ **Correlation coefficient.** You will need to use this test to consider the relationship between two sets of data. This will determine if and how the data is correlated.

☐ **Simpson's Index** biodiversity in a h Diversity, the mor

☐ **Proportion of poly** measure genetic b

220

Introduction at the beginning of each module summarises what you will cover.

A checklist helps you assess your knowledge from KS4, before starting work on the module. There is also a maths skills checklist to demonstrate the skills you will learn in that module.

Visual summaries show how some of the key concepts of that module interlink with other modules, across the entire A Level course.

Application task brings together some of the key concepts of the module in a new context.

Extension task bring together some key concepts of the module and develop them further, leading you towards greater understanding and further study.

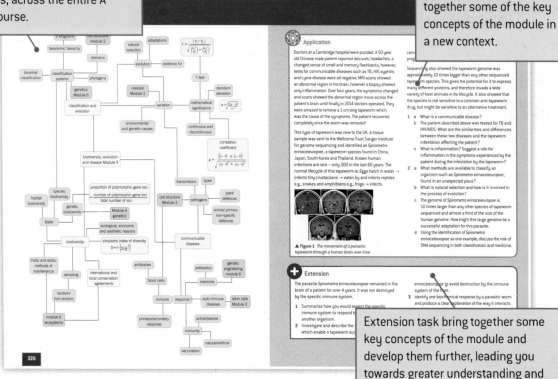

Application

Doctors at a Cambridge hospital were puzzled. A 50 year old Chinese male patient reported seizures, headaches, a changed sense of smell and memory flashbacks, however, tests for communicable diseases such as TB, HIV, syphilis and Lyme disease were all negative. MRI scans showed an abnormal region in his brain, however a biopsy showed only inflammation. Over four years, the symptoms changed and scans showed the abnormal region move across the patient's brain until finally in 2014 doctors operated. They were amazed to remove a 1 cm long tapeworm which was the cause of the symptoms. The patient recovered completely once the worm was removed!

This type of tapeworm was new to the UK. A tissue sample was sent to the Wellcome Trust Sanger Institute for genome sequencing and identified as *Spirometra erinaceieuropae*, a tapeworm species found in China, Japan, South Korea and Thailand. Known human infections are rare – only 300 in the last 60 years. The normal lifecycle of this tapeworm is: Eggs hatch in water → Infects tiny crustaceans → eaten by and infects reptiles e.g., snakes and amphibians e.g., frogs → infects

▲ **Figure 1** *The movement of a parasitic tapeworm through a human brain over time*

carni prop

Sequencing also showed the tapeworm genome was approximately 10 times bigger than any other sequenced tapeworm species. This gives the potential for it to express many different proteins, and therefore invade a wide variety of host animals in its lifecycle. It also showed that the species is not sensitive to a common anti-tapeworm drug, but might be sensitive to an alternative treatment.

1 a What is a communicable disease?
  b The patient described above was tested for TB and HIV/AIDS. What are the similarities and differences between these two diseases and the tapeworm infestation affecting the patient?
  c What is inflammation? Suggest a role for inflammation in the symptoms experienced by the patient during the infestation by the tapeworm?

2 a What methods are available to classify an organism such as *Spirometra erinaceieuropae*, found in an unexpected place?
  b What is natural selection and how is it involved in the process of evolution?
  c The genome of *Spirometra erinaceieuropae* is 10 times larger than any other species of tapeworm sequenced and almost a third of the size of the human genome. How might this large genome be a successful adaptation for this parasite?
  d Using the identification of *Spirometra erinaceieuropae* as one example, discuss the role of DNA sequencing in both classification and medicine.

Extension

The parasite *Spirometra erinaceieuropae* remained in the brain of a patient for over 4 years. It was not destroyed by the specific immune system.

1 Summarise how you would expect the specific immune system to respond t another organism.

2 Investigate and describe the which enable a tapeworm su

erinaceieuropae to avoid destruction by the immune system of the host.

3 Identify one biochemical response by a parasitic worm and produce a clear explanation of the way it interacts

326

vii

a Calculate the strain of the wire. *(3 marks)*

b Design a simple electrical circuit to determine the resistivity of the metal used to make the wire. *(4 marks)*

c Figure 10 shows an electrical circuit. Calculate the total resistance between **A** and **B**. *(3 marks)*

▲ Figure 10

the top of a smooth ramp.

▲ Figure 8

The ramp makes an angle of 10° to the horizontal. The block is released and it slides down the ramp.

a Calculate the acceleration of the block along the length of the ramp. *(2 marks)*

b The block travels a total distance of 45 cm down the ramp. Calculate the time it takes to reach the bottom of the ramp. *(3 marks)*

c The speed of the block at the bottom of the ramp is v. Describe a simple experiment a student can carry out to determine an approximate value of the speed v. The student only has a metre rule and a stopwatch. *(3 marks)*

13 Figure 9 shows a metal wire stretched horizontally between two supports on a laboratory bench.

▲ Figure 9

The tension in the wire is 15 N and it has a cross-sectional area $3.1 \times 10^{-7} \, m^2$. The Young modulus of the wire is 4.2×10^{10} Pa.

14 a Explain what is meant by coherent waves. *(1 mark)*

b State two ways in which a stationary waves differs from a progressive wave. *(2 marks)*

c Figure 11 shows a stationary pattern on a length of stretched string.

▲ Figure 11

The distances shown in Figure 11 are **drawn to scale**. The frequency of vibration of the string is 110 Hz.

(i) By taking measurements from Figure 11, determine the wavelength of the progressive waves on the string. *(2 marks)*

(ii) Calculate the speed of the progressive waves on the string. *(2 marks)*

15 a Sketch a graph of energy E of a photon against frequency f of the electromagnetic radiation. *(1 mark)*

b Electromagnetic waves of frequency 8.93×10^{14} Hz are incident on the surface of metal. The work function of the metal is 3.20×10^{-19} J.

(i) Calculate the energy of the photons. *(2 marks)*

(ii) Calculate the maximum speed v_{max} of the photoelectrons emitted from the metal surface. *(3 marks)*

16 a Define *refractive index* of a material. *(1 mark)*

b Figure 12 shows the path of a ray of light as it crosses the boundary between two materials A and B.

▲ Figure 12

The refractive index of material A is n_1 and the angle of incidence of the ray of light is θ_1. The angle of refraction in material B is θ_2 and the refractive index of material B is n_2. Write an equation that relates n_1, n_2, θ_1 and θ_2. *(1 mark)*

c A student is investigating the refraction of light by a transparent material by measuring the angles of incidence i and refraction r. Figure 13 show the results from the experiment.

▲ Figure 13

Use Figure 13 to determine

(i) the refractive index of the material *(2 marks)*

(ii) the critical angle for this material. *(2 marks)*

d You are provided with a semi-circular glass block and a ray-box with a suitable supply.

Design a laboratory experiment to determine the critical angle of the glass of the semi-circular block and hence the refractive index of the glass. You may use other equipment available in the laboratory. In your description pay particular attention to

• how the apparatus is used
• what measurements are taken
• how the data is analysed. *(4 marks)*

17 a State *Ohm's law*. *(1 mark)*

b The *I-V* characteristic of a particular component is shown in Figure 14.

▲ Figure 14

(i) Use Figure 14 to describe how the resistance of this component depends on the potential difference (p.d.) across it. You may do calculations to support your answer. *(3 marks)*

(ii) Draw a circuit diagram for an arrangement that could be used to collect results to plot the graph shown in Figure 14. *(3 marks)*

c Figure 15 shows an electrical circuit.

▲ Figure 15

The e.m.f. of the battery is 6.0 V and it has negligible internal resistance.

Calculate

(i) the current in the 36Ω resistor *(2 marks)*

(ii) the potential difference across the 12Ω resistor *(1 mark)*

(iii) the potential difference between points **P** and **Q**. *(2 marks)*

Paper 1 style questions

1 Which of the following statements is/are correct with respect to phylogenetic trees?

Statement 1: Phylogenetic trees depict the evolutionary relationships among groups of organisms

polymer, is not usually deposited uniformly but laid down in rings or spirals.

a Describe the role of lignin in xylem vessels. *(4 marks)*

b Suggest the benefit to a plant of way that lignin is deposited in the walls of xylem vessels. *(2 marks)*

4 Consider the statement:

'evolution can be summarised as change over time'

a Discuss why this statement is an over simplification as a description of evolution. *(4 marks)*

b Explain, using the finch populations observed by Darwin in the Galapagos islands the process of disruptive selection. *(6 marks)*

5 a Describe the difference between conservation and preservation. *(3 marks)*

b Describe the difference between 'in situ' and 'ex situ' conservation. *(4 marks)*

6 Fill in the missing words.

T memory cells live for a long time and are part of the .... If they meet an .... for a second time, they undergo .... to form a large .... of T .... cells that destroy the pathogen. *(5 marks)*

7 Figure 2 shows diagrams of four cells that have been placed in different solutions.

▲ Figure 2

a In the table below write the letter **K, L, M** or **N** next to the description that best matches the diagram. One has been done for you.

description	letter
an animal cell that has been placed in distilled water	
an animal cell that has been placed in a concentrated sugar solution	
a plant cell that has been placed in distilled water	
a plant cell that has been placed in concentrated sugar solution	**M**

(3 marks)

b Explain, using the term **water potential**, what has happened to cell **M**. *(3 marks)*

c Small non-polar substances enter cells in different ways to large or polar substance.

Outline the ways in which the substances below, can enter a cell through the plasma (cell surface) membrane.

Small, non-polar substances

Large substances

Polar substances *(5 marks)*

OCR F211 2009

8 a The structure of cell membranes can be described as 'proteins floating in a sea of lipids'. This membrane structure allows certain substances to pass through freely whereas other substances cannot.

State the term used to describe a membrane through which some substances can pass freely but others cannot. *(1 mark)*

b Copy and complete the following paragraph about cell membranes through which some substances can pass freely but others cannot.

The model of cell membrane structure is called the .... model. Phospholipid bilayers with specific membrane proteins account for the ability of the membrane to allow both passive and .... transport mechanisms. Ions and most polar molecules are insoluble in the phospholipid bilayer. However, the bilayer allows diffusion of most non-polar molecules such as ....

Protein channels, which may be gated, and .... proteins enable the cell to control the movement of most polar substances. *(4 marks)*

c One function of membranes that is not mentioned in (b) is cell signalling.

(i) State what is meant by *cell signalling*. *(1 mark)*

(ii) Explain how cell surface membranes contribute to the process of cell signalling. In your answer you should use appropriate technical terms, spelled correctly. *(4 marks)*

OCR F211/01 2013

9 Various measurements of lung function are used to help diagnose lung disease and to monitor its treatment.

b State what is meant by the following terms:

(i) vital capacity; *(1 mark)*

(ii) forced expiratory volume 1 (FEV1). *(1 mark)*

One measure of lung function is:

$$\text{Percentage lung function} = \frac{\text{FEV 1}}{100} \times 100$$

This is particularly useful in identifying possible obstructive disorders of the airways and lungs, such as asthma or chronic obstructive pulmonary disease (COPD).

• Asthma is a condition that responds to, and can be controlled by, the use of bronchodilators. These are drugs that dilate the airways and improve airflow.

• COPD lasts for a long period of time and is caused by progressive and permanent damage to the lung tissue.

When the value calculated for the percentage lung function is less than or equal to 70%, this indicates an obstructive disorder. A 'normal' value is approximately 80%.

Table 1 shows data relating to three patients, **C, D** and **E**, before and after treatment with a bronchodilator drug.

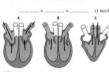

▲ Figure 1

c Describe, using the diagrams in Figure 1, what is happening at each stage in the cardiac cycle shown. *(6 marks)*

3 Xylem vessels have two walls, a primary wall made of cellulose and a secondary wall composed of lignin. Lignin, a strong inflexible

.... × .... = ... *(1 mark)*

Practice questions at the end of each chapter and the end of each module, including questions that cover practical and math skills.

Practice questions at the end of the book, with multiple choice questions and synoptic style questions, also covering the practical and math skills.

This book is supported by next generation Kerboodle, offering unrivalled digital support for independent study, differentiation, assessment, and the new practical endorsement.

If your school subscribes to Kerboodle, you will also find a wealth of additional resources to help you with your studies and with revision.

- Study guides
- Maths skills boosters and calculation worksheets
- On your marks activities to help you achieve your best
- Practicals and follow up activities to support the practical endorsement
- Interactive objective tests that give question-by-question feedback
- Animations and revision podcasts
- Self-assessment checklists

Revise with ease using the study guides to guide you through each chapter and direct you towards the resources you need.

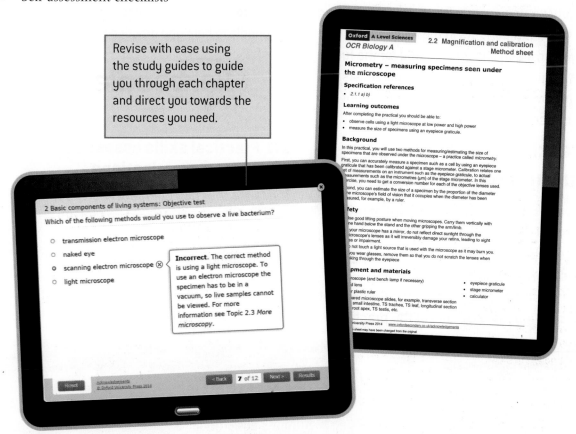

For teachers, Kerboodle also has plenty of further assessment resources, answers to the questions in the book, and a digital markbook along with full teacher support for practicals and the worksheets, which include suggestions on how to support and stretch students. All of the resources are pulled together into teacher guides that suggest a route through each chapter.

MODULE 1
Development of practical skills in biology

Developing your practical skills is a fundamental part of a complete education in science. A good grounding of practical skills will help your understanding of Biology and help to prepare you for studying beyond A level.

You will carry out a number of practicals in your AS year of study. Practical skills knowledge will account for 15% of the marks in your written exams. There are no assessed practical components in the AS qualification, however, if you choose to study for the full A level qualification then these practicals will count towards the practical endorsement in your second year of study.

Practical coverage throughout this book

It is a good idea to keep a record of your practical work during your AS year of study, this will be very useful when you come to revise for your exams, and you can use it later as part of your practical endorsement. You can find more details of the practical endorsement from your teacher or from the specification.

In this book and its supporting materials, practical skills are covered in a number of ways – look out for the conical flask symbol. By studying **Application boxes** and **practice questions** in this student book you will have many opportunities to learn about scientific method and how to carry out practical activities. There are also more resources available on Kerboodle to help you practise your skills.

1.1 Practical skills assessed in written examinations

There are four key components of practical skill assessments. These are laid out here with a skills checklist to help you as you study.

1.1.1 Planning

- Experimental design
- Identification of variables
- Evaluation of experimental method

Skills checklist

- [ ] Forming a hypothesis
- [ ] Selecting suitable equipment
- [ ] Considering accuracy and precision
- [ ] Identifying dependent and independent variables
- [ ] Identifying variables which need to be controlled

1.1.2 Implementing

- Use of practical apparatus
- Appropriate units for measurement
- Presenting observations and data

Skills checklist

- ☐ Confidence using apparatus and techniques correctly
- ☐ Understanding S.I. units and prefixes
- ☐ Results table design
- ☐ Presenting data in the most suitable way:
    - Scatter graph
    - Line graph
    - Bar chart
    - Pie chart

1.1.3 Analysis

- Processing, analysing, and interpreting results
- Appropriate mathematical skills for data analysis
- Use of appropriate number of significant figures
- Plotting and interpreting graphs

Skills checklist

- ☐ Understanding results and using this to reach valid conclusions
- ☐ Using mathematical skills to process results
- ☐ Using significant figures correctly
- ☐ Plotting and interpreting graphs
    - labelling axes correctly
    - using appropriate scales
    - reading intercepts and gradients from graphs

1.1.4 Evaluation

- Evaluate results to draw conclusions
- Identify anomalies
- Explain limitations in method
- Precision and accuracy of measurements
- Uncertainties and errors
- Suggest improvements to help improve experimental design

Skills checklist

- ☐ Evaluate results to draw sound conclusions
- ☐ Understand and explain any limitations in the method, and make suggestions on how the method could be improved
- ☐ Understand accuracy of measurements and margins of error (including percentage error) and uncertainty in apparatus

Maths skills and How Science Works across AS Biology

Maths is a vital tool for scientists, and throughout this course you will become familiar with maths techniques and equations that support the development of your science knowledge. Each **module opener** in this book has an overview of the maths skills that relate to the theory in the chapter. There are also **questions using maths skills** throughout the book that will help you practice.

How Science Works are skills that will help you to apply your knowledge in a wider context, and the relevance of what you have learnt in the real world. This includes developing your critical and creative skills to help you solve problems in many different contexts. How Science Works is embedded throughout this book, particularly in **application boxes** and **practice questions**.

1.2 Practical skills assessed in practical endorsement

* A level qualification only *

You are required to carry out 12 assessed practical activities over both years of your A level studies. The practical endorsement does not count towards your final A level grade but is reported alongside it as a pass or a fail. Universities and employers will look for this as evidence that you have a good level of practical competence. These practicals will be assessed by your teacher and will help to develop your skills and confidence.

The practicals you carry out should be recorded in a lab book or practical portfolio where the hypothesis, method, results, and conclusion are clearly displayed. The information below details the types of skills and equipment you should become familiar with.

1.2.1 Practical skills

Independent thinking
- Investigating and analysing the methods used in practicals in order to solve problems

Use and application of scientific methods and practices
- using practical equipment correctly and safely
- following written instructions, recording observations, taking measurements, and presenting data scientifically
- Using appropriate software and technology throughout.

Research and referencing
Using information available from a variety of different sources including websites, scientific journals, and textbooks to help provide context and background for the practical. It is important to use many sources of information as you can and to cite these correctly.

Instruments and equipment
Correct and appropriate use of a wide range of equipment, instruments, and techniques.

1.2.2 Use of apparatus and techniques

- Apparatus for quantitative measuring,
- Use of glassware apparatus
- Use of a light microscope at high and low power and the use of stage graticule
- Producing clear and well labelled scientific drawings
- Use of qualitative reagents
- Experience of carrying out electrophoresis or chromatography
- Ethical and safe use of organisms
- Microbial aseptic techniques
- Use of sampling techniques for fieldwork
- Use of IT to collect and process data.

A level PAG overview and Application features

The practical activity requirements (PAGs) for the OCR A Biology practical endorsement are listed below. You should take all opportunities to develop you practical skills and techniques in your first year of study to help build a greater understanding.

The table below shows the practical activity requirements, and where these are covered throughout the AS course.

Specification reference	Topic reference
PAG1 Microscopy	2.1, 2.2, 2.3
PAG2 Dissection	7.4, 8.2, 8.5, 9.1
PAG3 Sampling techniques	11.3
PAG4 Enzyme controlled reactions	4.2
PAG5 Colorimeter or potometer	3.4
PAG6 Chromatography or electrophoresis	3.6
PAG7 Microbial technique	A level only
PAG8 Transport in and out of cells	5.2, 5.3, 5.5
PAG9 Qualitative testing	3.4
PAG10 Investigation using a data logger or computer modelling	7.3
PAG11 Investigation into the measurement of plant or animal responses	9.3, 12
PAG12 Research skills	A level only

MODULE 2
Foundations in Biology

Chapters in this module

2 Basic components of living systems

3 Biological molecules

4 Enzymes

5 Plasma membranes

6 Cell division

Introduction

Biology is the study of living organisms. Every living organism is made up of one or more cells, therefore understanding the structure and function of the cell is a fundamental concept in the study of biology. Since Robert Hooke coined the phrase 'cells' in 1665, careful observation using microscopes has revealed details of cell structure and ultrastructure and provided evidence to support hypotheses regarding the roles of cells and their organelles.

Basic components of living systems provides an introduction to cells and microscopy techniques. An understanding of cell biology is essential for most onward routes for biologists.

Biological molecules will begin to explore the biochemistry you need for your A Level course. An understanding of biochemistry provides a firm grounding for the study of key biological disciplines, such as medicine and disease research.

Enzymes are vital for many biological processes. In this chapter you will learn how they are structured and how they function.

Plasma membranes control the substances that move in and out of cells, and so a knowledge of how they work is essential for all areas of biology involving cellular processes.

Cell division explores the two processes by which cells divide – mitosis and meiosis. An understanding of these processes and how they can go wrong will help you understand health and disease, as well as explore new technologies in genetics and cloning.

Knowledge and understanding checklist

From your Key Stage 4 study you should be able to answer the following questions. Work through each point, using your Key Stage 4 notes and other resources. There is also support available on Kerboodle.

- [ ] Describe a cell as the basic structural unit of all organisms.
- [ ] Describe the main sub-cellular structures of eukaryotic and prokaryotic cells.
- [ ] Relate sub-cellular structures to their functions.
- [ ] Describe the cell cycle.
- [ ] Describe cell differentiation.
- [ ] Relate the adaptation of specialised cells to their function.
- [ ] Explain the mechanism of enzyme action.
- [ ] Recall the difference between intracellular enzymes and extracellular enzymes.
- [ ] Describe some anabolic and catabolic processes in living organisms including the importance of sugars, amino acids, fatty acids, and glycerol in the synthesis and breakdown of carbohydrates, lipids, and proteins.

Maths skills checklist

All biologists need to use maths in their studies and field of work. In this module, you will need to use the following maths skills.

- [ ] **Changing the subject of an equation.** You will need to be able to do this when working with microscopy.
- [ ] **Converting units.** You will need to be able to do this when working with microscopy.
- [ ] **Working with negative numbers.** You will need to be able to do this when calculating water potential when studying osmosis.
- [ ] **Working in standard form.** You will need to be able to do this throughout this chapter.

MyMaths.co.uk
Bringing Maths Alive

2 BASIC COMPONENTS OF LIVING SYSTEMS

2.1 Microscopy

Specification reference: 2.1.1

Learning outcomes

Demonstrate knowledge, understanding, and application of:

→ the use of **light microscopy**

→ the preparation of microscope slides for use in light microscopy

→ the use of staining in light microscopy

→ the representation of cell structure seen under light microscope using scientific annotated drawings.

Before the invention of microscopes, we knew nothing of bacteria, cells, sperm, pollen grains, chromosomes – the list is endless. Microscopes have given us the power to understand disease, see how a new life is formed, watch the dance of the chromosomes as cells divide, and manipulate the processes of life itself.

Seeing is believing

A microscope is an instrument which enables you to magnify an object hundreds, thousands and even hundreds of thousands of times. We can see many large organisms with the naked eye, but microscopes open up a whole world of unicellular organisms. By making visible the individual cells which make up multicellular organisms, microscopes allow us to discover how details of their structures relate to their functions.

The first types of microscopes to be developed were **light microscopes** in the 16th to 17th century. Since then they have continued to be developed and improved.

By the mid-19th century, scientists, for the first time, had access to microscopes with a high enough level of magnification to allow them to see individual cells. Cell theory was developed. It states that:

● both plant and animal tissue is composed of cells

● cells are the basic unit of all life

● cells only develop from existing cells.

Light microscopy continues to be important – it is easily available, relatively cheap and can be used out in the field, and it can be used to observe living organisms as well as dead, prepared specimens.

 History of the light microscope and the development of cell theory

Late in the Roman Empire the Romans began to develop and experiment with glass. They noted how objects looked bigger when viewed through pieces of glass that were thicker in the middle than at the edges.

There was little further development of glass lenses until around the 13th century and the invention of spectacles or eye glasses.

The credit for the invention of the light microscope is much disputed. Some accredit it to two Dutch spectacle makers who invented the telescope when experimenting with multiple glass lenses in a tube in the late 15th century. Others claim it was Galileo Galilei in 1609 who developed the first true or compound microscope (Table 1). Galileo's instrument was the first to be given the name 'microscope'.

Cell theory

The development of cell theory is a good example of how scientific theories change over time as new evidence is gained and as knowledge increases. Theories are proposed, accepted and can then be later disproved as new evidence comes to light. New evidence can arise in a number of ways, including as technology develops. This is the case with cell theory - as microscopes with higher magnification and resolution were developed, cells could be observed for the first time. Table 1 summarises some of the developments in cell theory.

▼ Table 1 *Cell theory timeline*

Timeline	Development of cell theory
1665	**Cell first observed** Robert Hooke, an English scientist, observed the structure of thinly sliced cork using an early light microscope. He described the compartments he saw as 'cells' – coining the term we still use today. As this was dead plant tissue he was observing only cell walls.
1674–1683	**First living cells observed** Anton van Leeuwenhoek, a Dutch biologist, developed a technique for creating powerful glass lenses and used his handcrafted microscopes to examine samples of pond water. He was the first person to observe bacteria and protoctista and described them as 'little animals' or 'animalcules' – today we call them microorganisms. He went on to observe red blood cells, sperm cells, and muscle fibres for the first time.
1832	**Evidence for the origin of new plant cells** Barthélemy Dumortier, a Belgian botanist, was the first to observe cell division in plants providing evidence against the theories of the time, that new cells arise from *within* old cells or that cells formed spontaneously from non-cellular material. However it was several more years until cell division as the origin of all new cells became the accepted theory.
1833	**Nucleus first observed** Robert Brown, an English botanist, was the first to describe the nucleus of a plant cell.
1837–1838	**The birth of a universal cell theory** Matthias Schleiden, a German botanist, proposed that *all* plant tissues are composed of cells. Jan Purkyně, a Czech scientist, was the first to use a microtome to make ultra-thin slices of tissue for microscopic examination. Based on his observations, he proposed that not only are animals composed of cells but also that the "basic cellular tissue is clearly analogous to that of plants". Not long after this, and independently, Theodor Schwann, a German physiologist, made a similar observation and declared that "all living things are composed of cells and cell products". He is the scientist credited with the 'birth' of cell theory.
1844 (1855)	**Evidence for the origin of new animal cells** Robert Remak, a Polish/German biologist, was the first to observe cell division in animal cells, disproving the existing theory that new cells originate from *within* old cells. He was not believed at the time however, and Rudolf Virchow, a German biologist, published these findings as his own a decade later in 1855.
1860	**Spontaneous generation disproved** Louis Pasteur disproved the theory of spontaneous generation of cells by demonstrating that bacteria would only grow in a sterile nutrient broth after it had been exposed to the air.

1   Outline the importance of microscopes in the study of living organisms.
2   Suggest, with reasons, why cell theory was not fully developed before the mid-19th century.

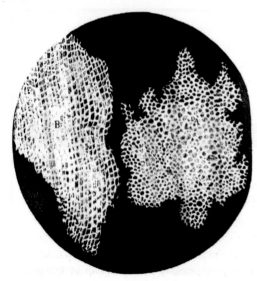

▲ **Figure 1** *Drawing of cork from 1663 seen under an early microscope. Robert Hooke described the pores as cells, thus coining the term. He prepared the specimen by making thin slices with a razor blade, inventing the technique of sectioning*

▲ **Figure 2** *Robert Hooke's drawing of his own compound microscope, which he used to see the 'cells' in a sample of cork*

How a light microscope works

A **compound light microscope** has two lenses – the objective lens, which is placed near to the specimen, and an eyepiece lens, through which the specimen is viewed. The objective lens produces a magnified image, which is magnified again by the eyepiece lens. This objective/eyepiece lens configuration allows for much higher magnification and reduced chromatic aberration than that in a simple light microscope.

Illumination is usually provided by a light underneath the sample. Opaque specimens can be illuminated from above with some microscopes.

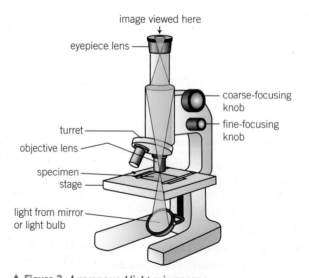

▲ **Figure 3** *A compound light microscope*

Sample preparation

There are a number of different ways in which samples and specimens can be prepared for examination by light microscopy. The method chosen will depend on the nature of the specimen and the resolution that is desired.

- **Dry mount** – Solid specimens are viewed whole or cut into very thin slices with a sharp blade, this is called *sectioning*. The specimen is placed on the centre of the slide and a cover slip is placed over the sample. For example hair, pollen, dust and insect parts can be viewed whole in this way, and muscle tissue or plants can be sectioned and viewed in this way.

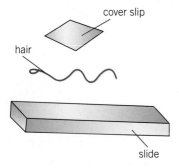

- **Wet mount** – Specimens are suspended in a liquid such as water or an immersion oil. A cover slip is placed on from an angle, as shown. For example, aquatic samples and other living organisms can be viewed this way.

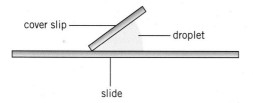

- **Squash slides** – A wet mount is first prepared, then a lens tissue is used to gently press down the cover slip. Depending on the material, potential damage to a cover slip can be avoided by squashing the sample between two microscope slides. Using squash slides is a good technique for soft samples. Care needs to be taken that the cover slip is not broken when being pressed. For example, root tip squashes are used to look at cell division.

- **Smear slides** – The edge of a slide is used to smear the sample, creating a thin, even coating on another slide. A cover slip is then placed over the sample. An example of a smear slide is a sample of blood. This is a good way to view the cells in the blood.

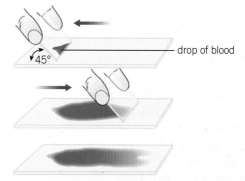

1 Suggest reasons for the following, with reference to slide preparation:
  a Specimens must be thin.
  b When preparing a wet mount the refractive index (ability to bend light) of the medium should be roughly the same as glass.
  c A cover slip must be placed onto a wet mount at an angle.

Using staining

In basic light microscopy the sample is illuminated from below with white light and observed from above (brightfield microscopy). The whole sample is illuminated at once (wide-field microscopy). The images tend to have low contrast as most cells do not absorb a lot of light. Resolution is limited by the wavelength of light and diffraction of light as it passes through the sample. Diffraction is the bending of light as it passes close to the edge of an object.

The cytosol (aqueous interior) of cells and other cell structures are often transparent. Stains increase contrast as different components within a cell take up stains to different degrees. The increase in contrast allows components to become visible so they can be identified (Figure 4).

To prepare a sample for staining it is first placed on a slide and allowed to air dry. This is then heat-fixed by passing through a flame. The specimen will adhere to the microscope slide and will then take up stains.

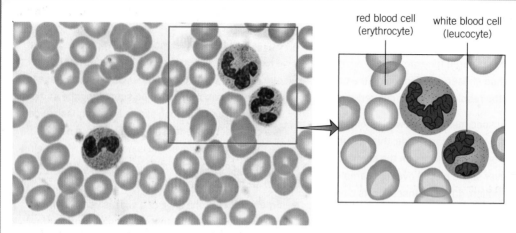

▲ **Figure 4** *Light micrograph and annotated diagram of a blood sample that has been stained using Wright's stain (a mixture of eosin red and methylene blue dyes). The nuclei of the white blood cells are stained purple*

Crystal violet or methylene blue are positively charged dyes, which are attracted to negatively charged materials in cytoplasm leading to staining of cell components.

Dyes such as nigrosin or Congo red are negatively charged and are repelled by the negatively charged cytosol. These dyes stay outside cells, leaving the cells unstained, which then stand out against the stained background. This is a negative stain technique.

Differential staining can distinguish between two types of organisms that would otherwise be hard to identify. It can also differentiate between different organelles of a single organism within a tissue sample.

● **Gram stain technique** is used to separate bacteria into two groups, Gram-positive bacteria and Gram-negative bacteria (Figure 5). Crystal violet is first applied to a bacterial specimen on a slide, then iodine, which fixes the dye. The slide is then washed with alcohol. The Gram-positive bacteria retain the crystal violet stain and will appear blue or purple under a microscope. Gram-negative bacteria have thinner cell walls and therefore lose the stain. They are then stained with safranin dye, which is called a **counterstain**. These bacteria will then appear red. Gram-positive bacteria are susceptible to the antibiotic penicillin, which inhibits the formation of cell walls. Gram-negative bacteria have much thinner cell walls that are not susceptible to penicillin.

● **Acid-fast technique** is used to differentiate species of *Mycobacterium* from other bacteria. A lipid solvent is used to carry carbolfuchsin dye into the cells being studied. The cells are then washed with a dilute acid-alcohol solution. Mycobacterium are not affected by the acid-alcohol and retain the carbolfuchsin stain,

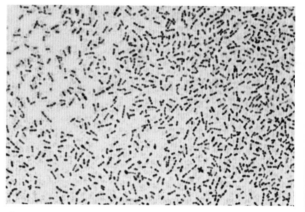

▲ **Figure 5** *Gram stain of Streptococcus pneumoniae, a Gram-positive bacteria which cause pneumonia (×200 magnification). Gram-positive bacteria retain a crystal violet dye during the Gram stain process and appear blue or violet under a microscope*

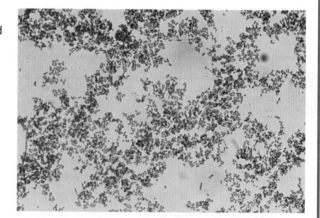

▲ **Figure 6** *Yersinia pestis, the Gram-negative bacteria which cause the bubonic plague infecting humans and animals (×1450 magnification). Gram-negative bacteria appear red or pink under a microscope*

which is bright red. Other bacteria lose the stain and are exposed to a methylene blue stain, which is blue.

You will often look at slides that have been pre-prepared. A number of stages are involved in the production of these slides:

- Fixing – chemicals like formaldehyde are used to preserve specimens in as near-natural a state as possible.
- Sectioning – specimens are dehydrated with alcohols and then placed in a mould with wax or resin to form a hard block. This can then be sliced thinly with a knife called a microtome.
- Staining – specimens are often treated with multiple stains to show different structures.
- Mounting – the specimens are then secured to a microscope slide and a cover slip placed on top.

Risk management

Many of the stains used in the preparation of slides are toxic or irritants. A risk assessment must carried out before any practical is started to identify any procedures involved that may result in harm.

CLEAPSS (Consortium of Local Education Authorities for the Provision of Science Services) is the organisation that provides support for the practical work carried out in schools. One of the main areas covered is health and safety, including risk assessment. Advice and support is provided to all types of educational establishments and their employees about all aspects of practical work such as the use of chemicals or living organisms, laboratory design, and even where to obtain the right equipment.

CLEAPSS provide student safety sheets that identify specific risks, advise on the measures to be taken to reduce these risks and the action to be taken in any emergency.

In fact, in schools many of the microscopy slides that are used are bought in ready-prepared and pre-stained, not only because of the harmful nature of the stains but also because of the long complex process needed to produce high quality sections.

1   Use your knowledge of the staining technique used to distinguish Gram-positive from Gram-negative bacteria and information from the paragraph above to answer the following question:
    Suggest why Gram-negative infections are more difficult to treat than Gram-positive infections.

2   Crystal violet and potassium iodide are chemicals classed as irritants. Crystal violet is also toxic. Describe the precautions you should take when using these chemicals.

Less is more

Scientific drawings are line drawings not pictures. They are used to highlight particular features and should not include unnecessary detail. The focus can be changed to help draw selected features.

The following is a list of rules for producing good scientific drawings:

- include a title
- state magnification
- use a sharp pencil for drawings and labels
- use white, unlined paper
- use as much of the paper as possible for the drawing
- draw smooth, continuous lines
- do not shade
- draw clearly defined structures
- ensure proportions are correct
- label lines should not cross and should not have arrow heads
- label lines should be parallel to the top of the page and drawn with a ruler

The light micrograph (Figure 7) shows a layer of onion cells as seen under a light microscope. Next to the micrograph is an example of a good scientific drawing of this image.

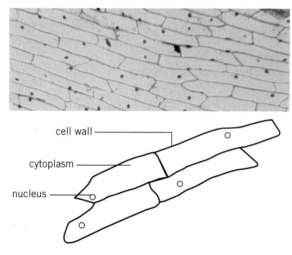

cell wall

cytoplasm

nucleus

× 18 magnification

▲ **Figure 7** *Top: Light micrograph of a layer of onion cuticle, showing the bands of large, rectangular cells. The dark spot in the centre of each cell is its nucleus. × 18 magnification. Bottom: A scientific drawing from the micrograph*

Below is an example of a poor scientific drawing:

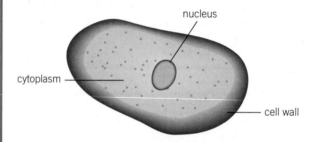

nucleus

cytoplasm

cell wall

1   Describe how this diagram is incorrect as a scientific drawing.

Summary questions

1 Outline the basic concepts of cell theory.   *(3 marks)*

2 Explain why staining is used in microscopy. *(2 marks)*

3 Compound microscopes led to new discoveries essential for cell theory to be fully explained. Explain the benefit of having two lenses in a microscope.   *(4 marks)*

4 a Calculate the low and high power magnifications of a microscope with the following lenses:
   i   Eyepiece lens ×10
   ii Objective lenses ×10 and ×40.   *(2 marks)*

   b You are observing a specimen of squamous tissue under high power. The individual cells have an average diameter of 60 μm and the diameter of the field of view of the objective lens is 2 mm. Calculate the maximum number of whole cells that are visible in the field of view.   *(3 marks)*

2.2 Magnification and calibration

Specification reference: 2.1.1

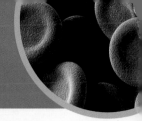

Magnification, resolution, and the magnification formula

Magnification is how many times larger the image is than the actual size of the object being viewed. Interchangeable objective lenses on a compound light microscope allow a user to adjust the magnification.

Simply magnifying an object does not increase the amount of detail that can be seen. The resolution also needs to be increased. The resolution of a microscope determines the amount of detail that can be seen – the higher the resolution the more details are visible.

Resolution is the ability to see individual objects as separate entities. Imagine a car coming towards you at night with its headlights on. When it is a long way off you will only see one light but as the car gets closer you eventually see that there are, in fact, two headlights – they have been resolved.

Resolution is limited by the diffraction of light as it passes through samples (and lenses). Diffraction is the tendency of light waves to spread as they pass close to physical structures such as those present in the specimens being studied. The structures present in the specimens are very close to each other and the light reflected from individual structures can overlap due to diffraction. This means the structures are no longer seen as separate entities and detail is lost. In optical microscopy structures that are closer than half the wavelength of light cannot be seen separately (resolved).

Resolution can be increased by using beams of electrons which have a wavelength thousands of times shorter than light (Topic 2.3, More microscopy). Electron beams are still diffracted but the shorter wavelength means that individual beams can be much closer before they overlap. This means objects which are much smaller and closer together can be seen separately without diffraction blurring the image.

Calculation for magnification

The magnification of an object can be calculated using the magnification formula:

$$\text{magnification} = \frac{\text{size of image}}{\text{actual size of object}}$$

In practice, the size of the image refers to the length of the image as measured, for example with a ruler. You may need to change the units of measurement to that of the actual size of the object. The magnification formula, like all mathematical equations, can be rearranged to find any of the unknowns, where the remaining values are known. To help you, you can imagine this three-part formula in a standard formula triangle (Figure 2).

Learning outcomes

Demonstrate knowledge, understanding, and application of:

→ slide preparation and examination
→ the magnification formula
→ the difference between magnifcation and resolution.

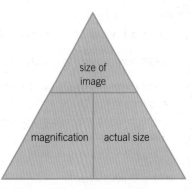

▲ Figure 2 *Formula triangle for magnification calculations*

Study tip

When solving magnification problems, carry out your calculations with all measurements having the same units.

Measure the image size in mm and convert it to the smallest unit present in the problem, usually micrometres.

1000 nanometres (nm) = 1 micrometre (µm)

1000 micrometres (µm) = 1 millimetre (mm)

1000 millimetres (mm) = 1 metre (m)

Magnification itself does not have units of measurement.

Students should also practise rearranging the magnification formula.

So, if the actual size of the object isn't known but the magnification is known, the actual size can be calculated by rearranging the formula to give:

$$\text{actual size of object} = \frac{\text{size of image}}{\text{magnification}}$$

Or, the size of the image can be calculated by rearranging the formula as below:

$$\text{size of image} = \text{magnification} \times \text{actual size of object}$$

 Worked example: Magnification calculation

Calculate the magnification of the image of the nuclear pore shown here.

To calculate the magnification you first convert all figures to the smallest unit, in this case nm.

24 millimetres is equal to $(24 \times 1000 \times 1000)$ nanometres or 24 000 000 nanometres.

120 nm

$$\text{Magnification} = \frac{\text{size of image}}{\text{actual size of object}}$$

$$= \frac{24\,000\,000\,\text{nm}}{120\,\text{nm}}$$

$$= 200\,000 \text{ or 2 hundred thousand times.}$$

 Using a graticule to calibrate a light microscope

To measure the size of a sample under a microscope you use an eyepiece graticule. The true magnification of the different lenses of a microscope can vary slightly from the magnification stated so every microscope, and every lens, has to be calibrated individually using an eyepiece graticule and a slide micrometer.

- An *eyepiece graticule* is a glass disc marked with a fine scale of 1 to 100. The scale has no units and remains unchanged whichever objective lens is in place. The relative size of the divisions, however, increases with each increase in magnification. You need to know what the divisions represent at the different magnifications so you can measure specimens. The scale on the graticule at each magnification is calibrated using a stage micrometer.

- A *stage micrometer* is a microscope slide with a very accurate scale in micrometres (µm) engraved on it.

The scale marked on the micrometer slide is usually 100 divisions = 1 mm, so 1 division = 10 µm.

You calibrate the eyepiece graticule scale for each objective lens separately. Once all three lenses are calibrated, if you measure the same cell using the three different lenses you should get the same actual measurement each time.

For example:

Calibrating a ×4 objective lens:

1 Put the stage micrometer in place and the eyepiece graticule in the eyepiece.

2 Get the scale on the micrometer slide in clear focus.

3 Align the micrometer scale with the scale in the eyepiece. Take a reading from the two scales – see next page:

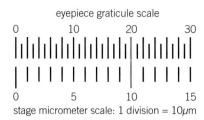

eyepiece graticule scale

stage micrometer scale: 1 division = 10μm

20 divisions on the eyepiece graticule = 10 divisions on the stage micrometer

Use these readings to calculate the calibration factor for the ×4 objective lens.

100 micrometer divisions = 1 mm

So each small division is 1/100 mm = 0.01 mm or 10.0 μm

20 graticule divisions = 10 micrometer divisions

10 micrometer divisions = 10 × 10 = 100 μm

$$1\ \text{graticule division} = \frac{\text{number of micrometres}}{\text{number of graticule divisions}}$$

20 graticule divisions = 100 μm so 1 graticule division = 100/20 = 5.0 μm

The magnification factor is 5.0

To use this magnification factor remove the stage micrometer and place a prepared slide on the stage. Measure the size of an object in graticule units. To find the actual size multiply the number of graticule units measured by the magnification factor to give you the length in μm

graticule divisions × magnification factor = measurement (μm)

e.g., the diameter of a cell seen using the ×4 objective lens measures 10 graticule divisions. Each graticule division = 5.0 μm so the cell diameter = 10 × 5.0 = 50.0 μm.

Calibration example

A student was asked to calibrate the ×10 lens of a light microscope and then to determine the diameter of a pollen grain from a sample slide provided.

1   The student placed a scale on the stage of the microscope and focused on it with the ×10 lens. 100 small divisions of the scale are 1 mm long, so 1 division is 10 μm.

2   The student then aligned the scale in the eyepiece with the scale on the microscope stage and calibrated it.

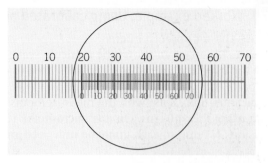

3   The student replaced the micrometer slide with pollen sample slides and used the calibrated scale in the eyepiece to measure the diameter of the pollen grains.

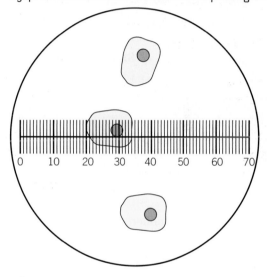

Results

The student decided to use the ×10 objective lens.

Result	1	2	3
Diameter of pollen grain/divisions	11	16	

1   State the correct names of the scale used on the stage of the microscope and the scale in the eyepiece.

2   Calculate the calibration factor of the ×10 objective lens.

3   a   Fill in the missing reading in the table for the pollen grain shown in the artwork above.

    b   Using the table calculate the diameter of the different pollen grains using the calibration factor you have calculated in question 2.

    c   Calculate the mean diameter of the pollen grains.

    d   Suggest why it is important to calibrate the lens that is going to be used to view the pollen grains.

Worked example: Using calibrated scales to measure specimens

Using the ×4 objective lens

Step 1: The diagram shows the graticule scale and the eyepiece micrometer readings used to calibrate the ×4 lens described in the previous application box, along with the measurements seen using the ×40 lens. On this micrometer 100 divisions = 1 mm so each graticule unit = 10 μm.

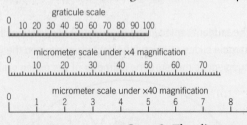

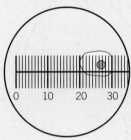 **Step 2:** The diameter of this pollen grain seen using the ×4 objective lens measures 10 graticule divisions. The calibration calculations on the previous page tell you that the magnification factor for the ×4 lens is 5.0

the diameter of the pollen grain = graticule units x magnification factor = 10 × 5.0 = 50 μm

Using the ×40 objective lens:

Step 1: 20 divisions on the eyepiece graticule = 1 division on the stage micrometer
Each division of the micrometer is 1/100 mm = 10 μm

Step 2: Using our observations
20 graticule divisions = 1.0 micrometer division
1.0 micrometre division = 1 × 10 μm = 10 μm
20 graticule divisions = 10 μm
1 graticule division = 10/20 = 0.5 μm
The magnification factor is 0.5

Step 3: The stage micrometer is removed and the same prepared slide placed on the stage. In this example you measure the same pollen grain: Calibrate the eyepiece graticule using the ×40 objective lens – see previous diagram.

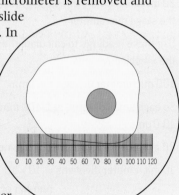

Step 4: The diameter of the pollen grain seen using the ×40 objective lens measures 100 graticule divisions

the pollen grain diameter = graticule units × magnification factor = 100 × 0.5 = 50 μm

The diameter of the pollen grain is the same measured using lenses of two very different magnifications. This is what you would expect but it is reassuring to see the theory confirmed and gives confidence in any measurements you take, regardless of which objective lens is in place.

Summary questions

1 Suggest why you should put all measurements into the same units before carrying out calculations.
(*2 marks*)

2 Calculate how many nanometres are present in 3846 centimetres.
Give your answer in standard form.

3 Explain the difference between contrast and resolution.
(*2 marks*)

4 Calculate the magnification of the micrograph showing human cheek cells (Figure 3).
The average diameter of a cheek cell is 60 μm.
(*2 marks*)

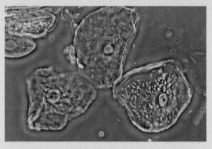
▲ **Figure 3** *Light micrograph of squamous epithelium cells from the inside of a human cheek*

5 Explain how diffraction limits resolution.   (*5 marks*)

6 Explain why eyepiece graticules do not have units.
(*2 marks*)

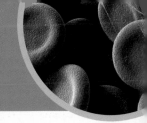

2.3 More microscopy

Specification reference: 2.1.1

Light microscopy started the science of cell biology, but it has limitations. In the middle of the 20th century a new invention, the electron microscope, revolutionised the study of cells and enabled biologists to see deep inside structures that were invisible under a light microscope.

Electron microscopy

In light microscopy, increased magnification can be achieved easily using the appropriate lenses, but if the image is blurred no more detail will be seen. Resolution is the limiting factor.

In **electron microscopy**, a beam of electrons with a wavelength of less than 1 nm is used to illuminate the specimen. More detail of cell **ultrastructure** can be seen because electrons have a much smaller wavelength than light waves. They can produce images with magnifications of up to ×500 000 and still have clear resolution.

Electron microscopes have changed the way we understand cells but there are some disadvantages to this technique. They are very expensive pieces of equipment and can only be used inside a carefully controlled environment in a dedicated space. Specimens can also be damaged by the electron beam and because the preparation process is very complex, there is a problem with **artefacts** (structures that are produced due to the preparation process). However, as techniques improve a lot of theses artefacts can be eliminated.

There are two types of electron microscope:

- In a **transmission electron microscope (TEM)** a beam of electrons is transmitted through a specimen and focused to produce an image. This is similar to light microscopy. This has the best resolution with a resolving power of 0.5 nm (Figure 1).

- In a **scanning electron microscope (SEM)** a beam of electrons is sent across the surface of a specimen and the reflected electrons are collected. The resolving power is from 3–10 nm, so the resolution is not as good as with transmission electron microscopy but stunning three-dimensional images of surfaces are produced, giving us valuable information about the appearance of different organisms (see Figure 2).

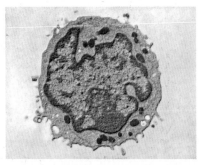

▲ Figure 1 *Coloured transmission electron micrograph of a lymphocyte (white blood cell). Magnification ×1600*

▲ Figure 2 *Coloured scanning electron micrograph of a lymphocyte (white blood cell). Magnification × 2000*

Sample preparation for electron microscopes ⚙️

The inside of an electron microscope is a vacuum to ensure the electron beams travel in straight lines. Because of this, samples need to be processed in a specific way.

Specimen preparation involves fixation using chemicals or freezing, staining with heavy metals and dehydration with solvents. Samples for a TEM will then be set in resin and may be stained again. Samples for a SEM may be fractured to expose the inside and will then need to be coated with heavy metals.

1 Suggest reasons for the following steps in the preparation of samples for electron microscopy:
- fixation
- dehydration
- embedding in resin
- staining with heavy metals.

Table 1 summarises the differences between light and electron microscopy.

▼ Table 1 *A comparison of light and electron microscopy*

Light microscope	Electron microscope
inexpensive to buy and operate	expensive to buy and operate
small and portable	large and needs to be installed
simple sample preparation	complex sample preparation
sample preparation does not usually lead to distortion	sample preparation often distorts material
vacuum is not required	vacuum is required
natural colour of sample is seen (or stains are used)	black and white images produced (but can be coloured digitally)
up to ×2000 magnification	over ×500 000 magnification
resolving power is 200 nm	resolving power of transmission electron microscope is 0.5 nm and a scanning electron microscope is 3–10 nm
specimens can be living or dead	specimens are dead

 ## Scientific drawings from electron micrographs

Electron microscopes produce images with much greater resolution than light microscopes and therefore much more detail can be seen.

Producing good scientific drawings from electron micrographs takes practice. The same rules used in producing drawings from light micrographs must still be observed (Topic 2.1, Microscopy).

a
starch granules — chloroplasts

cell wall — nucleus

b
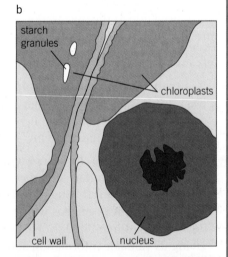
starch granules

chloroplasts

cell wall   nucleus

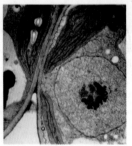

◀ **Figure 3** *Transmission electron micrograph of a section through two leaf cells at their junction. Their cell walls run from top centre to lower left. A nucleus is seen at lower right. Starch granules (pale ovals) can be seen in the chloroplast (dark grey, upper left and right). ×18 700 magnification*

1   Study the two drawings above and state, with reasons, which of them represents the best scientific drawing.

Creation of artefacts

An artefact is a visible structural detail caused by processing the specimen and not a feature of the specimen. Artefacts appear in both light and electron microscopy. The bubbles that get trapped under the cover slip as you prepare a slide for light microscopy are artefacts. When preparing specimens for electron microscopy, changes in the ultrastructure of cells are inevitable during the processing that the samples must undergo. They are seen as the loss of continuity in membranes, distortion of organelles and empty spaces in the cytoplasm of cells

Experience enables scientists to distinguish between an artefact and a true structure.

Identifying artefacts

Identifying artefacts in microscopy preparation can cause much discussion and controversy. 'Mesosome' was the name given to invaginations (inward foldings) of cell membranes that were observed using electron microscopes after bacterial specimens had been chemically fixed. They were thought to be a normal structure, or organelle, found within prokaryotes. The large surface area of the folded membrane was considered to be an important site for the process of oxidative phosphorylation. However, when specimens were fixed by the more recently developed, non-chemical technique called cryofixation, the mesosomes were no longer visible.

It is now widely thought that the majority of mesosomes observed are actually artefacts produced by the chemicals used in the fixation process in electron microscopy preparation, which damage bacterial cell membranes. However, there are still a number of scientists who believe

that some species of bacteria do have mesosomes as part of their normal structure, but this is not the general consensus.

This is a good example of how the scientific community accepted an idea based on the evidence available at the time and as techniques improved and more evidence became available, the collective knowledge and understanding developed and changed. New evidence can either provide further support for a theory or disprove an earlier theory. Scientific knowledge is constantly developing.

1 Structures that look similar to mesosomes have recently been observed in bacteria after treatment with certain types of antibiotics.
Suggest, with reasons, whether this information is evidence to support the current theory that mesosomes are artefacts or the theory that they are, in fact, organelles.

Laser scanning confocal microscopy

Light microscopy has also continued to develop. Some of the latest technology produces images that are very different from electron micrographs but are just as useful.

Conventional optical microscopes use visible light to illuminate specimens and a lens to produce a magnified image. In fluorescent microscopes a higher light intensity is used to illuminate a specimen that has been treated with a fluorescent chemical (a fluorescent 'dye'). Fluorescence is the absorption and re-radiation of light. Light of a longer wavelength and lower energy is emitted and used to produce a magnified image.

A **laser scanning confocal microscope** moves a single spot of focused light across a specimen (point illumination). This causes fluorescence from the components labelled with a 'dye'. The emitted light from the specimen is filtered through a pinhole aperture. Only light radiated from very close to the focal plane (the distance that gives the sharpest focus) is detected (Figure 4).

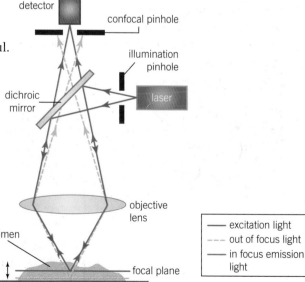

▲ **Figure 4** The light rays from the laser and the fluorescing sample follow the same path and have the same focal plane

Light emitted from other parts of the specimen would reduce the resolution and cause blurring. This unwanted radiation does not pass through the pinhole and is not detected. A laser is used instead of light to get higher intensities, which improves the illumination.

As very thin sections of specimen are examined and light from elsewhere is removed, very high resolution images can be obtained.

The spot illuminating the specimen is moved across the specimen and a two dimensional image is produced. A three dimensional image can be produced by creating images at different focal planes.

Laser scanning confocal microscopy is non-invasive and is currently used in the diagnosis of diseases of the eye and is also being developed for use in endoscopic procedures. The fact that it can be used to see the distribution of molecules within cells means it is also used in the development of new drugs.

The future uses for advanced optical microscopy include virtual biopsies, particularly in cases of suspected skin cancer.

The beamsplitter is a dichroic mirror, which only reflects one wavelength (from the laser) but allows other wavelengths (produced by the sample) to pass through.

The positions of the two pinholes means the light waves from the laser (illuminating the sample) follow the same path as the light waves radiated when the sample fluoresces. This means they will both have the same focal plane, hence the term *confocal*.

Synoptic link

You will learn more about antibodies in Topic 12.6, The specific immune system.

Fluorescent tags

By using antibodies with fluorescent 'tags', specific features can be targeted and therefore studied by confocal microscopy with much more precision than when using staining and light microscopy.

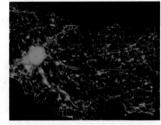

Green fluorescent protein (GFP) is produced by the jellyfish *Aequorea victoria*. The protein emits bright green light when illuminated by ultraviolet light. GFP molecules have been engineered to fluoresce different colours, meaning different components of a specimen can be studied at the same time. The gene for this protein has been isolated and can be attached, by genetic engineering, to genes coding for proteins under investigation. The fluorescence indicates that a protein is being made and is used to see where it goes within the cell or organism. Bacterial, fungal, plant, and human cells have all been modified to express this gene and fluoresce. The use of these fluorescing proteins provides a non-invasive technique to study the production and distribution of proteins in cells and organisms.

a   Define the term resolution with reference to microscopy.
b   Suggest whether fluorescent microscopy has a higher resolution than normal light microscopy. Explain your answer.

➕ Atomic force microscopy

The atomic force microscope (AFM) gathers information about a specimen by 'feeling' its surface with a mechanical probe. These are scanning microscopes that generate three-dimensional images of surfaces.

An AFM consists of a sharp tip (probe) on a cantilever (a lever supported at one end) that is used to scan the surface of a specimen. When this is brought very close to a surface, forces between the tip and the specimen cause deflections of the cantilever. These deflections are measured using a laser beam reflected from the top of the cantilever into a detector.

Fixation and staining are not required and specimens can be viewed in almost normal cell conditions without the damage caused during the preparation of specimens for electron microscopy. Living systems can even be examined.

The resolution of AFM is very high, in the order of 0.1 nm. Information can be gained at the atomic level, even about the bonds within molecules.

The pharmaceutical industry in particular uses AFM to identify potential drug targets on cellular proteins and DNA. These microscopes can lead to a better understanding of how drugs interact with their target molecule or cell.

AFM is also being employed to identify new drugs. Finding and identifying new chemical compounds from the natural world, which may have medical applications, takes a long time, and is expensive. The molecular structures need to be understood before their potential use in medicine is known. Atomic force microscopes can speed up this process, saving money and, potentially, lives. The case study below is a good example of the importance of AFM.

Case Study: Deep sea molecules

In 2010, scientists working on a species of bacterium from a mud sample taken from the Mariana Trench — the deepest place on the planet located nearly 11 000 metres beneath the Pacific Ocean, found that the bacteria produced an unknown chemical compound.

The chemical composition (the number and type of atoms present) was easily determined. However,

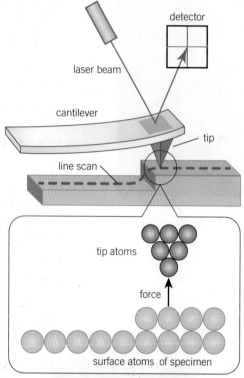

▲ **Figure 5** *Top: The principle of atomic force microscopy. Bottom: A nuclear pore as shown by atomic force microscopy*

the molecular structure, the way in which the atoms were joined together, was not so easy to work out and would have taken months using conventional techniques.

Using atomic force microscopy the scientists were able to image the molecules at very high, atomic level resolution within one week, giving them the molecular structure they needed.

This was the first time this method had been used in this way. This new approach could lead to much faster identification of unknown compounds and ultimately speed up the process of the development of new medicines.

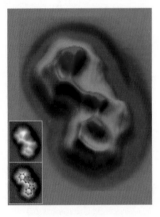

◀ **Figure 6** *Atomic force microscopy unveiled the previously unknown structure of cephalandole A, a chemical compound that could lead to the development of new drugs*

▲ **Figure 7** *Atomic force microscopy image of* Staphylococcus aureus *bacteria, commonly known for causing MRSA infections*

1   Explain why atomic force microscopy has a greater resolution than traditional light microscopy.

2   AFM is capable of producing magnifications equal or better than electron microscopes (this is demonstrated in Figure 7). Explain why.

3   Discuss why, despite the comparable magnification, atomic force microscopy could not have resulted in the same advances in the study of cell function as electron microscopy.

➕ Super resolved fluorescence microscopy

Electron microscopes cannot be used to examine living cells and it was always believed that the maximum resolution for light microscopes was 0.2 µm, about half the wavelength of light. This limits the detail that can be seen in living cells. In 2014 Eric Betzig, Stefan W. Hell, and William E. Moerner were awarded the Nobel Prize in Chemistry for achieving resolutions greater than 0.2 µm using light microscopy.

Two principles were involved, both forms of super resolution fluorescent microscopy (SRFM). One involved building up a very high resolution image by combining many very small images. The other involved superimposing many images with normal resolution to create one very high resolution image.

Stefan Hell developed stimulated emission depletion (STED) which involves the use of two lasers which are slightly offset. The first laser scans a specimen causing fluorescence, followed by the second laser which negates the fluorescence from all but a molecular sized area. A picture is built up with a resolution much greater than

that produced normally in light microscopy. In this way, individual strands of DNA become visible.

Eric Betzig and William E. Moerner independently developed the second principle which relies on the ability to control the fluorescence of individual molecules. Specimens are scanned multiple times but each time different molecules are allowed to fluoresce. The images are then superimposed and the resolution of the combined image is at the molecular level, much greater than 0.2 µm.

It is now possible to follow individual molecules during cellular processes. Proteins involved in Parkinson's and Alzheimer's diseases can be observed interacting and fertilised eggs dividing into embryos can be studied at a molecular level.

1   **a**   Explain why electron microscopes cannot be used to examine living cells.

     b   Describe how the ability to control the fluorescence of individual molecules helped uncover cell processes.

Summary questions

1 Explain why you would see more detail with an electron microscope than with a light microscope.   (*2 marks*)

2 **a** Define the term artefact with reference to microscopy. (*2 marks*)
   b Explain why artefacts are more likely to be produced when preparing samples for electron microscopy than for light microscopy.   (*3 marks*)

3 Study the two images below.

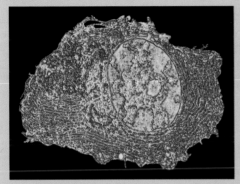

   a Suggest which form of microscopy was used to produce each image.   (*1 mark*)
   b Explain the reasons for your choices. (*3 marks*)
   c Outline the advantages and disadvantages of each technique.   (*6 marks*)

4 Confocal microscopy is used in medicine to study the cornea of the eye and the progression of skin cancer.
   a Explain the meaning of the term fluorescence. (*2 marks*)
   b State why lasers are used to provide illumination. (*1 mark*)
   c Explain the purpose of the pinhole aperture in confocal microscopy.   (*3 marks*)
   d One limitation of confocal microscopy is that it can not be used for deep tissue imaging. Suggest why. (*1 mark*)

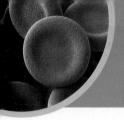

2.4 Eukaryotic cell structure

Specification reference: 2.1.1

Learning outcomes

Demonstrate knowledge, understanding, and application of:

→ the ultrastructure and function of eukaryotic cellular components

→ the importance of the cytoskeleton. The interrelationship between the organelles involved in the production and secretion of proteins.

Microscopes not only make cells visible – they also enable us to look deep inside individual cells. Using different types of microscopes you can discover how cells are organised and investigate the ways in which the structures you can see relate to their function. Microscopy allows you to see what goes on in a healthy cell, and to observe some of the changes which take place if the cell is attacked or diseased.

Relative sizes of molecules, organelles and cells

The diagram below demonstrates how the development of microscopes has allowed biologists to discover increasing amounts of detail of cell ultrastructure. The increased knowledge of structure has led to a better understanding of cell function.

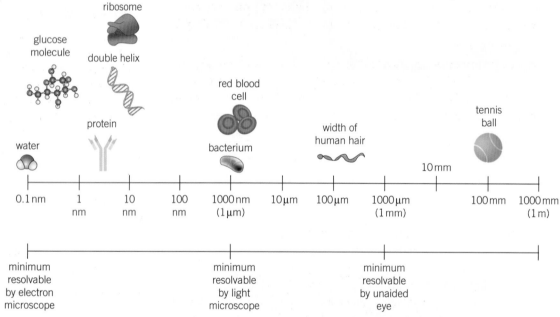

▲ **Figure 1** *This diagram illustrates the relative sizes of the different components of living organisms*

Cells

The basic unit of all living things is the cell – but not all cells are the same. There are two fundamental types of cell – **prokaryotic** and **eukaryotic**. Prokaryotes are single-celled organisms with a simple structure of just a single undivided internal area called the **cytoplasm** (composed of cytosol, which is made up of water, salts and organic molecules). Eukaryotic cells make up multicellular organisms like animals, plants, and fungi. Eukaryotic cells have a much more complicated internal structure, containing a membrane-bound nucleus (nucleoplasm) and cytoplasm, which contains many membrane-bound cellular components. You will learn more about the differences between prokaryotic and eukaryotic cells in Topic 2.6.

Synoptic link

You will learn about the role of enzymes in cellular metabolism and how they are affected by cellular conditions in Chapter 4, Enzymes.

In this topic you will learn about the ultrastructure of eukaryotic cells. The ultrastructure of a cell is those features that can be seen using an electron microscope.

Compartments for life

Chemical reactions are the fundamental processes of life and in cells they require both enzymes and specific reaction conditions. **Metabolism** involves both the synthesis (building up) and the breaking down of molecules. Different sets of reactions take place in different regions of the ultrastructure of the cell.

The reactions take place in the cytoplasm. The cell cytoplasm is separated from the external environment by a cell-surface membrane. In eukaryotic cells the cytoplasm is divided into many different membrane-bound compartments, known as **organelles**. These provide distinct environments and therefore conditions for the different cellular reactions.

Membranes are selectively permeable and control the movement of substances into and out of the cell and organelles. Membranes are effective barriers in controlling which substances enter and exit cells but they are fragile.

There are a number of organelles that are common to all eukaryotic cells. Each type has a distinct structure and function. They are clearly seen in animal cells, the focus of this topic. The ultrastucture specific to plant cells is discussed in the next topic.

Synoptic link

You will learn about the building blocks of membranes, and phospholipids in Topic 3.5, Lipids. The structure and function of cell membranes and the transport of substances across membranes are discussed in Chapter 5, Plasma membranes.

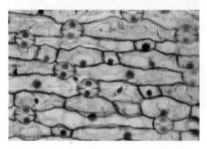

▲ **Figure 2** *The detail you can see inside a cell depends on the type of microscope used to produce the image. This photomicrograph is of onion cells as seen under a light microscope. Only the nuclei can be seen inside the cells (x 200 magnification)*

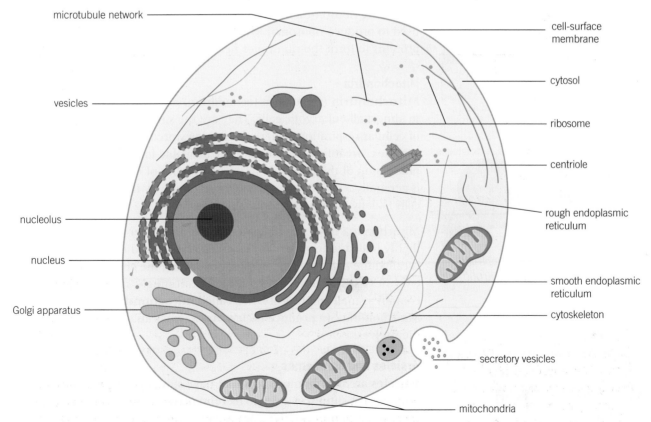

▲ **Figure 3** *A drawing of a eukaryotic animal cell showing the many other components that are not visible with a light microscope*

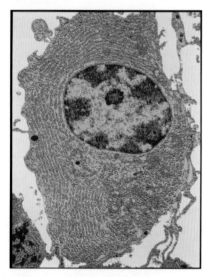

▲ **Figure 4** *Coloured transmission electron micrograph of a human cell showing the nucleus (large oval) and endoplasmic reticulum. Magnification: ×10 000*

Synoptic link

You will learn more about DNA, RNA and protein synthesis in Chapter 3, Biological molecules.

Synoptic link

You will learn more about ATP in Topic 3.11, ATP.

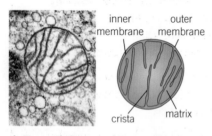

inner membrane   outer membrane

crista   matrix

▲ **Figure 5** *Electron micrograph and drawing of a mitochondrion x30 000 magnification*

Synoptic link

You will learn about phagocytic cells in Topic 12.6, The specific immune system.

Nucleus

The **nucleus** (plural nuclei) contains coded genetic information in the form of DNA molecules. DNA directs the synthesis of all proteins required by the cell (although this protein synthesis occurs outside of the nucleus at ribosomes). In this way the DNA controls the metabolic activities of the cell, as many of these proteins are the enzymes necessary for metabolism to take place. Not surprisingly, the nucleus is often the biggest single organelle in the cell (Figure 4).

DNA is contained within a double membrane called a *nuclear envelope* to protect it from damage in the cytoplasm. The nuclear envelope contains *nuclear pores* (Topic 2.2, Magnification and calibration) that allow molecules to move into and out of the nucleus. DNA itself is too large to leave the nucleus to the site of protein synthesis in the cell cytoplasm. Instead it is transcribed into smaller RNA molecules, which are exported via the nuclear pores.

DNA associates with proteins called **histones** to form a complex called **chromatin**. Chromatin coils and condenses to form structures known as **chromosomes**. These only become visible when cells are preparing to divide.

Nucleolus

The nucleolus is an area within the nucleus and is responsible for producing ribosomes. It is composed of proteins and RNA. RNA is used to produce ribosomal RNA (rRNA) which is then combined with proteins to form the ribosomes necessary for protein synthesis.

Mitochondria

Mitochondria (singular mitochondrion) are essential organelles in almost all eukaryotic cells. They are the site of the final stages of cellular respiration, where the energy stored in the bonds of complex, organic molecules is made available for the cell to use by the production of the molecule ATP. The number of mitochondria in a cell is generally a reflection of the amount of energy it uses, so very active cells usually have a lot of mitochondria.

Mitochondria have a double membrane. The inner membrane is highly folded to form structures called **cristae** and the fluid interior is called the **matrix**. The membrane forming the cristae contains the enzymes used in aerobic respiration. Interestingly, mitochondria also contain a small amount of DNA, called **mitochondrial (mt)DNA**. Mitochondria can produce their own enzymes and reproduce themselves.

Vesicles and lysosomes

Vesicles are membranous sacs that have storage and transport roles. They consist simply of a single membrane with fluid inside. Vesicles are used to transport materials inside the cell.

Lysosomes are specialised forms of vesicles that contain hydrolytic enzymes. They are responsible for breaking down waste material in cells, including old organelles. They play an important role in the immune system as they are responsible for breaking down pathogens ingested by phagocytic cells. They also play an important role in programmed cell death or apoptosis.

The cytoskeleton

The **cytoskeleton** is present throughout the cytoplasm of all eukaryotic cells. It is a network of fibres necessary for the shape and stability of a cell. Organelles are held in place by the cytoskeleton and it controls cell movement and the movement of organelles within cells.

The cytoskeleton has three components:

- Microfilaments – contractile fibres formed from the protein **actin**. These are responsible for cell movement and also cell contraction during cytokinesis, the process in which the cytoplasm of a single eukaryotic cell is divided to form two daughter cells.

- Microtubules – globular tubulin proteins polymerise to form tubes that are used to form a scaffold-like structure that determines the shape of a cell. They also act as tracks for the movement of organelles, including vesicles, around the cell. Spindle fibres, which have a role in the physical segregation of chromosomes in cell division, are composed of microtubules.

- Intermediate fibres – these fibres give mechanical strength to cells and help maintain their integrity.

> ### Synoptic link
>
> You will learn about the role of spindle fibres in cell division in Chapter 6, Cell division.

Cell movement

The movement of cells like phagocytes depends on the activity of the actin filaments in the cytoskeleton. The filament lengths change with the addition and removal of monomer subunits. The rate at which these subunits are added is different at each end of a filament. The subunits are not symmetrical and can only be added if they are in the correct orientation.

The subunits have to change shape before they are added to one end (the minus end) of the filament but not the other end (the plus end). This means that the subunits are added at a faster rate at the plus end. The filaments therefore increase in length at a faster rate in one particular direction.

Whether subunits are added or removed, at either end, is determined by the concentration of subunits in the cytoplasm. Due to the different rates of addition at either end, at certain concentrations subunits will be added at one end and removed at the other. This called treadmilling.

The increasing length of the filaments at one edge of a cell, the leading edge, leads to cells such as phagocytes moving in a particular direction.

> 1 Suggest, giving your reasons, which components of the cytoskeleton undergo treadmilling and which components do not.

Centrioles

Centrioles are a component of the cytoskeleton present in most eukaryotic cells with the exception of flowering plants and most fungi. They are composed of microtubules. Two associated centrioles form the *centrosome*, which is involved in the assembly and organisation of the spindle fibres during cell division.

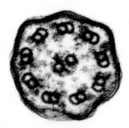

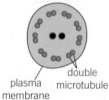

plasma
membrane
double
microtubule

▲ **Figure 6** *TEM of a cross-section through a single cilium from a protozoan (× 70 000 magnification)*

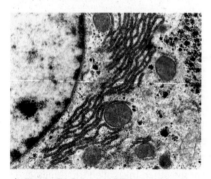

▲ **Figure 7** *Coloured TEM showing the rough endoplasmic reticulum (folds, centre). The cell nucleus is partially seen to the left. The round structures are vesicles that are being used to transport proteins from the rough endoplasmic reticulum to elsewhere in the cell. Magnification × 20 000*

In organisms with flagella and cilia, centrioles are thought to play a role in the positioning of these structures.

Flagella and cilia

Both flagella (whip-like) and cilia (hair-like) are extensions that protrude from some cell types. Flagella are longer than cilia but cilia are usually present in much greater numbers.

Flagella are used primarily to enable cells motility. In some cells they are used as a sensory organelle detecting chemical changes in the cell's environment.

Cilia can be mobile or stationary. Stationary cilia are present on the surface of many cells and have an important function in sensory organs such as the nose. Mobile cilia beat in a rhythmic manner, creating a current, and cause fluids or objects adjacent to the cell to move. For example, they are present in the trachea to move mucus away from the lungs (helping to keep the air passages clean), and in fallopian tubes to move egg cells from the ovary to the uterus.

Each cilium contains two central microtubules (black circles) surrounded by nine pairs of microtubules arranged like a "wheel". This is known as the 9+2 arrangement (Figure 6). Pairs of parallel microtubules slide over each other causing the cilia to move in a beating motion.

Organelles of protein synthesis

A key function of a cell is to synthesise proteins (including enzymes) for internal use and for **secretion** (transport out of the cell). A significant proportion of the internal structure of a cell is required for this process. The ribosomes, the endoplasmic reticulum, and the Golgi apparatus are all closely linked and coordinate the production of proteins and their preparation for different roles within the cell. The cytoskeleton plays a key role in coordinating protein synthesis.

Endoplasmic reticulum

The **endoplasmic reticulum (ER)** is a network of membranes enclosing flattened sacs called cisternae. It is connected to the outer membrane of the nucleus. There are two types:

- **Smooth endoplasmic reticulum** is responsible for lipid and carbohydrate synthesis, and storage.
- **Rough endoplasmic reticulum** has ribosomes bound to the surface and is responsible for the synthesis and transport of proteins.

Secretory cells, which release hormones or enzymes, have more rough endoplasmic reticulum than cells that do not release proteins.

Ribosomes

Ribosomes can be free-floating in the cytoplasm or attached to endoplasmic reticulum, forming rough endoplasmic reticulum. They are not surrounded by a membrane. They are constructed of RNA molecules made in the nucleolus of the cell. Ribosomes are the site of protein synthesis.

Mitochondria and chloroplasts also contain ribosomes, as do prokaryotic cells.

Golgi apparatus

The **Golgi apparatus** is similar in structure to the smooth endoplasmic reticulum. It is a compact structure formed of cisternae and does not contain ribosomes. It has a role in modifying proteins and 'packaging' them into vesicles. These may be secretory vesicles, if the proteins are destined to leave the cell, or lysosomes, which stay in the cell.

Protein production

Proteins are synthesised on the ribosomes bound to the endoplasmic reticulum (1). They then pass into its cisternae and are packaged into transport vesicles (2). Vesicles containing the newly synthesised proteins move towards the Golgi apparatus via the transport function of the cytoskeleton (3). The vesicles fuse with the cis face of the Golgi apparatus and the proteins enter. The proteins are structurally modified before leaving the Golgi apparatus in vesicles from its trans face (4).

Secretory vesicles carry proteins that are to be released from the cell. The vesicles move towards and fuse with the cell-surface membrane, releasing their contents by exocytosis. Some vesicles form lysosomes – these contain enzymes for use in the cell (5).

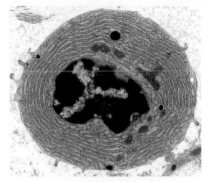

▲ Figure 8 *Transmission electron micrograph of a plasma cell with a large central nucleus surrounded by large amounts of rough endoplasmic reticulum. Plasma cells, which are found in the blood and lymph, produce and secrete antibodies (which are made of protein) during an immune response. × 6000 magnification*

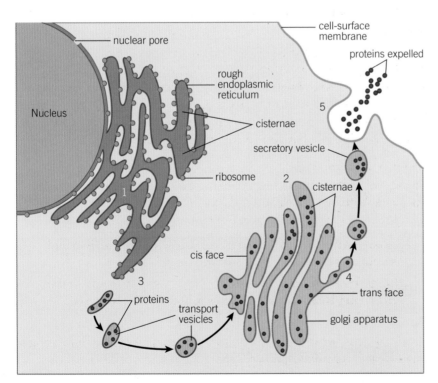

▲ Figure 10 *The ribosomes, endoplasmic reticulum, and Golgi apparatus work together to synthesise, modify and then transport proteins, including enzymes and hormones, out of the cell*

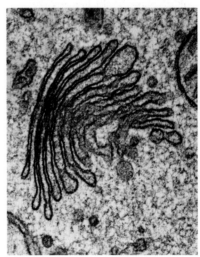

▲ Figure 9 *Transmission electron micrograph of the Golgi apparatus. Golgi are membrane-bound organelles that modify and package proteins for onward transport. × 8000 magnification*

Synoptic link

You will learn more about the details of protein synthesis at the ribosome in Topic 3.10, Protein synthesis.

Summary questions

1 What is a lysosome and why is the membrane that surrounds it so important? *(3 marks)*

2 Explain why cells need to be compartmentalised, and describe three examples of compartmentalisation within an animal cell. *(4 marks)*

3 Compare the structure and function of the rough and smooth endoplasmic reticulum. *(3 marks)*

4 Describe the structure and function of the cytoskeleton. *(5 marks)*

5 Given the following information about a eukaryotic cell

 7×10^7 base pairs of DNA per chromosome 0.34×10^{-9} m per base pair diploid number is 46

 a Calculate the length of DNA in a single cell. Give your answer in metres.

 b Suggest how this DNA is packed into a cell only 50 μm in diameter.

6 Discuss how the structure of microfilaments and microtubules means these components of the cytoskeleton are involved in the movement of cells but the intermediate fibres are not. *(6 marks)*

2.5 The ultrastructure of plant cells

Specification reference: 2.1.1

Plant cells have all of the cellular components you have just seen in animal cells. However, there are some structures that are only seen in plant cells, that carry out photosynthesis.

Cellulose cell wall

Plant cells, unlike animal cells, are rigid structures. They have a cell wall surrounding the cell-surface membrane.

Plant cell walls are made of cellulose, a complex carbohydrate. They are freely permeable so substances can pass into and out of the cell through the cellulose wall. The cell walls of a plant cell give it shape. The contents of the cell press against the cell wall making it rigid. This supports both the individual cell and the plant as a whole. The cell wall also acts as a defence mechanism, protecting the contents of the cell against invading pathogens. All plant cells have cellulose cell walls.

Synoptic link

You will learn about carbohydrates, including cellulose, in Topic 3.3, Carbohydrates.

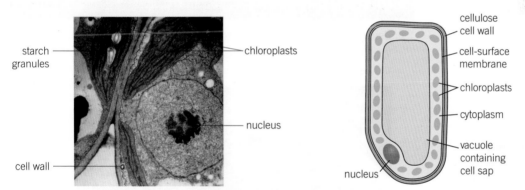

▲ Figure 1 *Left: A transmission electron micrograph of the junction of two leaf cells × 18 700 magnification. Right: A representation of a plant cell as seen under a light microscope*

Plant cell organelles

Plant cells, unlike animal cells, are rigid structures. Structures which are unique to plant cells include:

Vacuoles

Vacuoles are membrane lined sacs in the cytoplasm containing cell sap. Many plant cells have large permanent vacuoles which are very important in the maintenance of turgor, so that the contents of the cell push against the cell wall and maintain a rigid framework for the cell. The membrane of a vacuole in a plant cell is called the **tonoplast**. It is selectively permeable, which means that some small molecules can pass through it but others cannot. If vacuoles appear in animal cells, they are small and transient (not permanent).

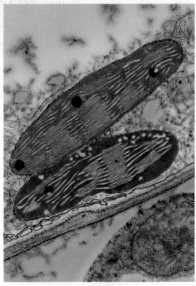

Chloroplasts

Chloroplasts are the organelles responsible for photosynthesis in plant cells. They are found in the cells in the green parts of plants such as the leaves and the stems but not in the roots. They have a double membrane structure, similar to mitochondria. The fluid enclosed in the chloroplast is called the **stroma**. They also have an internal network of membranes, which form flattened sacs called thylakoids. Several thylakoids stacked together are called a **granum** (plural grana). The grana are joined by membranes called lamellae. The grana contain the chlorophyll pigments, where light-dependent reactions occur during photosynthesis. Starch produced by photosynthesis is present as starch grains. Like mitochondria, chloroplasts also contain DNA and ribosomes. Chloroplasts are therefore able to make their own proteins.

The internal membranes provide the large surface area needed for the enzymes, proteins and pigment molecules necessary in the process of photosynthesis.

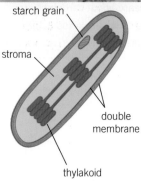

▲ **Figure 2** *Transmission electron micrograph (top) and drawing of chloroplasts seen in the leaf of a pea plant (bottom). The chloroplasts are seen cut lengthways so the grana are visible. Starch produced during photosynthesis is seen as dark circles (starch grains) within each chloroplast.* ×13 000 *magnification*

Summary questions

1 Using your knowledge of cell ultrastructure, identify the structures visible in the micrograph below. State, with reasons, whether the cell is a plant or animal cell *(4 marks)*

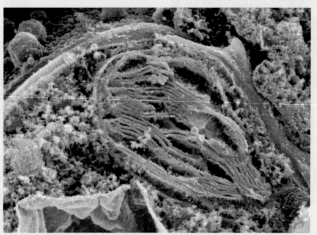

2 a Many different organisms have cell walls including fungi and bacteria. What is unique about plant cell walls? *(1 mark)*
  b Give three functions of plant cell walls *(3 marks)*

3 Describe the similarities and differences between a human cell and a plant root cell *(3 marks)*

2.6 Prokaryotic and eukaryotic cells

Specification reference: 2.1.1

Animals, plants, and fungi are all complex multicellular organisms. The cells making up these organisms are eukaryotic. There is a lot of evidence that suggests that eukaryotic cells evolved from less complex prokaryotic cells. These prokaryotic cells, present in great numbers, live in an incredibly diverse range of habitats. These unicellular organisms can be classed into two evolutionary domains – Archaea and Bacteria, which evolved from an ancient common ancestor.

Prokaryotic cells

Prokaryotic cells may have been among the earliest forms of life on Earth. They first appeared around 3.5 billion years ago when the surface of the Earth was a very hostile environment. Scientists believe that these early cells were adapted to living in extremes of salinity, pH and temperature.

These organisms are known as extremophiles and they still exist today. They can be found in hydrothermal vents and salt lakes – similar environments to those believed to have made up the early Earth. They are usually of the domain Archaea and more recently they have been found in more hospitable environments such as soil and the human digestive system.

Prokaryotic organisms are always unicellular with a relatively simple structure. Their DNA is not contained within a nucleus, they have few organelles and the organelles they do have are not membrane-bound.

DNA

The structure of the DNA contained within prokaryotes is fundamentally the same as in eukaryotes but it is packaged differently. Prokaryotes generally only have one molecule of DNA, a chromosome, which is supercoiled to make it more compact. The genes on the chromosome are often grouped into operons, meaning a number of genes are switched on or off at the same time.

Ribosomes

The ribosomes in prokaryotic cells are smaller than those in eukaryotic cells. Their relative size is determined by the rate at which they settle, or form a sediment, in solution. The larger eukaryotic ribosomes are designated 80S and the smaller prokaryotic ribosomes, 70S. They are both necessary for protein synthesis, although the larger 80S ribosomes are involved in the formation of more complex proteins.

Cell wall

Prokaryotic cells have a cell wall made from peptidoglycan, also known as murein. It is a complex polymer formed from amino acids and sugars.

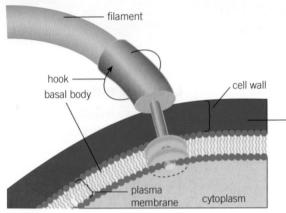

▲ **Figure 1** *A prokaryotic flagellum*

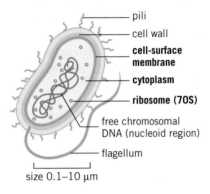

size 0.1–10 μm

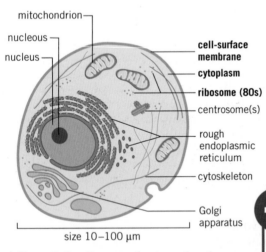

size 10–100 μm

▲ **Figure 2** *Top: Features of prokaryotic cell (bacterium). Bottom: features of a typical eukaryotic cell (animal). Common features are highlighted in* **bold**

Flagella

The flagella of prokaryotes is thinner than the equivalent structure of eukaryotes and does not have the 9 + 2 arrangement. The energy to rotate the filament that forms the flagellum is supplied from the process of chemiosmosis, not from ATP as in eukaryotic cells.

The flagellum is attached to the cell membrane of a bacterium by a basal body and rotated by a molecular motor.

The basal body attaches the filament comprising the flagellum to the cell-surface membrane of a bacterium. A molecular motor causes the hook to rotate giving the filament a whip like movement, which propels the cell.

A comparison with eukaryotic cells

The first eukaryotic cells appeared about 1.5 billion years ago. As you have learned eukaryotic cells are much more complex than prokaryotic cells. Their DNA is present within a nucleus and exists as multiple **chromosomes**, which are supercoiled, and each one wraps around a number of proteins called **histones**, forming a complex for efficient packaging. This complex is called **chromatin** and chromatin coils and condenses to form chromosomes. Eukaryotic genes are generally switched on and off individually.

As you learnt earlier, eukaryotic cells have membrane-bound organelles including mitochondria and chloroplasts (Topic 2.4, Eukaryotic cell structure and Topic 2.4, The ultrastructure of plant cells).

Organisms from the plant, animal, fungi, and protoctista kingdoms are all composed of eukaryotic cells. Many are multicellular.

➕ Endosymbiosis

The theory of endosymbiosis is that mitochondria and chloroplasts, and possibly other eukaryotic organelles, were formerly free-living bacteria, that is, prokaryotes. The theory is that these prokaryotes were taken inside another cell as an endosymbiont – an organism that lives within the body or cells of another organism. This eventually led to the evolution of eukaryotic cells.

1   Discuss, using information from this topic, any evidence that supports the endosymbiotic theory.

The similarities and differences between prokaryotic and eukaryotic cells

The similarities and differences between prokaryotic and eukaryotic cells are summarised in Table 1.

▼ **Table 1** *Prokaryotic and eukaryotic cells compared*

Feature	Prokaryotic	Eukaryotic
nucleus	not present	present
DNA	circular	linear
DNA organisation	proteins fold and condense DNA	associated with proteins called histones
extra chromosomal DNA	circular DNA called plasmids	only present in certain organelles such as chloroplasts and mitochondria
organelles	non membrane-bound	both membrane-bound and non membrane-bound
cell wall	peptidoglycan	chitin in fungi, cellulose in plants, not present in animals
ribosomes	smaller, 70S	larger, 80S
cytoskeleton	present	present, more complex
reproduction	binary fission	asexual or sexual
cell type	unicellular	unicellular and multicellular
cell-surface membrane	present	present

➕ Prokaryotic cell study

Here is a transmission electron micrograph image of a slice through a rod-shaped Gram-negative *Escherichia coli* bacterium. The cell wall can be seen as a double line around the cell. The darker area inside is the nucleoid which contains the DNA. Look closely at the photo and answer the questions below

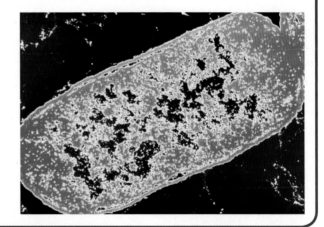

1 a Use the micrograph shown here to produce a scientific drawing of the bacterium.
   b Describe the differences you would see if you were observing a eukaryotic cell with the same microscope, at the same magnification.

Summary questions

1 List three structural differences between prokaryotic cells and eukaryotic cells.   (*3 marks*)

2 Suggest why the lack of membrane-bound organelles does not stop prokaryotic cells making proteins.   (*4 marks*)

3 Some antibiotics kill bacteria by disrupting the formation of peptidoglycan molecules. Explain why these antibiotics kill bacteria and why they do not have any effect on eukaryotic cells.   (*4 marks*)

Practice questions

1 The cytoskeleton is present throughout the cytoplasm of all eukaryotic cells.

Which of the following statements is/are correct with respect to the structure of the cytoskeleton?

Statement 1: Intermediate fibres - these fibres give mechanical strength to cells and help maintain their integrity.

Statement 2: Microtubules - contractile fibres formed from the protein actin. Responsible for cell movement

Statement 3: Microfilaments - formed from the cylindrical-shaped protein tubulin. They form a scaffold-like structure determining the shape of a cell.

A 1, 2 and 3 are correct

B Only 1 and 2 are correct

C Only 2 and 3 are correct

D Only 1 is correct *(1 mark)*

2 Serous cells are present in the salivary glands of animals. They are responsible for the production of the enzyme amylase which begins the breakdown of starch.

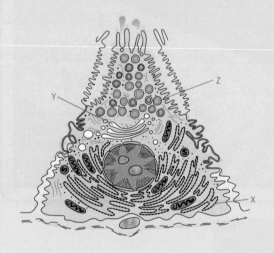

a (i) identify the structures labelled in the diagram.

x *(1 mark)*

y *(1 mark)*

z *(1 mark)*

(3 marks)

(ii) State whether the cell is eukaryotic or prokaryotic giving the reason for your decision. *(2 marks)*

b (i) State which group of enzymes contains amylase. *(1 mark)*

(ii) Outline the stages and organelles involved in the production and release of amylase. *(5 marks)*

c Explain the process of exocytosis. *(3 marks)*

d Discuss the different roles of vesicles, vacuoles and lysosomes. *(4 marks)*

3 The photo below shows a transmission electron micrograph of plankton. These single-celled marine micro-organisms are thought to be the most abundant photosynthetic organisms on Earth.

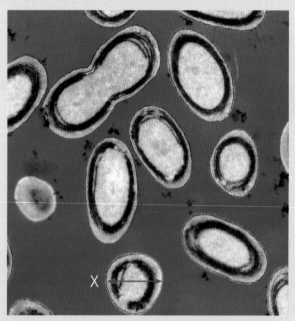

a Calculate the magnification of the Plankton labelled X. The actual diameter of the plankton is 2.6 μm. *(2 marks)*

The amount of detail that can be seen with a microscope depends on both the magnification and resolution possible with the microscope being used. Any increase in magnification beyond the limit of resolution results in 'empty magnification'.

b Define the following terms

   (i)   *resolution*   (*2 marks*)

   (ii) *magnification*   (*1 mark*)

   (iii) Suggest what is meant by the term 'empty magnification' (*1 mark*)

c Outline how a compound light microscope magnifies an image of a specimen.
                                         (*4 marks*)

d Describe three different ways of preparing microscope slides for light microscopy.
                                         (*6 marks*)

4 a Explain the meaning of the term artefact with reference to microscopy.   (*2 marks*)

b Discuss the advantages and disadvantages of using an electron microscope to study the ultrastructure of cells.   (*4 marks*)

c Outline how laser scanning confocal microscopes produce an image.   (*4 marks*)

5 a Complete and complete the table below.

The first row has been done for you.
                                         (*5 marks*)

Feature	Prokaryotic	Eukaryotic
DNA	*circular*	*linear*
Extra chromosomal DNA		only present in certain organelles such as chloroplasts and mitochondria
Organelles	non membrane bound	
Cell wall	peptidoglycan	
Ribosomes		large, 80 s
Cell surface membrane		present

b Define the term 'cell ultrastructure'.
                                         (*2 marks*)

6 Human genomes contain many more genes than bacterial genomes, and they are much longer.

Discuss the way in which this affects the packing of DNA in eukaryotes and prokaryotes.   (*6 marks*)

3 BIOLOGICAL MOLECULES
3.1 Biological elements
Specification reference: 2.1.2

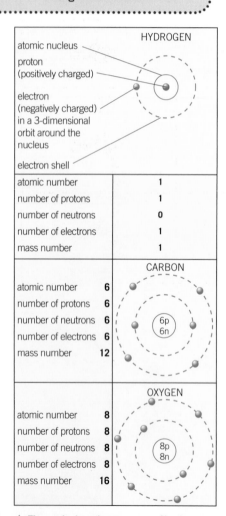

▲ Figure 1 *Atomic structure of hydrogen, carbon and oxygen which are common biological elements*

The building blocks of life

In section 2 you looked at cells and cellular components, however, these too are composed from smaller components called molecules. Molecules are built from even smaller components called atoms. In fact, atoms are built from yet smaller components including protons, neutrons, and electrons, which you will have learned about in your GCSE science or chemistry studies.

Knowledge of biochemistry is essential, as it underpins a proper understanding of the metabolic processes, structures, and limitations of biology, for example the complex series of reactions and molecules involved in cellular respiration. You do not need to be a chemist to understand basic biochemistry, as you will see in coming topics, but you do need to understand some essential chemical concepts and rules, which will be explored in this chapter.

Elements

Different types of atoms are called **elements**. Elements are distinguished by the number of protons in their atomic nuclei. There are over a hundred known elements in the universe but only a small percentage of these are present in the living world.

If you have ever built a model using interlocking bricks you will know how useful the bricks that make the most connections are, at the start of a new build.

In the same way that complex models can be built from a small range of simple bricks, all living things are made primarily from four key elements – carbon (C), hydrogen (H), oxygen (O) and nitrogen (N). In addition, phosphorus (P) and sulfur (S) also have important roles in the biochemistry of cells. These six elements are by far the most abundant elements present in biological molecules.

Other elements, including sodium (Na), potassium (K), calcium (Ca), and iron (Fe), also have important roles in biochemistry. You will learn about some of the roles of these elements later in this chapter.

Bonding

Atoms connect with each other by forming bonds. Atoms can bond to other atoms of the same element, or atoms of different elements, provided this follows the 'bonding rules' (described on the next page). When two or more atoms bond together the complex is called a molecule.

A covalent bond occurs when two atoms share a pair of electrons. The electrons used to form bonds are unpaired and present in the outer orbitals of the atoms.

Bonding follows some simple rules, determined by the number of unpaired electrons present in the outer orbitals of different elements:

- Carbon atoms can form four bonds with other atoms.
- Nitrogen atoms can form three bonds with other atoms.
- Oxygen atoms can form two bonds with other atoms.
- Hydrogen atoms can only form one bond with another atom.

▼ Table 1 *Hydrogen atoms can only form one bond with other atoms. Carbon atoms can form four bonds, nitrogen three and oxygen two. The dots and crosses represent electrons and which atom they belong to*

Molecule	Electron diagram	Displayed formula	'Ball and stick' model
Hydrogen (H_2)		$H \text{---} H$	
Water (H_2O)			
Carbon dioxide (CO_2)		$O = C = O$	
Methane (CH_4)			
Ammonia (NH_3)			

The number of bonds formed by these elements can be no more or less than stated. There are, however, exceptions to this rule, which you will learn about in later sections. Life on this planet is often referred to as being 'carbon-based' because carbon, which can form four bonds, forms the backbone of most biological molecules.

Ions

An atom or molecule in which the total number of electrons is not equal to the total number of protons is called an ion. If an atom or molecule loses one or more electrons it has a net positive charge and is known as a cation. If an atom or molecule gains electrons, it has a net negative charge and is known as an anion.

In ionic bonds, one atom in the pair donates an electron and the other receives it. This forms positive and negative ions that are held together by the attraction of the opposite charges.

Ions in solution are called electrolytes. The following tables list *some* of the important roles of ions in living organisms.

▼ Table 2 *Roles of cations*

Cations	Necessary for
calcium ions (Ca^{2+})	nerve impulse transmission
	muscle contraction
sodium ions (Na^+)	nerve impulse transmission
	kidney function
potassium ions (K^+)	nerve impulse transmission
	stomatal opening
hydrogen ions (H^+)	catalysis of reactions
	pH determination
ammonium ions (NH_4^+)	production of nitrate ions by bacteria

▼ Table 3 *Roles of anions*

Anions	Necessary for
nitrate ions (NO_3^-)	nitrogen supply to plants for amino acid and protein formation
hydrogen carbonate ions (HCO_3^-)	maintenance of blood pH
chloride ions (Cl^-)	balance positive charge of sodium and potassium ions in cells
phosphate ions (PO_4^{3-})	cell membrane formation
	nucleic acid and ATP formation
	bone formation
hydroxide ions (OH^-)	catalysis of reactions
	pH determination

Biological molecules

Below is a summary of the elements present in some of the key biological molecules. You will learn more about each of these classes of molecule in the coming topics of this chapter.

- **Carbohydrates** – carbon, hydrogen, and oxygen, usually in the ratio $C_x(H_2O)_x$.
- **Lipids** – carbon, hydrogen, and oxygen.
- **Proteins** – carbon, hydrogen, oxygen, nitrogen, and sulfur.
- **Nucleic acids** – carbon, hydrogen, oxygen, nitrogen, and phosphorus.

Polymers

Biological molecules are often **polymers**. Polymers are long-chain molecules made up by the linking of multiple individual molecules (called **monomers**) in a repeating pattern. In carbohydrates the monomers are sugars (saccharides) and in proteins the monomers are amino acids.

Summary questions

1 Explain how atoms join together to form molecules. (*2 marks*)

2 Explain the difference between a cation and an anion. (*4 marks*)

3 Explain how the bonds between the atoms in both water *and* carbon dioxide molecules fulfil the 'bonding rules'. (*4 marks*)

4 The image below, obtained in 1953, helped confirm the recently proposed structure of DNA. The equipment that was used to obtain the image is also shown, X-ray diffraction photograph of DNA (deoxyribonucleic acid). This image was obtained in 1953 and results from a beam of X-rays being scattered onto a photographic plate by the DNA. Various features about the structure of the DNA can be determined from the pattern of spots and bands. The cross of bands indicates the helical nature of DNA.

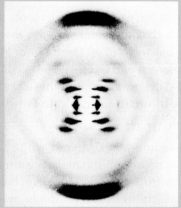

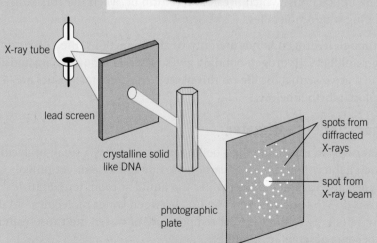

a Suggest why the x-ray diffraction technique used to produce this image was not considered a form of microscopy but the use of electrons to produce images is called electron microscopy. (*3 marks*)

b Explain why cells are visible with light microscopes but electron microscopes are needed to see ribosomes. (*3 marks*)

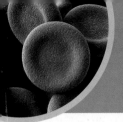

3.2 Water

The bonds of life

Atoms join together to form molecules by making bonds with each other. As you learnt in the previous topic, in ionic bonds, atoms give or receive electrons. They form negative or positive ions that are held together by the attraction of the opposite charges. Covalent bonds occur when atoms *share* electrons. However, the negative electrons are not always shared equally by the atoms of different elements. In many covalent bonds, the electrons will spend more time closer to one of the atoms than to the other. The atom with the greater share of negative electrons will be slightly negative (δ^-) compared with the other atom in the bond, which will therefore be slightly positive (δ^+) (Figure 1).

Molecules in which this happens are said to be **polar** – they have regions of negativity and regions of positivity.

Oxygen and hydrogen are examples of elements that do not share electrons equally in a covalent bond. Oxygen always has a much greater share of the electrons in an O—H bond. Many organic molecules contain oxygen and hydrogen bonded together in what are called hydroxyl (OH) groups and so they are slightly polar. Water (H_2O) is an example of such a molecule, in fact, water contains two of these hydroxyl groups (Figure 2).

Polar molecules, including water, interact with each other as the positive and negative regions of the molecule attract each other and form bonds, called hydrogen bonds. Hydrogen bonds are relatively weak interactions, which break and reform between the constantly moving water molecules.

Although hydrogen bonds are only weak interactions, they occur in high numbers. Hydrogen bonding gives water its unique characteristics, which are essential for life on this planet. These characteristics are explored further below.

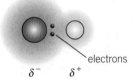

▲ Figure 1 *Covalent bond between oxygen and hydrogen. The unequal sharing of the electrons leads to oxygen being more negative compared with hydrogen*

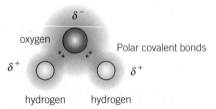

▲ Figure 2 *The polar covalent bonds make water a polar molecule*

Characteristics of water

Water has an unusually high boiling point. Water is a small molecule, much lighter than the gases carbon dioxide or oxygen, yet unlike oxygen and carbon dioxide, water is a liquid at room temperature. This is due to the hydrogen bonding between water molecules. It takes a lot of energy to increase the temperature of water and cause water to become gaseous (evaporate).

When water freezes it turns to ice. Most substances are more dense in their solid state than in their liquid state, but when water turns to ice it becomes less dense. This is because of the hydrogen bonds formed. As water is cooled below 4 °C the hydrogen bonds fix the positions of the polar molecules slightly further apart than the average distance in the liquid state. This produces a giant, rigid but open structure, with every oxygen atom at the centre of a tetrahedral arrangement

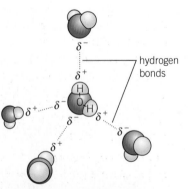

▲ Figure 3 *Five water molecules interacting via hydrogen bonds*

of hydrogen atoms, resulting in a solid that is less dense than liquid water. For this reason, ice floats.

Water therefore has *cohesive* properties. It moves as one mass because the molecules are attracted to each other (cohesion). It is in this way that plants are able to draw water up their roots and how you are able to drink water through a straw. Water also has *adhesive* properties – this is where water molecules are attracted to other materials. For example, when you wash your hands your hands become wet, the water doesn't run straight off.

Water molecules are more strongly cohesive to each other than they are to air, this results in water having a 'skin' of surface tension. (Figure 4)

Water for life

The characteristics and properties of water are critical in sustaining life. In this way, water is unique. Some of the ways in which water is vital for life are summarised below.

Because it is a polar molecule, water acts as a *solvent* in which many of the solutes in an organism can be dissolved. The cytosol of prokaryotes (bacterial) and eukaryotes is mainly water. Many solutes are also polar molecules, amino acids, proteins (Topic 3.6, Structure of proteins) and nucleic acids (Topic 3.8, Nucleic acids). Water acts as a medium for chemical reactions and also helps transport dissolved compounds into and out of cells.

Water makes a very efficient *transport medium* within living things. Cohesion between water molecules means that when water is transported though the body, molecules will stick together. Adhesion occurs between water molecules and other polar molecules and surfaces. The effects of adhesion and cohesion result in water exhibiting **capillary action**. This is the process by which water can rise up a narrow tube against the force of gravity.

Water acts as a coolant, helping to buffer temperature changes during chemical reactions in prokaryotic and eukaroytic cells because of the large amounts of energy required to overcome hydrogen bonding. Maintaining constant temperatures in cellular environments is important as enzymes are often only active in a narrow temperature range.

Many organisms, such as fish, live in water and cannot survive out of it. Water is stable – it does not change temperature or become a gas easily, therefore providing a constant environment. Because ice floats, it forms on the surface of ponds and lakes, rather than from the bottom up. This forms an insulating layer above the water below. Aquatic organisms would not be able to survive freezing temperatures if their entire habitat froze solid. Some organisms also inhabit the surface of water. Surface tension is strong enough to support small insects such as pond skaters.

▲ **Figure 4** *This pond skater (Gerris lacustris) inhabits the surface of water, supported by surface tension*

Study tip

Do not confuse polarity with 'charged' or 'ionic'. In polar bonds electrons are shared, albeit unequally, but if atoms actually lose an electron to another atom they both become charged and are called ions. In ionic bonding, the atom that gains the electron becomes a negative ion and the atom that loses the electron becomes a positive ion.

Synoptic link

You will learn more about how temperature affects enzyme activity in Topic 4.2, Factors affecting enzyme activity.

Summary questions

1 Explain how hydrogen bonds form. (*3 marks*)

2 Explain why water is a polar molecule. (*2 marks*)

3 Suggest, with reasons, which properties of water make it such an important component of blood. (*5 marks*)

4 Water forms the basis of the stroma in chloroplasts and the matrix in mitochondria.

   Describe which properties of water make it such an important component of these particular organelles. (*5 marks*)

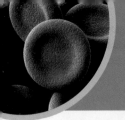

3.3 Carbohydrates

Specification reference: 2.1.2

Carbohydrates are molecules that only contain the elements carbon, hydrogen, and oxygen. Carbohydrate literally means 'hydrated carbon' (carbon and water). The elements in carbohydrates usually appear in the ratio $Cx(H20)y$. This is known as the general formula of carbohydrates.

Carbohydrates are also known as saccharides or sugars. A single sugar unit is known as a **monosaccharide**, examples include glucose, fructose, and ribose. When two monosaccharides link together they form a disaccharide, for example lactose and sucrose. When two or more (usually many more) monosaccharides are linked they form a polymer called a **polysaccharide**. Glycogen, cellulose, and starch are examples of polysaccharides.

Glucose

The basic building blocks, or monomers, of some biologically important large carbohydrates are **glucose** molecules, which have the chemical formula $C_6H_{12}O_6$. Glucose is a monosaccharide composed of six carbons and therefore is a **hexose monosaccharide** (hexose sugar) (Figure 1).

In molecular structure diagrams, the carbons are numbered clockwise, beginning with the carbon to the right (clockwise) of the oxygen atom within the ring.

There are two structural variations of the glucose molecule, alpha (α) and beta (β) glucose, in which the OH (hydroxyl) group on carbon 1 is in opposite positions, as shown in Figure 1.

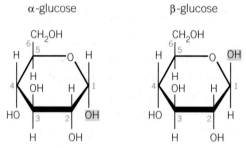

▲ Figure 1 *The sugars alpha and beta glucose, which are examples of monosaccharides (single sugar units). Note the different position of the OH group on carbon 1*

Glucose molecules are polar and soluble in water. This is due to the hydrogen bonds that form between the hydroxyl groups and water molecules. This solubility in water is important, because it means glucose is dissolved in the cytosol of the cell.

Condensation reactions

When two alpha glucose molecules are side by side, two hydroxyl groups interact (react). When this happens bonds are broken and new bonds reformed in different places producing new molecules.

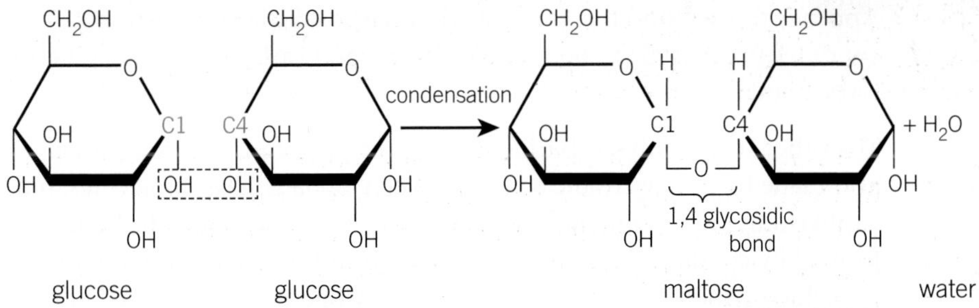

CH$_2$OH   CH$_2$OH   CH$_2$OH   CH$_2$OH

condensation

C1   C4

OH   OH

C1   C4

+ H$_2$O

1,4 glycosidic
bond

glucose   glucose   maltose   water

▲ **Figure 2** *As the two OH are so close they react, forming a covalent bond called a glycosidic bond between the two glucose molecules*

As you can see in Figure 2, two hydrogen atoms and an oxygen atom are removed from the glucose monomers and join to form a water molecule. A bond forms between carbons 1 and 4 on the glucose molecules and the molecules are now joined.

A covalent bond called a **glycosidic bond** is formed between two glucose molecules. The reaction is called a **condensation reaction** because a water molecule is formed as one of the products of the reaction. Because in this reaction carbon 1 of one glucose molecule is joined to carbon 4 of the other glucose molecule, the bond is known as a 1,4 glycosidic bond. In this reaction the new molecule is called **maltose**. This is an example of a **disaccharide** (a molecule made up of two monosaccharides).

Other sugars
Fructose and galactose are also hexose monosaccharides. Fructose naturally occurs in fruit, often in combination with glucose forming the disaccharide **sucrose**, commonly known as cane sugar or just sugar.

Galactose and glucose form the disaccharide **lactose**. Lactose is commonly found in milk and milk products.

Fructose is sweeter than glucose and glucose is sweeter than galactose.

Pentose monosaccharides are sugars that contain five carbon atoms. Two pentose sugars are important components of biological molecules – **ribose** is the sugar present in RNA nucleotides and deoxyribose is the sugar present in DNA nucleotides.

Starch and glycogen
Many alpha glucose molecules can be joined by glycosidic bonds to form two slightly different polysaccharides known collectively as **starch**. Glucose made by photosynthesis in plant cells is stored as starch. It is a chemical energy store.

One of the polysaccharides in starch is called amylose. Amylose is formed by alpha glucose molecules joined together only by 1–4 glycosidic bonds. The angle of the bond means that this long chain of glucose twists to form a helix which is further stabilised by hydrogen bonding within the molecule. This makes the polysaccharide more compact, and much less soluble, than the glucose molecules used to make it.

> **Synoptic link**
>
> You will find out more about the structure of nucleotides in Topic 3.8, Nucleic acids.

▲ **Figure 3** *The characteristic helix shape of amylose*

Another type of starch is formed when glycosidic bonds form in condensation reactions between carbon 1 and carbon 6 on two glucose molecules.

The other starch polysaccharide is called amylopectin. Amylopectin is also made by 1-4 glycosidic bonds between alpha glucose molecules, but (unlike amylose) in amylopectin there are also some glycosidic bonds formed by condensation reactions between carbon 1 and carbon 6 on two glucose molecules. this means that amylopectin has a branched structure, with the 1-6 branching points occurring approximately once in every 25 glucose subunits.

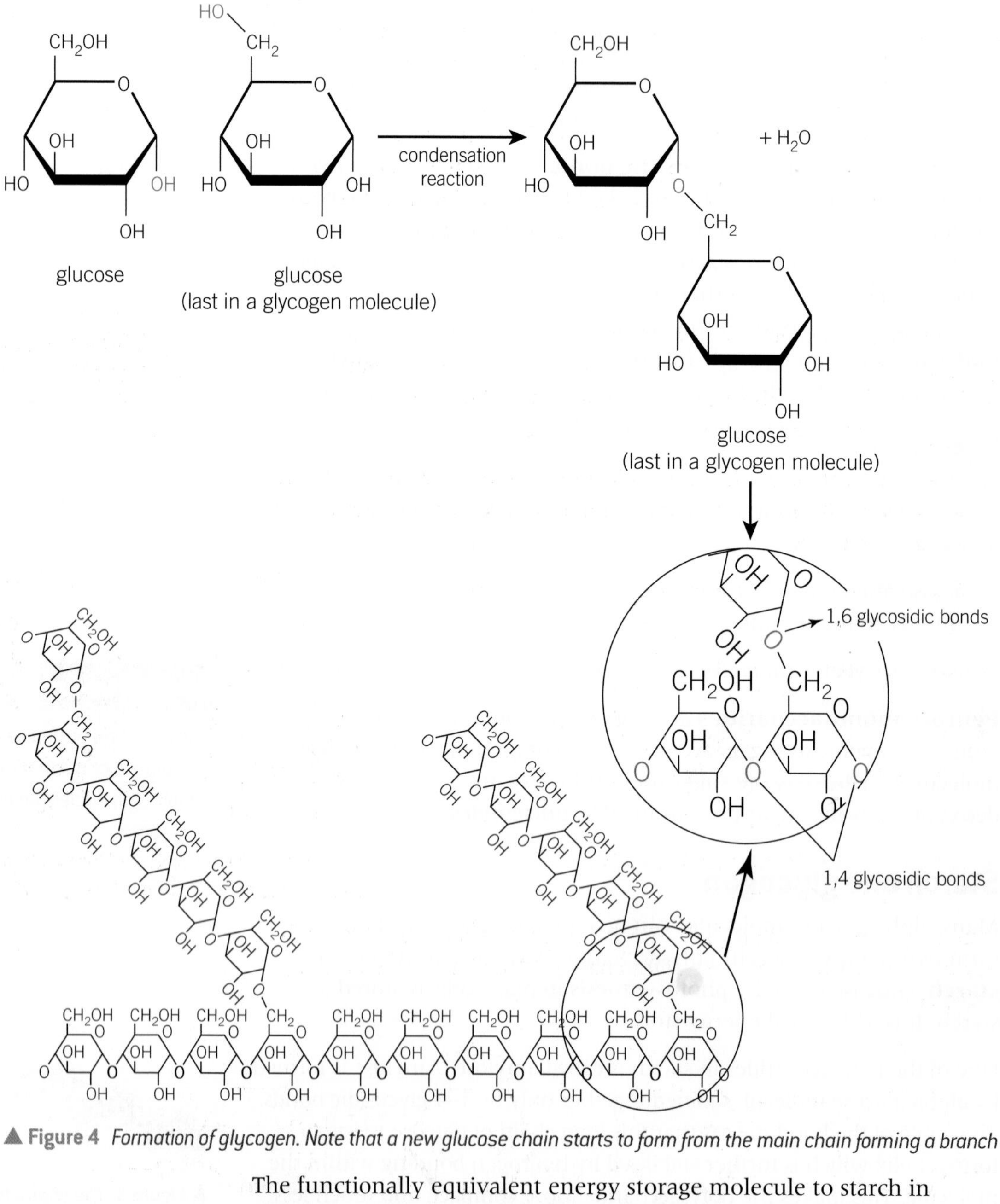

▲ Figure 4 Formation of glycogen. Note that a new glucose chain starts to form from the main chain forming a branch

The functionally equivalent energy storage molecule to starch in animals and fungi is called **glycogen** (Figure 4). Glycogen forms more branches than amylopectin, which means it is more compact and less space is needed for it to be stored. This is important as animals are mobile, unlike plants. The coiling or branching of these polysaccharides makes them very compact, which is ideal for storage. The branching also means there are many free ends where glucose molecules can be added or removed. This speeds up the processes of storing or releasing glucose molecules required by the cell.

So, the key properties of amylopectin and glycogen are that they are insoluble, branched, and compact. These properties mean they are ideally suited to the storage roles that they carry out.

Hydrolysis reactions

Glucose is stored as starch by plants or glycogen by animals and fungi, until it is needed for respiration – the process in which biochemical energy in these stored nutrients is converted into a useable energy source for the cell.

To release glucose for respiration, starch or glycogen undergo **hydrolysis reactions**, requiring the addition of water molecules. The reactions are catalysed by enzymes. These are the reverse of the condensation reactions that form the glycosidic bonds.

Cellulose

Beta glucose molecules are unable to join together in the same way that alpha glucose molecules can. As you can see in Figure 5, the hydroxyl groups on carbon 1 and carbon 4 of the two glucose molecules are too far from each other to react.

The only way that beta glucoses molecules can join together and form a polymer is if alternate beta glucose molecules are turned upside down as in Figure 6.

Synoptic link

You will learn more about ATP in Topic 3.11, ATP.

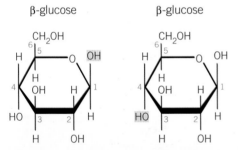

▲ Figure 5 *Note how far apart the OH groups are on these two β-glucose molecules*

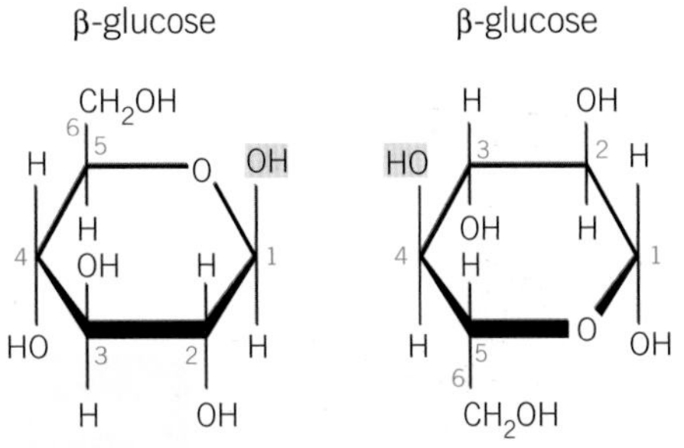

▲ Figure 6 *The OH groups of the two β-glucoses are now close enough to react and a 1,4 glycosidic bond is formed*

When a polysaccharide is formed from glucose in this way it is unable to coil or form branches. A straight chain molecule is formed called **cellulose** (Figure 7).

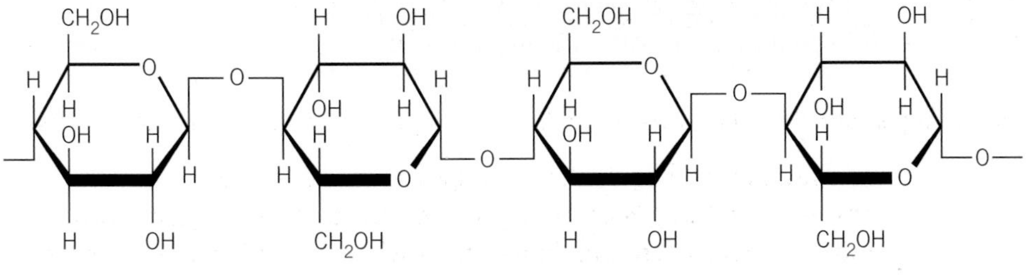

▲ Figure 7 *The cellulose molecule is straight and unbranched*

Cellulose molecules make hydrogen bonds with each other forming microfibrils. These microfibrils join together forming macrofibrils, which combine to produce fibres (Figure 8). These fibres are strong and insoluble and are used to make cell walls. Cellulose is an important part of our diet, it is very hard to break down into its monomers and forms the 'fibre' or 'roughage' necessary for a healthy digestive system.

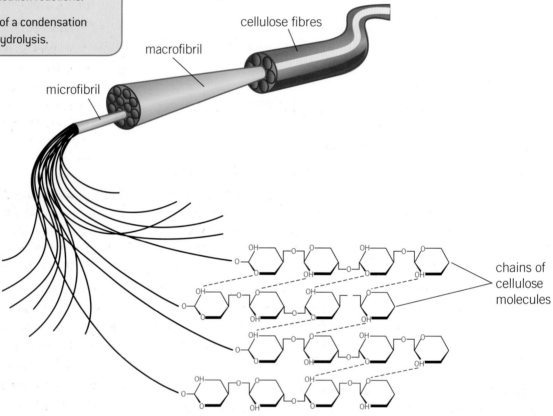

▲ **Figure 8** *Formation of cellulose fibres*

Summary questions

1 Describe the difference between alpha and beta glucose. (*2 marks*)

2 Describe the formation of a glycosidic bond. (*4 marks*)

3 Explain how the structure of cellulose is related to its function. (*4 marks*)

4 Explain why beta glucose, when polymerised, leads to the production of cellulose instead of starch. (*6 marks*)

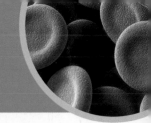

3.4 Testing for carbohydrates
Specification reference: 2.1.2

Chemical tests

Benedict's test for reducing sugars

In chemistry reduction is a reaction involving the gain of electrons. All monosaccharides and some disaccharides (for example maltose and lactose) are **reducing sugars**. This means that they can donate electrons, or reduce another molecule or chemical.

In the chemical test for a reducing sugar, this chemical is **Benedict's reagent**, an alkaline solution of copper(II)sulfate.

The test is carried out as follows:

1   Place the sample to be tested in a boiling tube. If it is not in liquid form, grind it up or blend it in water.

2   Add an equal volume of Benedict's reagent.

3   Heat the mixture gently in a boiling water bath for five minutes.

Reducing sugars will react with the copper ions in Benedict's reagent. This results in the addition of electrons to the blue Cu^{2+} ions, reducing them to brick red Cu^+ ions. When a reducing sugar is mixed with Benedict's reagent and warmed, a brick-red precipitate is formed indicating a positive result.

The more reducing sugar present, the more precipitate formed and the less blue Cu^{2+} ions are left in solution, so the actual colour seen will be a mixture of brick-red (precipitate) and blue (unchanged copper ions) and will depend on the concentration of the reducing sugar present (Figure 1). This makes the test qualitative.

Using Benedict's test for non-reducing sugars

Non-reducing sugars do not react with Benedict's solution and the solution will remain blue after warming, indicating a negative result. Sucrose is the most common non-reducing sugar.

If sucrose is first boiled with dilute hydrochloric acid it will then give a positive result when warmed with Benedict's solution. This is because the sucrose has been hydrolysed by the acid to glucose and fructose, both reducing sugars.

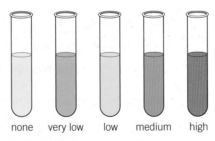

none   very low   low   medium   high

▲ **Figure 1** *Colour changes in a Benedict's test according to the concentration of reducing sugar present*

Iodine test for starch

The **iodine test** is used to detect the presence of starch. To carry out the test, a few drops of iodine dissolved in potassium iodide solution are mixed with a sample. If the solution changes colour from yellow/brown to purple/black starch is present in the sample.

▲ **Figure 2** *A positive test result for starch using iodine*

If the iodine solution remains yellow/brown it is a negative result and starch is not present.

Reagent strips

Manufactured reagent test strips can be used to test for the presence of reducing sugars, most commonly glucose. The advantage is that, with the use of a colour-coded chart, the concentration of the sugar can be determined.

Quantitative methods to determine concentration:

Colorimetry

In a Benedict's test, the colour produced is dependent on the concentration of reducing sugar present in the sample.

A colorimeter is a piece of equipment used to quantitatively measure the absorbance, or transmission, of light by a coloured solution. The more concentrated a solution is the more light it will absorb and the less light it will transmit. This can be used to calculate the concentration of reducing sugar present.

▲ Figure 3 *A colorimeter measures the colour of a liquid. It passes filtered light through the sample and the results can be transmitted to a graph plotter or computer*

A student was asked to determine the concentration of a solution of glucose.

The procedure was carried out as follows:

1   A filter was placed in the colorimeter.
2   The colorimeter was calibrated using distilled water.
3   Benedict's test was performed on a range of known concentrations of glucose.

4   The resulting solutions were filtered to remove the precipitate.
5   The % transmission of each of the solutions of glucose was measured using the colorimeter.
6   Using this information a calibration curve was plotted. Steps 3–6 were repeated using the solution with the unknown concentration of glucose.

▼ Table 1 *Shows the results of the experiment*

Absorbance / %	Concentration of glucose / mM
68	1.0
56	2.0
47	3.0
40	4.0
27	5.0
17	6.0
7	7.0
44	Unknown solution

1   Describe how you would calculate % absorbance from a % transmission reading.
2   Explain why it is important to use the correct filter (step 1).
3   Describe how you calibrate a colorimeter (step 2).
4   Describe what you have after the solutions have been filtered (step 4).
5   Plot a graph of the results from Table 1 and draw a calibration curve.
6   Estimate the concentration of glucose in the unknown solution.

Biosensors

Biosensors use biological components to determine the presence and concentration of molecules such as glucose.

The basic components of a biosensor are shown in Figure 4.

The analyte is the compound under investigation.

- Molecular recognition – a protein (enzyme or antibody) or single strand of DNA (ssDNA) is immobilised to a surface, for example a glucose test strip. This will interact with, or bind to, the specific molecule under investigation.

- Transduction – this interaction will cause a change in a transducer. A transducer detects changes, for example in pH, and produces a response such as the release of an immobilised dye on a test strip or an electric current in a glucose-testing machine.

- Display – this then produces a visible, qualitative or quantitative signal such as a particular colour on a test strip or reading on a test machine.

1 Canaries used to be used in the coal mining industry to detect the presence of harmful gases such as carbon monoxide. Miners would take the canaries, in cages, into the mines where they were working, and if the birds started to show signs of distress this signalled the presence of harmful gas. Discuss whether a canary in a cage is a biosensor and suggest a disadvantage of the use of this method to detect toxic gases.

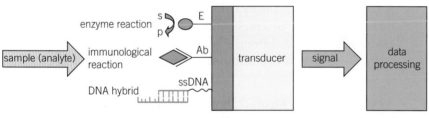

▲ Figure 4 *The main components in a biosensor*

Summary questions

1 Describe the feature of enzymes essential to their role as components in a biosensor. *(3 marks)*

2 Why does Benedict's reagent turn red when warmed with a reducing sugar? *(2 marks)*

3 Explain why an iodine test is used in experiments to show that plants require light for photosynthesis. *(3 marks)*

4 Suggest how reagent strips might be useful in the management of the medical condition diabetes, where a person's blood sugar level can become too high. *(4 marks)*

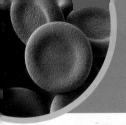

3.5 Lipids

Specification reference: 2.1.2

Learning outcomes

Demonstrate knowledge, understanding, and application of:

→ the structure of a triglyceride and a phospholipid

→ the synthesis and breakdown of triglycerides

→ properties of triglyceride, phospholipid, and cholesterol molecules

→ how to carry out and interpret the results of an emulsion test for lipids.

Lipids, commonly known as fats and oils, are molecules containing the elements carbon, hydrogen, and oxygen. Generally, fats are lipids that are solid at room temperature and oils are lipids that are liquid at room temperature.

Lipids are non-polar molecules as the electrons in the outer orbitals that form the bonds are more evenly distributed than in polar molecules. This means there are no positive or negative areas within the molecules and for this reason lipids are not soluble in water. Oil and water do not mix.

Lipids are large complex molecules known as **macromolecules**, which are not built from repeating units, or monomers, like polysaccharides. In this topic we will be looking at the lipids, triglycerides, phospholipids, and sterols.

Triglycerides

A **triglyceride** is made by combining one **glycerol** molecule with three **fatty acids**. Glycerol is a member of a group of molecules called alcohols. Fatty acids belong to a group of molecules called carboxylic acids – they consist of a carboxyl group (–COOH) with a hydrocarbon chain attached.

As you can see in Figure 1, both of these molecules contain hydroxyl (OH) groups. The hydroxyl groups interact, leading to the formation of three water molecules and bonds between the fatty acids and the glycerol molecule. These are called ester bonds and this reaction is called esterification. Esterification is another example of a condensation reaction, which you learnt about in Topic 3.3, Carbohydrates.

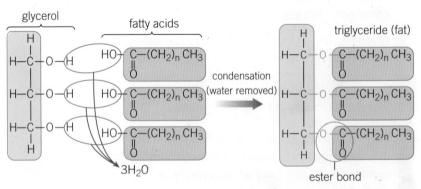

▲ Figure 1 *Synthesis of a triglyceride from glycerol and three fatty acids by the formation of three ester bonds, producing three water molecules*

When triglycerides are broken down, three water molecules need to be supplied to reverse the reaction that formed the triglyceride. This is another example of a hydrolysis reaction (Topic 3.3, Carbohydrates).

Saturated and unsaturated

Fatty acid chains that have *no* double bonds present between the carbon atoms are called saturated, because all the carbon atoms form the maximum number of bonds with hydrogen atoms (i.e., they are saturated with hydrogen atoms).

A fatty acid *with* double bonds between some of the carbon atoms is called unsaturated. If there is just one double bond it is called monounsaturated. If there are two or more double bonds it is called polyunsaturated. The presence of double bonds causes the molecule to kink or bend (Figure 2) and they therefore cannot pack so closely together. This makes them liquid at room temperature rather than solid, and they are therefore described as oils rather than fats.

Plants contain unsaturated triglycerides, which normally occur as oils, and tend to be more healthy in the human diet than saturated triglycerides, or (solid) fats. There has been some evidence that in excess, saturated fats can lead to coronary heart disease, however the evidence remains inconclusive. An excess of any type of fat can lead to obesity, which also puts a strain on the heart.

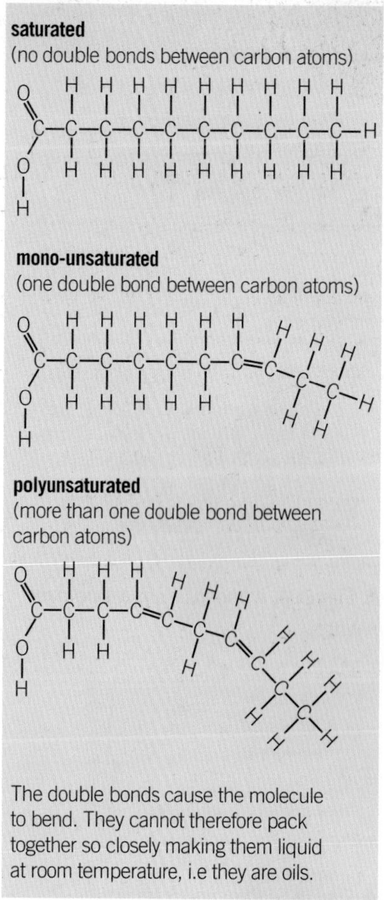

Phospholipids

Phospholipids are modified triglycerides and contain the element phosphorus along with carbon, hydrogen, and oxygen. Inorganic phosphate ions (PO_4^{3-}) are found in the cytoplasm of every cell. The phosphate ions have extra electrons and so are negatively charged, making them soluble in water.

One of the fatty acid chains in a triglyceride molecule is replaced with a phosphate group to make a phospholipid.

(a) chemical structure of a phospholipid

CH$_2$COO — fatty acid   non-polar long chain hydrocarbons

CHCOO — fatty acid   hydrophobic (repel water)

charged end of molecule hydrophilic (attracts water) ---CH$_2$

phosphate

(b) simplified way to draw a phospholipid

charged (hydrophilic) head   non-polar (hydrophobic) tails

▲ **Figure 3** *Structure of a phospholipid*

Phospholipids are unusual because, due to their length, they have a non-polar end or tail (the fatty acid chains) and a charged end or head (the phosphate group). The non-polar tails are repelled by water (but mix readily with fat). They are **hydrophobic**. The charged heads (often incorrectly called polar ends) will interact with, and are attracted to, water. They are **hydrophilic**.

saturated
(no double bonds between carbon atoms)

mono-unsaturated
(one double bond between carbon atoms)

polyunsaturated
(more than one double bond between carbon atoms)

The double bonds cause the molecule to bend. They cannot therefore pack together so closely making them liquid at room temperature, i.e they are oils.

▲ **Figure 2** *Saturated and unsaturated fatty acids*

Study tip

Polar or charged molecules and non-polar molecules should not be described as 'water loving' and 'water hating' but as hydrophilic and hydrophobic.

Synoptic link

You will learn about the role of surfactants in the lungs in Topic 6.2, The mammalian gaseous exchange system.

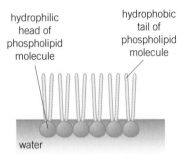

hydrophilic head of phospholipid molecule

hydrophobic tail of phospholipid molecule

water

▲ Figure 4 *A phospholipid monolayer in water*

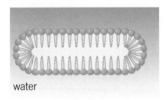

water

▲ Figure 5 *Phospholipid bilayer structure in water*

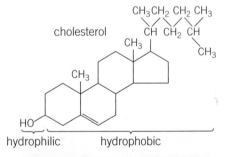

cholesterol

CH₃CH₂ CH₂ CH₃
CH CH₂ CH
CH₃
CH₃
CH₃
CH₃
HO

hydrophilic   hydrophobic

▲ Figure 6 *Cholesterol has a characteristic hydroxyl group and four carbon rings*

Synoptic link

You will learn about the roles of lipids in cell membranes in topic 5.1, The structure and function of membranes.

As a result of their dual hydrophobic/hydrophilic structure, phospholipids behave in an interesting way when they interact with water.

They will form a layer on the surface of the water with the phosphate heads in the water and the fatty acid tails sticking out of the water (Figure 4). Because of this they are called surface active agents or **surfactants** for short.

They can also form structures based on a two-layered sheet formation (a bilayer) with all of their hydrophobic tails pointing toward the centre of the sheet, protected from the water by the hydrophilic heads (Figure 5).

It is as a result of this bilayer arrangement that phospholipids play a key role in forming cell membranes. They are able to separate an aqueous environment in which cells usually exist from the aqueous cytosol within cells. It is thought that this is how the first cells were formed and, later on, membrane-bound organelles within cells.

Sterols

Sterols, also known as steroid alcohols, are another type of lipid found in cells. They are not fats or oils and have little in common with them structurally. They are complex alcohol molecules, based on a four carbon ring structure with a hydroxyl (OH) group at one end. Like phospholipids, however, they have dual hydrophilic/hydrophobic characteristics. The hydroxyl group is polar and therefore hydrophilic and the rest of the molecule is hydrophobic.

Cholesterol is a sterol. The body manufactures cholesterol primarily in the liver and intestines. It has an important role in the formation of cell membranes, becoming positioned between the phospholipids with the hydroxyl group at the periphery of the membrane. This adds stability to cell membranes and regulates their fluidity by keeping membranes fluid at low temperatures and stopping them becoming too fluid at high temperatures.

Vitamin D, steroid hormones, and bile are all manufactured using cholesterol.

Roles of lipids

Due to their non-polar nature, lipids have many biological roles. These include:

- membrane formation and the creation of hydrophobic barriers
- hormone production
- electrical insulation necessary for impulse transmission
- waterproofing, for example in birds' feathers and on plant leaves.

Lipids, triglycerides in particular, also have an important role in long-term energy storage. They are stored under the skin and around vital organs, where they also provide:

- thermal insulation to reduce heat loss for example, in penguins
- cushioning to protect vital organs such as the heart and kidneys
- buoyancy for aquatic animals like whales.

Identification of lipids

Lipids can be identified in the laboratory by a simple test known as the **emulsion test**. First, the sample is mixed with ethanol. The resulting solution is mixed with water and shaken. If a white emulsion forms as a layer on top of the solution this indicates the presence of a lipid. If the solution remains clear the text is negative.

Changing health advice

It can be confusing because health advice constantly changes. The way that new advice is issued in the media from new findings is partly responsible. The validity of the research has not usually been evaluated, the science is often not easy to explain, and as the majority of the general public do not have scientific background they are not aware of the fluid nature of scientific understanding. Scientific knowledge is also constantly changing as technology develops and so our understanding of biological processes increases.

It is often difficult to isolate the effect of just one nutrient and, in fact, it is now generally believed that nutrients do not work in isolation but as part of the combined effect of a whole range of nutrients. This is called food synergy. For example whole grains are believed to have a greater beneficial effect than any of their individual components and it is the combined effect of fish, fruit and vegetables that help prevent certain types of heart disease.

The data used in reports is often flawed, particularly where diet is concerned, as the subjects involved in studies often do not provide accurate information. People tend to underestimate what they eat, they forget what they have eaten and don't often know the exact ingredients of meals, particularly if they are eating out. People are also different due to their genetic make up and therefore respond differently to different nutrients.

The studies that catch the headlines often involve small numbers of subjects and these inherent differences distort the findings. The resulting headlines can be eye-catching, but not very accurate, and are often contradicted by the next study.

Fats in our diet

The presence of a double bond in a fatty acid leads to a kink in the chain causing the lipid to be more liquid in nature and, as you will discover in later sections, a more healthy component of the diet than saturated fats.

Plants contain unsaturated triglycerides, which normally occur as oils.

Animals (but generally not fish) contain saturated triglycerides, or (solid) fats. As mentioned, the evidence that saturated fats cause heart diseases is inconclusive.

Previously it was thought that saturated fats did cause heart disease, but more recent evidence has contradicted this.

Margarine versus butter

Butter is an animal fat made from cows' milk and is therefore high in saturated lipids. Various alternatives to butter have been developed over the last 200 years with the focus initially being to find cheaper or longer-lasting substitutes. More recently the aim has been to produce a more 'healthy' substitute for butter.

The main problem faced initially by food scientists was that the vegetable oils used to produce the 'substitute butters' are more liquid than the animal fats in milk.

This was overcome by using hydrogen to saturate, or removing the double bonds from, the unsaturated fatty acids in the vegetable oils. Solid hydrogenated fat was produced and the oil was said to be have been hardened.

The fat was then coloured and sometimes mixed with butter to improve the taste. Different degrees of hardening, colouring and mixing with butter gave rise to the many different margarines on the market.

An unwanted byproduct of the hardening process was the production of trans fats. These are unsaturated lipids in which the kinks that the double bonds naturally form in the fatty acid chains have been reversed. Trans fats, which actually increase the shelf life of baked products, have more recently been linked with the development of coronary heart disease and are now usually removed from foods.

With more focus on producing healthy alternatives to butter, and improvements in the manufacturing process, many spreads now contain less, if any, hydrogenated fats. Mono- and polyunsaturated plant oils are used instead and these have been shown to reduce high cholesterol levels, which are a factor in the development of coronary heart disease.

Reduced fat spreads
Lipids release the same quantity of energy gram for gram when respired whether saturated or unsaturated, so butter and margarine have always had the same calorific value. More recently the focus has been to reduce the overall fat content in such spreads.

1   Explain how hardening vegetable oils produces solid fats.
2   Explain why it is considered more healthy to have a low overall fat content as well as a low saturated fat content in a spread.

Summary questions

1 Using your knowledge of the structure of fatty acids, describe why oils are liquid and fats are solid at room temperature.　(4 marks)

2 Describe the formation and the hydrolysis of an ester bond.　(4 marks)

3 Some bacteria are extremophiles meaning they live in extreme environments that are very acidic or have very high temperatures. The phospholipids present in other bacteria or eukaryotic cells would be broken down in such extreme conditions.

   Extremophiles have membranes composed of modified phospholipids.

   a Identify which of the phospholipids in the diagram is present in the cell membrane of extremophiles.　(1 mark)
   b Outline the similarities and differences between the two types of phospholipid.　(3 marks)
   c Suggest why the phospholipids in the membranes of extremophiles can withstand extremes of temperature and pH.　(2 marks)

4 Read the following statements.

   Lipids are not soluble in water.
   Lipids and ethanol are soluble in water.
   Water is more soluble than lipids in ethanol.
   Use the information to explain how the emulsion test for lipids works.　(4 marks)

Peptides are polymers made up of **amino acid** molecules (the monomers). **Proteins** consist of one or more polypeptides arranged as complex macromolecules and they have specific biological functions. All proteins contain the elements carbon, hydrogen, oxygen, and nitrogen.

Amino acids

All amino acids have the same basic structure (Figure 1). Different **R-groups** (variable groups) result in different amino acids. Twenty different amino acids are commonly found in cells. Five of these are said to be non-essential as our bodies are able to make them from other amino acids. Nine are essential and can only be obtained from what we eat. A further six are said to be conditionally essential as they are only needed by infants and growing children.

Synthesis of peptides

Amino acids join when the amine and carboxylic acid groups connected to the central carbon atoms react. The R-groups are not involved at this point. The hydroxyl in the carboxylic acid group of one amino acid reacts with a hydrogen in the amine group of another amino acid. A **peptide bond** is formed between the amino acids and water is produced (this is another example of a condensation reaction, which you learnt about in Topic 3.3, Carbohydrates). The resulting compound is a dipeptide.

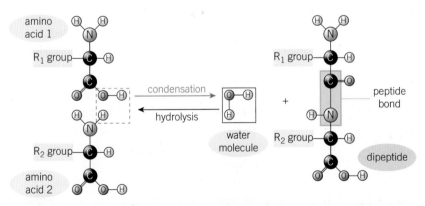

▲ Figure 2 *Condensation reaction to form a peptide bond*

When many amino acids are joined together by peptide bonds a **polypeptide** is formed. This reaction is catalysed by the enzyme peptidyl transferase present in ribosomes, the sites of protein synthesis.

The different R-groups of the amino acids making up a protein are able to interact with each other (R-group interactions) forming different types of bond. These bonds lead to the long chains of amino acids (polypeptides) folding into complex structures (proteins). The presence of different sequences of amino acids leads to different structures with

Learning outcomes

Demonstrate knowledge, understanding, and application of:

→ the general structure of an amino acid (monomer)

→ the synthesis and breakdown of dipeptides and polypeptides (polymers)

→ the levels of protein structure

→ how to carry out and interpret the results of the biuret test for proteins.

→ the principles of thin later chromatography.

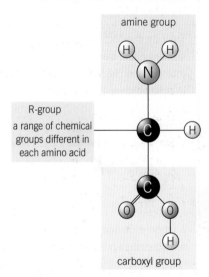

▲ Figure 1 *The general structure of an amino acid*

Study tip

Learn this structure and remember that the nitrogen-containing group is an amine group not an amino group.

Synoptic link

You will learn about protein synthesis in the cell in Topic 3.10, Protein synthesis.

different shapes being produced. The very specific shapes of proteins are vital for the many functions proteins have within living organisms.

If you look at the way protein structures are built up in stages, it is easier to understand what is happening.

Separating amino acids using thin layer chromatography

Thin layer chromatography (TLC) is a technique used to separate the individual components of a mixture. The technique can be used to separate and identify a mixture of amino acids in solution. There are two phases, the stationary phase and the mobile phase which involves an organic solvent. The mobile phase picks up the amino acids and moves through the stationary phase and the amino acids are separated.

In the stationary phase a thin layer of silica gel (or another adhesive substance) is applied to a rigid surface, for example a sheet of glass or metal. Amino acids are then added to one end of the gel. This end is then submerged in organic solvent. The organic solvent then moves through the silica gel, this is known as the mobile phase.

The rate at which the different amino acids in the organic solvent move through the silica gel depends on the interactions (hydrogen bonds) they have with the silica in the stationary phase, and their solubility in the mobile phase. This results in different amino acids moving different distances in the same time period resulting in them separating out from each other.

 Remember, when working with chemicals to take care, wear safety glasses and report any spillages/breakages to the teacher.

A student carried out the following procedure to separate and identify a mixture of amino acids in solution.

1   Wearing gloves, the student drew a pencil line on the chromatography plate about 2 cm from the bottom edge. The plate was only handled by the edges.
2   Four equally spaced points were marked at along the pencil line.
3   The amino acid solution was spotted onto the first pencil mark using a capillary tube. The spot was allowed to dry and then spotted again. The spot was labelled using a pencil.
4   The three remaining marks were spotted with solutions of three known amino acids.
5   The plate was then placed into a jar containing the solvent. The solvent was no more than 1 cm deep. The jar was then closed.

6   The plate was left in the solvent until it had reached about 2 cm from the top. The plate was then removed and a pencil line drawn along the solvent front. The plate was then allowed to dry.
7   The plate was then sprayed, in a fume cupboard, with ninhydrin spray. Amino acids react with ninhydrin and a purple/brown colour is produced. The centre of each spot present was then marked with a pencil.

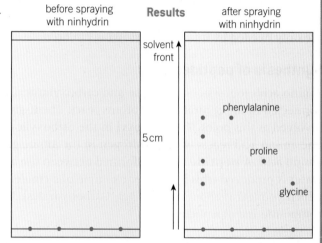

Here you can see the TLC plate showing the separated amino acids appearing purple after spraying with ninhydrin.

1   a   Suggest why gloves were worn by the student and the plate was only handled by the edges.
    b   A mixture of solvents (such as hexane, water, acetic acid, and butanol) is usually used as the mobile phase when separating an unknown mixture of amino acids. Suggest why.
    c   Explain why the solvent was no more than 1 cm deep.
    d   Suggest why the jar was sealed.
2   a   Using the information provided identify as many of the amino acids present in the solution as you can. The distance an amino acid travels is determined by the interactions it has with both the mobile phase and the stationary phase. Different amino acids will therefore move different distances in a set time. As long as the conditions are kept

the same, the same amino will always travel the same distance in the same time.

The retention value (Rf) for each amino acid is the distance travelled by the pigment divided by the distance travelled by the solvent, it can be calculated using the formula:

$$R_f = \frac{\text{distance travelled by component}}{\text{distance travelled by solvent}}$$

This will be constant for each amino acid tested under identical conditions.

b Calculate the Rf value for the two unidentified spots and, using the standard Rf values below, identify the amino acid.

Alanine 0.31   Cysteine 0.40
Aspartic acid 0.24   Methionine 0.49
Phenylalanine 0.59   Glutamine 0.13

Levels of protein structure

Primary structure – this is the *sequence* in which the amino acids are joined. It is directed by information carried within DNA (discussed further in Topic 3.8, Nucleic acids and Topic 3.9, DNA replication and the genetic code). The particular amino acids in the sequence will influence how the polypeptide folds to give the protein's final shape. This in turn determines its function. The only bonds involved in the primary structure of a protein are peptide bonds.

Secondary structure – the oxygen, hydrogen, and nitrogen atoms of the basic, repeating structure of the amino acids (the variable groups are not involved at this stage) interact. Hydrogen bonds may form within the amino acid chain, pulling it into a coil shape called an alpha helix (Figure 3a).

Polypeptide chains can also lie parallel to one another joined by hydrogen bonds, forming sheet-like structures. The pattern formed by the individual amino acids causes the structure to appear pleated, hence the name beta pleated sheet (Figure 3b).

Secondary structure is the result of hydrogen bonds and forms at regions along long protein molecules depending on the amino sequences.

Tertiary structure – this is the folding of a protein into its final shape. It often includes sections of secondary structure. The coiling or folding of sections of proteins into their secondary structures brings R-groups of different amino acids closer together so they are close enough to interact and further folding of these sections will occur. The following interactions occur between the R-groups:

- hydrophobic and hydrophilic interactions – weak interactions between polar and non-polar R-groups
- hydrogen bonds – these are weakest of the bonds formed
- ionic bonds – these are stronger than hydrogen bonds and form between oppositely charged R-groups
- disulfide bonds (also known as disulfide bridges) – these are covalent and the strongest of the bonds but only form between R-groups that contain sulfur atoms.

(a) alpha helix

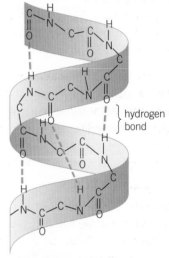

hydrogen bond

(b) beta pleated sheet

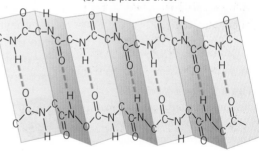

▲ Figure 3 *Depending on the amino acid composition, polypeptides initially form either (a) alpha helices or beta (b) pleated sheets– types of secondary structure*

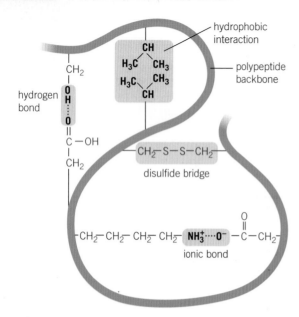

▲ **Figure 4** *The tertiary structures of proteins are very complex shapes involving multiple types of bonds and interactions between R-groups*

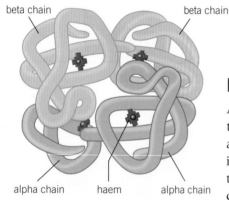

▲ **Figure 5** *Haemoglobin has a quaternary structure made up of four individual proteins (two alpha and two beta) as well as haem groups containing iron*

Synoptic link

You will learn more about enzymes in Chapter 4, Enzymes.

Synoptic link

You will learn more about haemoglobin in Topic 8.4, Transport of oxygen and carbon dioxide in the blood.

This produces a variety of complex-shaped proteins, with specialised characteristics and functions (Figure 4).

Quaternary structure – this results from the association of two or more individual proteins called subunits. The interactions between the subunits are the same as in the tertiary structure except that they are between different protein molecules rather than within one molecule.

The protein subunits can be identical or different. Enzymes often consist of two identical subunits whereas insulin (a hormone) has two different subunits. Haemoglobin, a protein required for oxygen transport in the blood, has four subunits, made up of two sets of two identical subunits (Figure 5).

Hydrophilic and hydrophobic interactions

Proteins are assembled in the aqueous environment of the cytoplasm. So the way in which a protein folds will also depend on whether the R-groups are hydrophilic or hydrophobic. Hydrophilic groups are on the outside of the protein while hydrophobic groups are on the inside of the molecule shielded from the water in the cytoplasm.

Breakdown of peptides

As you have learned, peptides are created by amino acids linking together in condensation reactions to form peptide bonds. Proteases are enzymes that catalyse the reverse reaction – turning peptides back into their constituent amino acids. A water molecule is used to break the peptide bond in a hydrolysis reaction, reforming the amine and carboxylic acid groups.

 Identification of proteins

Biuret test

Peptide bonds form violet coloured complexes with copper ions in alkaline solutions. This can be used as the basis of a test for proteins.

Remember, when working with chemicals:

⚠ Take care, wear safety glasses and report any spillages/breakages to the teacher.

A student carried out the following procedure to test a sample for the presence of protein.

1 $3\,cm^3$ of a liquid sample was mixed with an equal volume of 10% sodium hydroxide solution.
2 1% copper sulfate solution was then added a few drops at a time until the sample solution turned blue.
3 The solution was mixed and left to stand for five minutes.

This test is known as the biuret test. A mixture of an alkali and copper sulfate solution is called biuret reagent and can be used instead of adding the solutions individually.

1 State the colour you would expect to see on addition of the copper sulfate solution if protein is present in the sample.
2 State the colour you would expect to see if the sample contained amino acids instead of proteins.
  Explain the reason for this colour.
3 Suggest why this test is not used quantitatively.

Summary questions

1 Draw the structure of an amino acid. *(3 marks)*

2 Describe the formation of a peptide bond. *(3 marks)*

3 a Draw a box identifying the peptide bond in the diagram below.
   (1 mark)

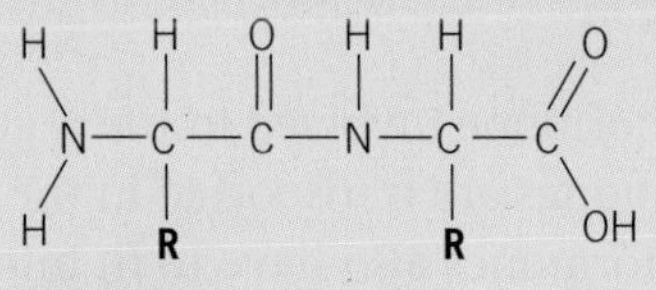

  b Describe how hydrogen bonds form within the secondary structure of proteins. *(2 marks)*
  c Alpha keratin, a protein found in sheep's wool, is primarily composed of alpha helices. Explain why alpha keratin has a more regular structure than the quaternary protein haemoglobin. *(3 marks)*

4 Compare and contrast the role of R-group interactions in the formation of the tertiary and quaternary structures of proteins. *(6 marks)*

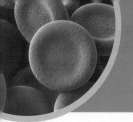

3.7 Types of proteins

Specification reference: 2.1.2

Learning outcomes

Demonstrate knowledge, understanding, and application of:

→ the structure and function of globular proteins including a conjugated protein

→ the properties and functions of fibrous proteins.

In the previous topic you saw how the complex tertiary and quaternary structures of proteins are built up. These structures determine the role the protein will play in the body. The two main groups are globular proteins and fibrous proteins.

Globular proteins

Globular proteins are compact, water soluble, and usually roughly spherical in shape. They form when proteins fold into their tertiary structures in such a way that the hydrophobic R-groups on the amino acids are kept away from the aqueous environment. The hydrophilic R-groups are on the outside of the protein. This means the proteins are soluble in water.

This solubility is important for the many different functions of globular proteins. They are essential for regulating many of the processes necessary to life. As you will see in later sections, these include processes such as chemical reactions, immunity, muscle contraction, and many more.

Insulin

Insulin is a globular protein. It is a hormone involved in the regulation of blood glucose concentration. Hormones are transported in the bloodstream so need to be soluble. Hormones also have to fit into specific receptors on cell-surface membranes to have their effect and therefore need to have precise shapes.

▲ Figure 1 *Insulin. The complex shape of a globular protein formed from the folding of the primary structure into secondary structures (helices), which are further folded into the tertiary structure*

Conjugated proteins

Conjugated proteins are globular proteins that contain a non-protein component called a **prosthetic group**. Proteins without prosthetic groups are called simple proteins.

There are different types of prosthetic groups. Lipids or carbohydrates can combine with proteins forming lipoproteins or glycoproteins. Metal ions and molecules derived from vitamins also form prosthetic groups.

Haem groups are examples of prosthetic groups. They contain an iron II ion (Fe^{2+}). Catalase and haemoglobin both contain haem groups.

Haemoglobin

Haemoglobin is the red, oxygen-carrying pigment found in red blood cells. It is a quaternary protein made from four polypeptides, two alpha and two beta subunits (Figure 5, Topic 3.6, Structure of proteins). Each subunit contains a prosthetic haem group. The iron II ions present in the haem groups are each able to combine reversibly with an oxygen molecule. This is what enables haemoglobin to transport oxygen around the body. It can pick oxygen up in the lungs and transport it to the cells that need it, where it is released.

▲ Figure 2 *Haemoglobin. Four subunits are each wrapped around a haem group (red), protecting it from being oxidised and destroyed by the oxygen it is intended to transport. The iron within each haem group reversibly bonds to oxygen in the blood. Four haemoglobin monomers (green, beige, purple, blue) usually bind together to form one large haemoglobin molecule*

Catalase

Catalase is an enzyme. Enzymes catalyse reactions, meaning they increase reaction rates, and each enzyme is *specific* to a particular reaction or type of reaction.

Catalase is a quaternary protein containing four haem prosthetic groups. The presence of the iron II ions in the prosthetic groups allow catalase to interact with hydrogen peroxide and speed up its breakdown. Hydrogen peroxide is a common byproduct of metabolism but is damaging to cells and cell components if allowed to accumulate. Catalase makes sure this doesn't happen.

Fibrous proteins

Fibrous proteins are formed from long, insoluble molecules. This is due to the presence of a high proportion of amino acids with hydrophobic R-groups in their primary structures. They contain a limited range of amino acids, usually with small R-groups. The amino acid sequence in the primary structure is usually quite repetitive. This leads to very organised structures reflected in the roles fibrous proteins often have. Keratin, elastin, and collagen are examples of fibrous proteins.

Fibrous proteins tend to make strong, long molecules which are *not* folded into complex three-dimensional shapes like globular proteins.

Keratin

Keratin is a group of fibrous proteins present in hair, skin, and nails. It has a large proportion of the sulfur-containing amino acid, cysteine. This results in many strong disulfide bonds (disulfide bridges) forming strong, inflexible, and insoluble materials. The degree of disulfide bonds determines the flexibility – hair contains fewer bonds making it more flexible than nails, which contain more bonds. The unpleasant smell produced when hair or skin is burnt is due to the presence of relatively large quantities of sulfur in these proteins.

Elastin

Elastin is a fibrous protein found in elastic fibres (along with small protein fibres). Elastic fibres are present in the walls of blood vessels and in the alveoli of the lungs – they give these structures the flexibility to expand when needed, but also to return to their normal size. Elastin is a quaternary protein made from many stretchy molecules called tropoelastin (see the Extension, The structure of fibrous proteins, for further detail of structure).

Collagen

Collagen is another fibrous protein. It is a connective tissue found in skin, tendons, ligaments and the nervous system. There are a number of different forms but all are made up of three polypeptides wound together in a long and strong rope-like structure. Like rope, collagen has flexibility (see the Extension, The structure of fibrous proteins, for further detail of structure).

Synoptic link

You will learn more about haemoglobin and its role in the transport of oxygen in Topic 8.4, Transport of oxygen and carbon dioxide in the blood.

Synoptic link

You will learn about enzymes in Chapter 4, Enzymes.

Study tip

You are not required to learn the detailed structure of fibrous proteins, but an overview is useful in understanding their properties and functions. Details of the structure of fibrous proteins are given in the Extension.

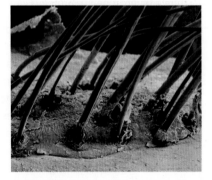

▲ **Figure 3** *A coloured SEM image of eyelash hairs growing from the surface of human skin. The shafts of hair are made up of the fibrous protein, keratin. × 50 magnification*

Synoptic link

You will learn more about the role of elastin in blood vessels in Topic 8.2, Blood vessels.

The structure of fibrous proteins

Elastin

Elastin is made by linking many soluble tropoelastin protein molecules to make a very large, insoluble, and stable, cross-linked structure (Figure 4).

Tropoelastin molecules are able to stretch and recoil without breaking, acting like small springs. They contain alternate hydrophobic and lysine-rich areas.

Elastin is formed when multiple tropoelastin molecules aggregate via interactions between the hydrophobic areas. The structure is stabilised by cross-linking covalent bonds involving the amino acid lysine, but the polypeptide structure still has flexibility.

Elastin confers strength and elasticity to the skin and other tissues and organs in the body.

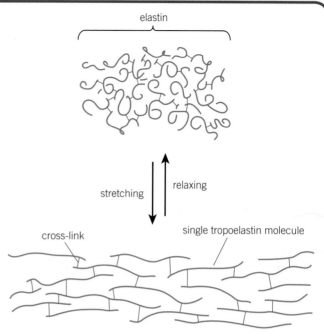

▲ **Figure 4** *The structure of elastin allows it to be stretched without breaking*

Collagen

Collagen molecules have three polypeptide chains wound around each other in a triple helix structure to form a tough, rope-like protein (Figure 5d and e).

Every third amino acid in the polypeptide chains is glycine, which is a small amino acid. Its small size allows the three protein molecules to form a closely packed triple helix. Many hydrogen bonds form between the polypeptide chains forming long quaternary proteins with staggered ends (Figure 5d). These allow the proteins to join end to end, forming long fibrils called tropocollagen (Figure 5c). The tropocollagen fibrils cross-link to produce strong fibres.

Collagen also contains high proportions of the amino acids proline and hydroxyproline. The R-groups in these amino acids repel each other and this adds to the stability of collagen.

In some tissues, multiple fibres of collagen aggregate into larger bundles (Figure 5a). This is the structure found in ligaments and tendons. In skin, collagen fibres form a mesh that is resistant to tearing.

▶ **Figure 5** *(a) A ligament is composed of a bundle of (b) collagen microfibrils. Each collagen microfibril is composed of (c) multiple triple-helix proteins wound together with staggered ends so that the individual polypeptides overlap, forming a very strong structure*

(a)

ligament tissue

(b)

multiple triple-helix proteins wound to form a microfibril

(c)

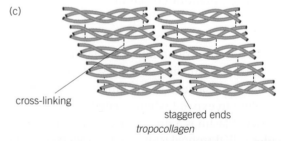

cross-linking

staggered ends
tropocollagen

(d)

three polypeptide chains form a triple helix – collagen

(e)

single polypeptide chain

1   Suggest what property the arrangement of collagen fibres into large bundles gives to tendons.

2   As we age the collagen in our skin starts to break down. This leads to the loss of skin structure and the formation of wrinkles. Many beauty products are available that contain collagen in the form of creams and capsules. Using your knowledge of the structure of collagen, suggest why these products are unlikely to have any beneficial effect in reducing or preventing wrinkles.

3   Which of the following sequences is correct in terms of increasing bond strength?
    a   ionic bonds, disulfide bonds, hydrogen bonds
    b   hydrogen bonds, ionic bonds, disulfide bonds
    c   disulfide bonds, ionic bonds, hydrogen bonds

Summary questions

1   Explain the difference between a simple protein and a conjugated protein.   (3 marks)

2   Describe the differences in properties and functions of insulin, a hormone, and keratin present in hair and nails.   (4 marks)

3   Describe why globular proteins are soluble in water but fibrous proteins are not.   (3 marks)

4   Myoglobin is an oxygen carrying molecule found primarily in muscle tissue. It is formed from a single polypeptide chain which is folded to form eight alpha helices. This chain is further folded around a central prosthetic group which binds reversibly with oxygen. The hydrophobic R groups of the amino acids are positioned towards the centre of the molecule.

    Discuss the similarities and differences in the structures of haemoglobin and myoglobin.   (6 marks)

Study tip

Make sure you are able to compare globular and fibrous proteins using named examples.

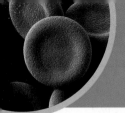

3.8 Nucleic acids

Specification reference: 2.1.2, 2.1.3

Learning outcomes

Demonstrate knowledge, understanding, and application of:

→ the structure of a nucleotide

→ the synthesis and breakdown of polynucleotides

→ the structure of DNA (deoxyribonucleic acid)

→ practical investigations into the purification of DNA by precipitation.

Nucleic acids are large molecules that were discovered in cell nuclei – hence their name. There are two types of nucleic acid – DNA and RNA, and both have roles in the storage and transfer of genetic information and the synthesis of polypeptides (proteins). They are the basis for heredity.

Nucleotides and nucleic acids

Nucleic acids contain the elements carbon, hydrogen, oxygen, nitrogen and phosphorus. They are large polymers formed from many **nucleotides** (the monomers) linked together in a chain.

An individual nucleotide is made up of three components, as shown in Figure 1:

- a pentose monosaccharide (sugar), containing five carbon atoms
- a phosphate group, $-PO_4^{2-}$, an inorganic molecule that is acidic and negatively charged
- a nitrogenous base – a complex organic molecule containing one or two carbon rings in its structure as well as nitrogen.

Nucleotides are linked together by condensation reactions to form a polymer called a polynucleotide. The phosphate group at the fifth carbon of the pentose sugar (5') of one nucleotide forms a covalent bond with the hydroxyl (OH) group at the third carbon (3') of the pentose sugar of an adjacent nucleotide. These bonds are called **phosphodiester bonds**. This forms a long, strong sugar-phosphate 'backbone' with a base attached to each sugar (Figure 2). The phosphodiester bonds are broken by **hydrolysis**, the reverse of condensation, releasing the individual nucleotides.

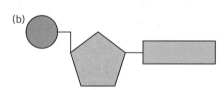

▲ Figure 1 (a) A nucleotide and its three component parts: a sugar, a phosphate and a base (the numbers in blue denote the standard numbering of the five carbons in the sugar). (b) A simple representation of a nucleotide

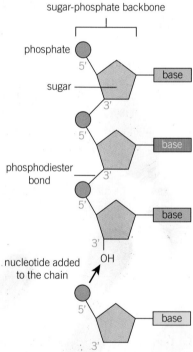

▲ Figure 2 Nucleotides (monomers) are linked together to form the polymer, nucleic acid (polynucleotide). Note the sugar-phosphate backbone with all the bases projecting from the opposite side

Study tip

You do **not** need to remember the *detailed* chemical structure of nucleotides or nucleic acids, but you should learn the basic structures of phosphate, sugars and bases.

Deoxyribonucleic acid (DNA)

As the name suggests, the sugar in **deoxyribonucleic acid (DNA)** is deoxyribose – a sugar with one fewer oxygen atoms than ribose, as shown in Figure 3.

The nucleotides in DNA each have one of four different bases. This means there are four different DNA nucleotides, see Figure 4. The four bases can be divided into two groups:

- **Pyrimidines** – the smaller bases, which contain single carbon ring structures – thymine (T) and cytosine (C)

- **Purines** – the larger bases, which contain double carbon ring structures – adenine (A) and guanine (G).

deoxyribose

ribose

◀**Figure 3** *The sugar in DNA is deoxyribose (top) and the sugar in RNA is ribose (bottom). Deoxyribose has one less oxygen atom than ribose*

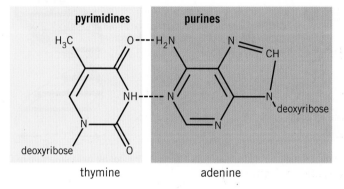

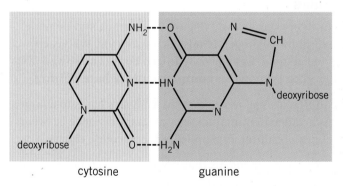

pyrimidines   purines

thymine   adenine

cytosine   guanine

▲ **Figure 4** *The chemical structure of the four different bases in DNA showing hydrogen bonding between complementary pairs*

The double helix

The DNA molecule varies in length from a few nucleotides to millions of nucleotides. It is made up of two strands of polynucleotides coiled into a helix, known as the DNA double helix, see Figure 6.

The two strands of the double helix are held together by hydrogen bonds between the bases, much like the rungs of a ladder. Each strand has a phosphate group (5′) at one end and a hydroxyl group (3′) at the other end. The two parallel strands are arranged so that they run in opposite directions (Figure 5) – they are said to be **antiparallel**.

The pairing between the bases allows DNA to be copied and transcribed – key properties required of the molecule of heredity.

Study tip

You do *not* need to remember the chemical structure of the bases, but you should remember that:

The comparative sizes of pyrimidines and purines is due to the presence of either a single ring or a double ring. The complementary pair thymine and adenine form *two* hydrogen bonds and the complementary pair cytosine and guanine form *three* hydrogen bonds *and* purines pair with pyrimidines.

Study tip

When referring to DNA, the molecule is the double helix composed of two antiparallel strands. Each strand is one chain of nucleotides (a polynucleotide).

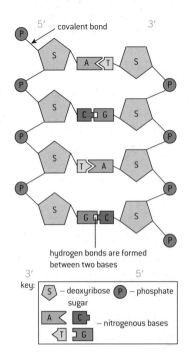

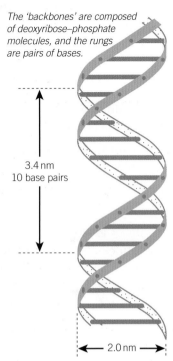

The 'backbones' are composed of deoxyribose–phosphate molecules, and the rungs are pairs of bases.

3.4 nm
10 base pairs

2.0 nm

▲ Figure 5 A simplified diagram of base pairing in DNA

▲ Figure 6 The double helix structure of DNA

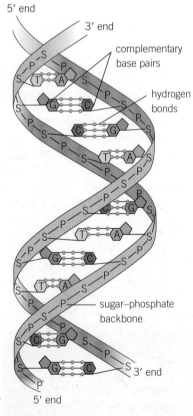

5′ end
3′ end
complementary base pairs
hydrogen bonds
sugar–phosphate backbone
3′ end
5′ end

▲ Figure 7 The DNA double helix – showing complementary base pairing and antiparallel strands

Base pairing rules

The bases bind in a very specific way (Figure 7). Adenine and thymine are both able to form *two* hydrogen bonds and always join with each other. Cytosine and guanine form *three* hydrogen bonds and so also only bind to each other. This is known as **complementary base pairing**.

These rules mean that a small pyrimidine base always binds to a larger purine base. This arrangement maintains a constant distance between the DNA 'backbones', resulting in parallel polynucleotide chains.

Complementary base pairing means that DNA always has equal amounts of adenine and thymine *and* equal amounts of cytosine and guanine. This was known long before the detailed structure of DNA was determined by Watson and Crick in 1953.

It is the *sequence* of bases along a DNA strand that carries the genetic information of an organism in the form of a code. In the next topic we will examine how the sequence of bases 'codes' for the sequences of amino acids that are needed to make different proteins.

Ribonucleic acid (RNA)

Ribonucleic acid (RNA) plays an essential role in the transfer of genetic information from DNA to the proteins that make up the enzymes and tissues of the body. DNA stores all of the genetic information needed by an organism, which is passed on from generation to generation. However, the DNA of each eukaryotic chromosome is a very long molecule, comprising many hundreds of genes, and is unable to leave the nucleus in order to supply the information directly to the sites of protein synthesis.

To get around this problem, the relatively short section of the long DNA molecule corresponding to a single gene is transcribed into a similarly short messenger RNA (mRNA) molecule. Each individual mRNA is therefore much shorter than the whole chromosome of DNA. It is a polymer composed of many nucleotide monomers.

RNA nucleotides are different to DNA nucleotides as the pentose sugar is ribose rather than deoxyribose (Figure 3) and the thymine base is replaced with the base uracil (U) (see Figure 9). Like thymine, uracil is a pyrimidine that forms two hydrogen bonds with adenine. Therefore the base pairing rules still apply when RNA nucleotides bind to DNA to make copies of particular sections of DNA.

The RNA nucleotides form polymers in the same way as DNA nucleotides – by the formation of phosphodiester bonds in condensation reactions. The RNA polymers formed are small enough to leave the nucleus and travel to the ribosomes, where they are central in the process of protein synthesis.

After protein synthesis the RNA molecules are degraded in the cytoplasm. The phosphodiester bonds are hydrolysed and the RNA nucleotides are released and reused.

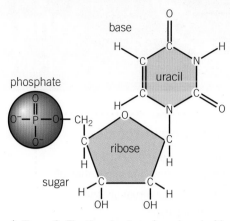

▲ Figure 8 *The thymine base is replaced with uracil in RNA nucleotides. The sugar is ribose not deoxyribose*

DNA extraction

DNA can be extracted from plant material using the following procedure:

- Grind sample in a mortar and pestle – this breaks down the cell walls.
- Mix sample with detergent – this breaks down the cell membrane, releasing the cell contents into solution.
- Add salt – this breaks the hydrogen bonds between the DNA and water molecules.
- Add protease enzyme – this will break down the proteins associated with the DNA in the nuclei.
- Add a layer of alcohol (ethanol) on top of the sample – alcohol causes the DNA to precipitate out of solution.
- The DNA will be seen as white strands forming between the layer of sample and layer of alcohol (Figure 9). The DNA can be picked up by 'spooling' it onto a glass rod.

▲ Figure 9 *DNA as a white precipitate*

1   The temperature should be kept low throughout this DNA extraction procedure. Suggest why.
2   Explain why detergent breaks down cell membranes.

Synoptic link

You will learn more about transcription and translation in protein synthesis in Topic 3.9, DNA replication and the genetic code and Topic 3.10, Protein synthesis.

Summary questions

1   Describe the differences between DNA and RNA nucleotides.   (*2 marks*)

2   Explain the base pairing rule.   (*3 marks*)

3   Explain how the structure of DNA is ideally suited to its role.   (*4 marks*)

4   A sample of DNA was tested and 17% of the total bases present were found to be adenine.

    Calculate the percentages of each of the other three bases present in this sample. Show all of your working.   (*3 marks*)

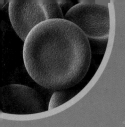

3.9 DNA replication and the genetic code

Specification reference: 2.1.3

Learning outcomes

Demonstrate knowledge, understanding, and application of:

→ semi-conservative DNA replication

→ the nature of the genetic code.

Synoptic link

You will learn more about cell division in Topic 6.2, Mitosis.

Synoptic link

You will learn more about enzymes in Chapter 4, Enzymes.

Cells divide to produce more cells needed for growth or repair of tissues. The two daughter cells produced as a result of cell division are genetically identical to the parent cell and to each other. In other words, they contain DNA with a base sequence identical to the original parent cell.

When a cell prepares to divide, the two strands of DNA double helix separate and each strand serves as a template for the creation of a new double-stranded DNA molecule. The complementary base pairing rules, which you learnt about in the previous topic, ensure that the two new strands are identical to the original. This process is called **DNA replication**.

Semi-conservative replication

For DNA to replicate, the double helix structure has to unwind and then separate into two strands, so the hydrogen bonds holding the complementary bases together must be broken (Figure 1c). Free DNA nucleotides will then pair with their complementary bases, which have been exposed as the strands separate. Hydrogen bonds are formed between them. Finally, the new nucleotides join to their adjacent nucleotides with phosphodiester bonds (Figure 1d).

In this way, two new molecules of DNA are produced. Each one consists of one old strand of DNA and one new strand. This is known as **semi-conservative** (meaning half the same) replication.

Roles of enzymes in replication

DNA replication is controlled by enzymes, a class of proteins that act as catalysts for biochemical reactions. Enzymes are only able to carry out their function by recognising and attaching to specific molecules or particular parts of the molecules.

Before replication can occur, the unwinding and separating of the two strands of the DNA double helix is carried out by the enzyme **DNA helicase**. It travels along the DNA backbone, catalysing reactions that break the hydrogen bonds between complementary base pairs as it reaches them. This can be thought of as the strand 'unzipping'.

Free nucleotides pair with the newly exposed bases on the template strands during the 'unzipping' process. A second enzyme, **DNA polymerase** catalyses the formation of phosphodiester bonds between these nucleotides.

a A representative portion of DNA, which is about to undergo replication.

b An enzyme, DNA helicase, causes the two strands of the DNA to separate.

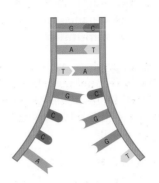

c DNA helicase completes the separation of the strand. Meanwhile, free nucleotides that have been activated are attracted to their complementary bases.

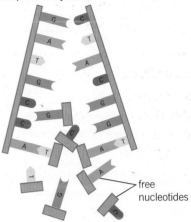

free nucleotides

d Once the activated nucleotides are lined up, they are joined together by DNA polymerase (bottom three nucleotides). The remaining unpaired bases continue to attract their complementary nucleotides.

e Finally, all the nucleotides are joined to form a complete polynucleotide chain using DNA polymerase. In this way, two identical molecules of DNA are formed. Each new molecule of DNA is composed of one original strand and one newly formed molecule – semi-conservative replication.

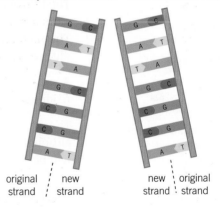

original strand | new strand   new strand | original strand

▲ **Figure 1** *The semi-conservative replication of DNA*

✚ Continuous and discontinuous replication

DNA polymerase always moves along the template strand in the same direction. It can only bind to the 3' (OH) end, so travels in the direction of 3' to 5'. As DNA only unwinds and unzips in one direction, DNA polymerase has to replicate each of the template strands in opposite directions. The strand that is unzipped from the 3' end can be continuously replicated as the strands unzip. This strand is called the *leading strand* and is said to undergo *continuous replication*.

The other strand is unzipped from the 5' end, so DNA polymerase has to wait until a section of the strand has unzipped and then work back along the strand.

This results in DNA being produced in sections (called *Okazaki fragments*), which then have to be joined. This strand is called the *lagging strand* and is said to undergo *discontinuous replication*.

1. Describe the difference between continuous and discontinuous replication. *(4 marks)*
2. Using your knowledge of enzymes, explain why DNA polymerase does not catalyse the joining of Okazaki fragments into a single strand but a different enzyme (DNA ligase) is used. *(3 marks)*

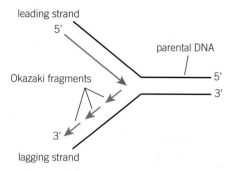

▲ **Figure 2** *As DNA polymerase only travels in the 3' to 5' direction, only the leading strand of DNA can be replicated continuously as the DNA unwinds but the lagging strand has to be replicated in the opposite direction in short sections called Okazaki fragments*

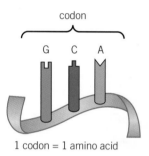

▲ **Figure 3** *A codon is a triplet of bases that code for an amino acid*

Replication errors

Sequences of bases are not always matched exactly, and an incorrect sequence may occur in the newly-copied strand. These errors occur randomly and spontaneously and lead to a change in the sequence of bases, known as a **mutation**.

Genetic code

DNA is contained within the cells of *all* organisms and scientists determined that this molecule was the means by which genetic information was passed from one generation to the next. But how does this happen? Scientists understood that DNA must carry the 'instructions' or 'blueprint' needed to synthesise the many different proteins needed by these organisms. Proteins are the foundation for the different physical and biochemical characteristics of living things. They are made up of a sequence of amino acids, folded into complex structures. Therefore DNA must *code* for a sequence of amino acids. This is called the **genetic code**.

A triplet code

The instructions that DNA carries are contained in the sequence of bases along the chain of nucleotides that make up the two strands of DNA. The code in the base sequences is a simple **triplet code**. It is a sequence of three bases, called a **codon**. Each codon codes for an amino acid.

A section of DNA that contains the complete sequence of bases (codons) to code for an entire protein is called a **gene**.

The genetic code is universal – all organisms use this same code, although the sequences of bases coding for each individual protein will be different.

Degenerate code

As you have learnt, there are four different bases, which means there are 64 different base triplets or codons possible (4^3 or $4 \times 4 \times 4$). This includes one codon that acts as the start codon when it comes at the beginning of a gene, signalling the start of a sequence that codes for a protein. If it is in the middle of a gene, it codes for the amino acid methionine. There are also three 'stop' codons that do not code for any amino acids and signal the end of the sequence.

Having a single codon to signal the start of a sequence ensures that the triplets of bases (codons) are read 'in frame'. In other words the DNA base sequence is 'read' from base 1, rather than base 2 or 3. So the genetic code is non-overlapping.

As there are only 20 different amino acids that regularly occur in biological proteins, there are a lot more codons than amino acids. Therefore, many amino acids can be coded for by more than one codon, see Figure 4. Due to this, the code is known as degenerate.

Study tip

Remember, the genetic code is a triplet code. Three bases = one codon. One codon codes for one amino acid.

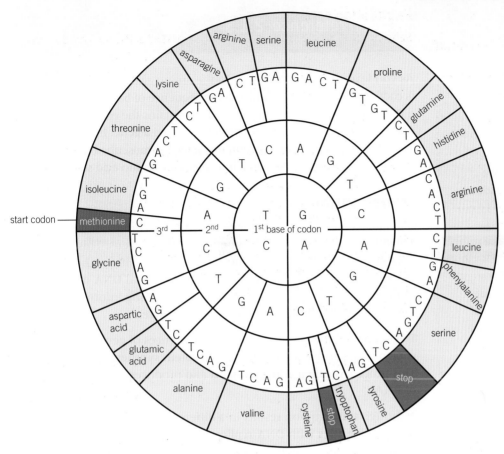

▲ **Figure 4** *The DNA code wheel shows how the genetic code is degenerate – different combinations of bases can code for the same amino acid*

Summary questions

1 Explain why DNA replication is described as semi-conservative.
(*3 marks*)

2 Explain what is meant by the triplet code. (*2 marks*)

3 Enzymes are cellular proteins that catalyse reactions – they have active sites in which specific substrates fit precisely. Suggest how a genetic mutation may result in an enzyme becoming non-functional. (*4 marks*)

4 The sequences of bases in specific sections of DNA and the sequences of amino acids in specific proteins are both used to compare how closely related different species are. The fewer differences in the sequences the more closely related the species. Explain why there are likely to be more differences, overall, between base sequences of DNA than between amino acid sequences of proteins. (*3 marks*)

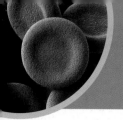

3.10 Protein synthesis

Specification reference: 2.1.3

Synoptic link

Refer back to Topic 2.4, Eukaryotic cell structure, to refresh your understanding of how some of the component structures of a eukaryotic cell are involved in protein synthesis.

As you learnt in the previous topic, the amino acid sequence of a protein is coded for by triplets (codons) in the sequence of bases along a strand of a DNA molecule. A section of DNA that contains the complete sequence of codons to code for an entire protein is called a gene. In this topic you will explore how the information in a gene is 'transcribed' into RNA molecules and then 'translated' into a specific amino acid sequence.

Transcription

In a eukaryotic cell, DNA is contained within a double membrane called the nuclear envelope that encloses the nucleus. This protects the DNA from being damaged in the cytoplasm. Protein synthesis occurs in the cytoplasm at ribosomes, but a chromosomal DNA molecule is too large to leave the nucleus to supply the coding information needed to determine the protein's amino acid sequence.

To get around this problem, the base sequences of genes have to be copied and transported to the site of protein synthesis, a ribosome. This process is called **transcription** and produces shorter molecules of RNA.

Although transcription results in a different polynucleotide, it has many similarities with DNA replication. The section of DNA that contains the gene unwinds and unzips under the control of a DNA helicase, beginning at a start codon. This involves the breaking of hydrogen bonds between the bases.

Only one of the two strands of DNA contains the code for the protein to be synthesised. This is the **sense strand** and it runs from 5' to 3'. The other strand (3' to 5') is a complementary copy of the sense strand and does not code for a protein. This is the **antisense strand** and it acts as the **template strand** during transcription, so that the complementary RNA strand formed carries the same base sequence as the sense strand.

Free RNA nucleotides will base pair with complementary bases exposed on the antisense strand when the DNA unzips. As you learnt in Topic 3.8 Nucleic acids, the thymine base in RNA nucleotides is replaced with the base uracil (U). So RNA uracil binds to adenine on the DNA template strand.

Phosphodiester bonds are formed between the RNA nucleotides by the enzyme **RNA polymerase**. Transcription stops at the end of the gene and the completed short strand of RNA is called **messenger (m)RNA**. It has the same base sequence as the sequence of bases making up the gene on the DNA, except that it has uracil in place of thymine.

The mRNA then detaches from the DNA template and leaves the nucleus through a nuclear pore. The DNA double helix reforms.

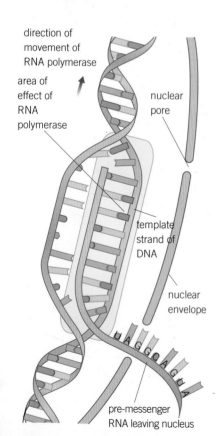

direction of movement of RNA polymerase

area of effect of RNA polymerase

nuclear pore

template strand of DNA

nuclear envelope

pre-messenger RNA leaving nucleus

▲ Figure 1 *Summary of transcription*

This mRNA molecule then travels to a ribosome in the cell cytoplasm for the next step in protein synthesis.

Translation

In eukaryotic cells, ribosomes are made up of two subunits, one large and one small. These subunits are composed of almost equal amounts of protein and a form of RNA known as **ribosomal (r)RNA**. rRNA is important in maintaining the structural stability of the protein synthesis sequence and plays a biochemical role in catalysing the reaction.

After leaving the nucleus, the mRNA binds to a specific site on the small subunit of a ribosome. The ribosome holds mRNA in position while it is decoded, or translated, into a sequence of amino acids. This process is called **translation**.

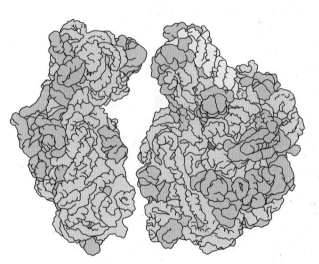

▲ Figure 2 *Large and small subunits of the ribosome with proteins shown in purple, ribosomal RNA in pink and yellow and the site that catalyses the formation of peptide bonds in green*

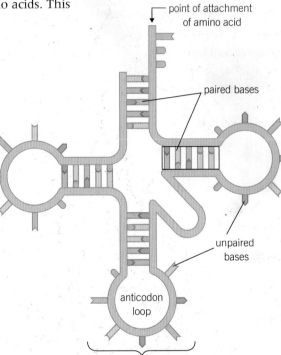

▲ Figure 3 *The structure of tRNA*

Transfer (t)RNA is another form of RNA, which is necessary for the translation of the mRNA. It is composed of a strand of RNA folded in such a way that three bases, called the anticodon, are at one end of the molecule (Figure 3). This anticodon will bind to a complementary codon on mRNA following the normal base pairing rules. The tRNA molecules carry an amino acid corresponding to that codon (Figure 4).

When the tRNA anticodons bind to complementary codons along the mRNA, the amino acids are brought together in the correct sequence to form the primary structure of the protein coded for by the mRNA.

This cannot happen all at once. Instead amino acids are added one at a time and the polypeptide chain (protein) grows as this happens. Ribosomes act as the binding site for mRNA and tRNA and catalyse the assembly of the protein. The sequence of events in translation is summarised on the next page and in Figure 5.

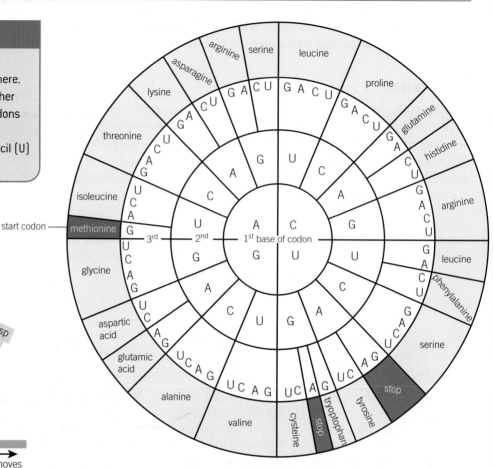

▲ Figure 4 *The codon-amino acid wheel, as seen in Topic 3.9, DNA replication and the genetic code. This time mRNA codons are shown*

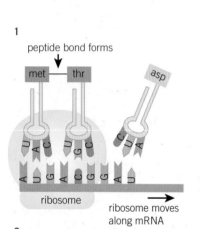

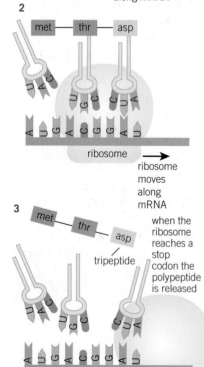

▲ Figure 5 *A summary of translation*

1 The mRNA binds to the small subunit of the ribosome at its start codon (AUG).

2 A tRNA with the complementary anticodon (UAC) binds to the mRNA start codon. This tRNA carries the amino acid methionine.

3 Another tRNA with the anticodon UGC and carrying the corresponding amino acid, threonine, then binds to the next codon on the mRNA (ACG). A maximum of two tRNAs can be bound at the same time.

4 The first amino acid, methionine, is transferred to the amino acid (threonine) on the second tRNA by the formation of a peptide bond. This is catalysed by the enzyme *peptidyl transferase*, which is an rRNA component of the ribosome.

5 The ribosome then moves along the mRNA, releasing the first tRNA. The second tRNA becomes the first.

Stages 3–5 are repeated, with another amino acid added to the chain each time. The process keeps repeating until the ribosome reaches the end of the mRNA at a stop codon and the polypeptide is released.

As the amino acids are joined together forming the primary structure of the protein, they fold into secondary and tertiary structures. This folding and the bonds that are formed are determined by the sequence of amino acids in the primary structure. The protein may undergo further modifications at the Golgi apparatus (Topic 2.4, Eukaryotic cell structure) before it is fully functional and and ready to carry out the specific role for which it has been synthesised.

Many ribosomes can follow on the mRNA behind the first, so that multiple identical polypeptides can be synthesised simultaneously (Figure 6).

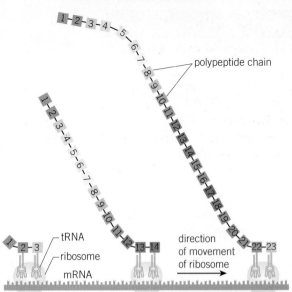

polypeptide chain

tRNA

ribosome

mRNA

direction of movement of ribosome

▶ **Figure 6** *The formation of multiple identical proteins simultaneously by translation of one mRNA by several ribosomes*

Summary questions

1 Copy and complete the diagram below to show the missing bases forming the tRNA anticodons. *(4 marks)*

DNA	TAC	CGG	AGT	GCA
mRNA	AUG	GCC	UCA	CGU
tRNA	A [met]	C [ala]	[ser]	[arg]

2 Describe the roles of mRNA, tRNA and rRNA in protein synthesis *(5 marks)*

3 a An enzyme forms part of the structure of ribosome. Suggest the role of this enzyme. *(2 marks)*

 b rRNA also forms part of the structure. Suggest why RNA needs to be present in a ribosome. *(2 marks)*

 c Ribosomes are either free floating within the cytoplasm or bound to endoplasmic reticulum. Suggest a reason for the different ribosomal sites. *(2 marks)*

4 Post-transcriptional modification of mRNA is carried out before it can leave the nucleus. This involves 'capping' each end to protect the mRNA from degradation in the cytoplasm and the removal of introns. Introns are non-coding sections of DNA and have no role in the formation of proteins.

 a Explain why unnecessary base sequences must be removed before protein synthesis begins. *(4 marks)*

 b Suggest an advantage of being able to edit mRNA *(2 marks)*

 c Suggest a reason for the presence of introns within genes. *(4 marks)*

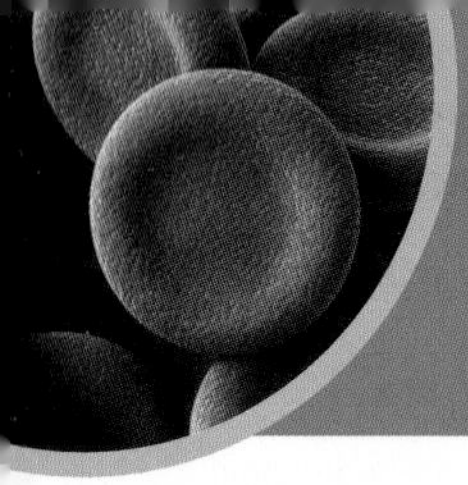

3.11 ATP

Specification reference: 2.1.3 and 2.1.2

Learning outcomes

Demonstrate knowledge, understanding, and application of:

→ the structure of ADP and ATP as nucleotides

→ the hydrolysis and condensation reactions involving ATP and ADP.

Study tip

You might be asked about the similarities and differences between ATP structure and DNA and RNA nucleotide structure. Make sure you understand these.

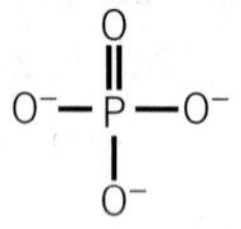

Phosphate group

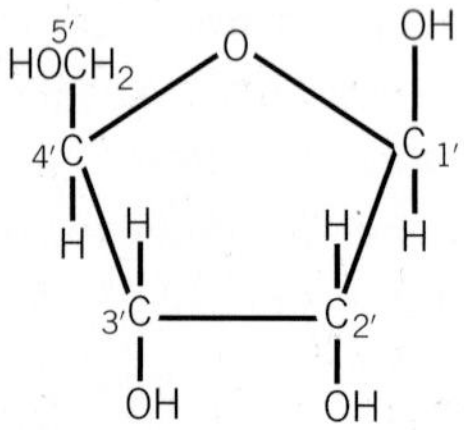

Ribose

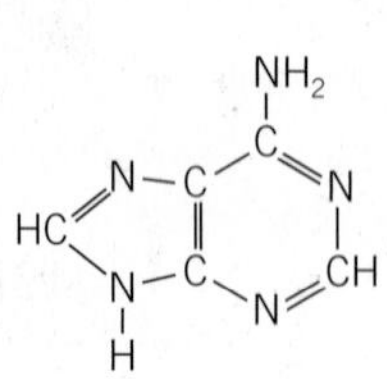

Adenine

▲ **Figure 2** *The structure of a phosphate group, the pentose sugar ribose, and the nitrogenous base adenine*

Universal energy currency

Muscle contraction, cell division, the transmission of nerve impulses, and even memory formation are just some of the many biological processes that require energy. Energy comes in many forms, such as heat, light, and the energy in chemical bonds. Energy has to be supplied in the right form and quantity to the processes that require it.

Cells require energy for three main types of activity:

● synthesis – for example of large molecules such as proteins
● transport – for example pumping molecules or ions across cell membranes by active transport
● movement – for example protein fibres in muscle cells that cause muscle contraction.

Inside cells, molecules of **adenosine triphosphate (ATP)** are able to supply this energy in such a way that it can be used.

An ATP molecule is composed of a nitrogenous base, a pentose sugar and three phosphate groups, as shown in Figure 1 – it is a nucleotide. You will notice that the structure of ATP is very similar to that of the nucleotides involved in the structure of DNA and RNA (Topic 3.8, Figure 1). However, in ATP the base is always adenine and there are three phosphate groups instead of one. The sugar in ATP is ribose, as in RNA nucleotides.

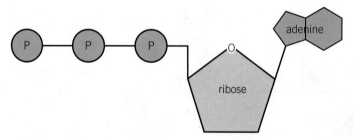

▲ **Figure 1** *The structure of an ATP molecule*

ATP is used for energy transfer in all cells of all living things. Hence it is known as the universal energy currency.

How ATP releases energy

Energy is needed to break bonds and is released when bonds are formed. A small amount of energy is needed to break the relatively weak bond holding the last phosphate group in ATP. However, a large amount of energy is then released when the liberated phosphate undergoes other reactions involving bond formation. Overall a lot more energy is released than used, approximately $30.6 \, \text{kJ} \, \text{mol}^{-1}$.

As water is involved in the removal of the phosphate group this is another example of a **hydrolysis reaction**.

$$ATP + H_2O \rightarrow ADP + P_i + energy$$

adenosine triphosphate   water   adenosine diphosphate   inorganic phosphate

The hydrolysis of ATP does not happen in isolation but in association with energy-requiring reactions. The reactions are said to be 'coupled' as they happen simultaneously.

ATP is hydrolysed into **adenosine diphosphate (ADP)** and a phosphate ion, releasing energy.

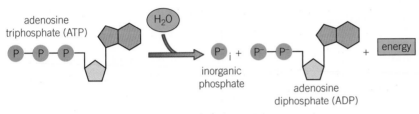

▲ **Figure 3** *The hydrolysis of ATP to ADP releases energy*

The instability of the phosphate bonds in ATP, however, means that it is not a good *long-term* energy store. Fats and carbohydrates are much better for this. The energy released in the breakdown of these molecules (a process called cellular respiration) is used to create ATP. This occurs by reattaching a phosphate group to an ADP molecule. The process is called *phosphorylation*. As water is removed in this process, the reaction is another example of a **condensation reaction**.

Due to the instability of ATP, cells do not store large amounts of it. However, ATP is rapidly reformed by the phosphorylation of ADP (Figure 4). This interconversion of ATP and ADP is happening constantly in all living cells, meaning cells do not need a large store of ATP. ATP is therefore a good *immediate* energy store.

Properties of ATP

The structure and properties of ATP mean that it is ideally suited to carry out its function in energy transfer. A summary of these properties is given below.

- Small – moves easily into, out of and within cells.
- Water soluble – energy-requiring processes happen in aqueous environments.
- Contains bonds between phosphates with intermediate energy: large enough to be useful for cellular reactions but not so large that energy is wasted as heat.
- Releases energy in small quantities – quantities are suitable to most cellular needs, so that energy is not wasted as heat.
- Easily regenerated – can be recharged with energy.

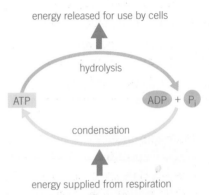

▲ **Figure 4** *Interconversion of ATP and ADP*

Study tip

Make sure you know the characteristics that make ATP so useful as the universal energy currency.

Summary questions

1 Describe the structure of ATP.
  (3 marks)

2 Describe why ATP is called the universal energy currency. *(3 marks)*

3 People and other animals store excess energy in the form of fat. Explain why fat is stored, not ATP? *(4 marks)*

4 a Outline how energy is transferred with reference to bond formation and cleavage. *(2 marks)*
  b Discuss the validity of the statement: ATP is the universal energy currency. *(4 marks)*

Practice questions

Water has a very simple molecular structure but is a vital component of living organisms. Water has many roles such as metabolite, solvent and reaction medium. It is also the environment in which many organisms live.

1 **a** Define the word 'molecule'.   (*2 marks*)

   b Explain what is meant by the term 'polar' with reference to water.   (*2 marks*)

   c Water has an essential role as a transport medium in both plants and animals.

      Explain why water makes an ideal transport medium using **two** examples.
(*4 marks*)

2 A colorimeter can be used to estimate the concentration of a solution of reducing sugar. This is done by carrying out Benedict's test on the solution, allowing the precipitate to settle and measuring the absorbance of the liquid portion or supernatant.

   Describe how you could estimate the concentration of a solution of glucose without using a colorimeter.   (*5 marks*)

3 Copy and fill in the table below.

   The first row has been done for you.
(*3 marks*)

Feature	Cellulose	Collagen
Fibrous	✓	✓
Monomers joined by condensation reactions		
Monomers identical		
Branching		

4 **a** Describe the difference between hydrophobic and hydrophilic using phospholipid molecules as an example.
(*4 marks*)

   b Describe how you would identify the presence of a lipid in an unknown sample.
(*4 marks*)

5 Explain the meaning of prosthetic group with reference to haemoglobin.   (*4 marks*)

6 **a** Identify A, B, C and D on the diagram below.   (*4 marks*)

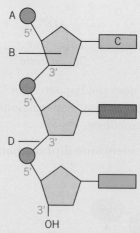

   b Explain why nucleic acids are described as polymers.   (*2 marks*)

   c Outline the differences between DNA and RNA.   (*2 marks*)

   d Chargaff's rule states that the ratios of adenine to thymine and guanine to cytosine in DNA are equal.

      (i) Explain, using your knowledge of the structure of DNA why the bases in DNA always obey this rule.
(*4 marks*)

      (ii) Suggest what has happened during replication if a section of DNA does not obey this rule.   (*3 marks*)

7 The graph below shows how the ratio of ribosome synthesis to total protein synthesis changes as the rate of cell division in a bacterial colony increases.

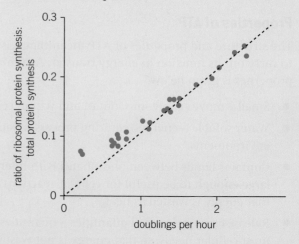

a Describe the relationship shown by the graph. *(4 marks)*

b Explain why ribosomes are necessary in protein synthesis. *(3 marks)*

c Assuming a constant rate of translation explain the trend shown by the graph. *(4 marks)*

8 The diagram below summarises the flow of information between DNA, RNA and protein.

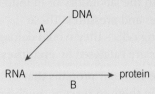

a Suggest which processes are happening at A and B on the diagram. *(2 marks)*

b Explain why there is no arrow between DNA and protein. *(3 marks)*

Viruses are often incorporated into DNA within the nucleus of infected cells.

c Suggest why an arrow could also be drawn going from RNA to DNA. *(2 marks)*

9 Compare the structure and function of ATP and a DNA nucleotide. *(5 marks)*

10 The diagram represents a water molecule

Water molecules are polar. As a result, they attract each other.

Draw a second water molecule on the diagram

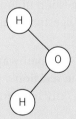

Your drawing should show:

- The bond(s) between the two molecules
- The name of the bond

The charges on each atom *(3 marks)*

OCR F212 2010

11 a Ponds provide a very stable environment for aquatic organisms.

Three properties that contribute to this stability are as follows:

- The density of water decreases as the temperature falls below 4 °C so ice floats on the top of the bond
- It acts as a solvent for ions such as nitrates (NO_3)
- A large quantity of energy is required to raise the temperature of water by 1 °C.

Explain how these three properties help organisms survive in the pond.

In you answer you should make clear the links between the behaviour of the water molecules and the survival of the organisms. *(8 marks)*

b Water is important in many biological questions.

Complete table 1 by writing an appropriate term next to each description.

description	term
The type of reaction that occurs when water is added to break a bond in a molecule	
The phosphate group of a phospholipid that readily attracts water molecules	

OCR F212 2010

4 ENZYMES
4.1 Enzyme action
Specification reference: 2.1.4

Learning outcomes

Demonstrate knowledge, understanding, and application of:

→ the role of enzymes in catalysing reactions that affect metabolism at a cellular and whole organism level

→ the role of enzymes in catalysing both intracellular and extracellular reactions

→ the mechanism of enzyme action.

Synoptic link

You learned about globular proteins in Topic 3.7, Types of proteins.

Study tip

Enzymes are often named for the reaction they are involved in, or the substrate that they act upon. Many enzyme names end in 'ase', but not all. Some enzyme names indicate both substrate and function. For example:

● Phosphorylases catalyse the addition of a phosphate groups to molecules.

● Lactate dehydrogenase catalyses the transfer of a hydrogen ion to and from lactate.

● Pepsins catalyse the breakdown of proteins (peptides) in the stomach.

● ATPase uses ATP as its substrate.

Why are enzymes important?

Most of the processes necessary to life involve chemical reactions, and these reactions need to happen very fast. In the laboratory or in industry this would demand very high temperatures and pressures. These extreme conditions are not possible in living cells – they would damage the cell components. Instead, the reactions are catalysed by **enzymes**.

Enzymes are biological catalysts. They are globular proteins that interact with **substrate** molecules causing them to react at much faster rates without the need for harsh environmental conditions. Without enzymes many of the processes necessary to life would not be possible.

The role of enzymes in reactions

Living organisms need to be built and maintained. This involves the synthesis of large polymer-based components. For example, cellulose forms the walls of plants cells and long protein molecules form the contractile filaments of muscles in animals. The different cell components are synthesised and assembled into cells, which then form tissues, organs, and eventually the whole organism. The chemical reactions required for growth are **anabolic** (building up) reactions and they are all catalysed by enzymes.

Energy is constantly required for the majority of living processes, including growth. Energy is released from large organic molecules, like glucose, in metabolic pathways consisting of many **catabolic** (breaking down) reactions. Catabolic reactions are also catalysed by enzymes.

These large organic molecules are obtained from the digestion of food, made up of even larger organic molecules, like starch. Digestion is also catalysed by a range of enzymes.

Reactions rarely happen in isolation but as part of multi-step pathways. Metabolism is the sum of all of the different reactions and reaction pathways happening in a cell or an organism, and it can only happen as a result of the control and order imposed by enzymes.

Just like reactions in a laboratory, the speed at which different cellular reactions proceed varies considerably and is usually dependent on environmental conditions. The temperature, pressure, and pH may all have an effect on the rate of a chemical reaction. Enzymes can only increase the rates of reaction up to a certain point called the V_{max} (maximum initial velocity or rate of the enzyme-catalysed reaction).

Mechanism of enzyme action

Molecules in a solution move and collide randomly. For a reaction to happen, molecules need to collide in the right orientation. When high temperatures and pressures are applied the speed of the molecules will increase, therefore so will the number of successful collisions and the overall rate of reaction.

Many different enzymes are produced by living organisms, as each enzyme catalyses one biochemical reaction, of which there are thousands in any given cell. This is termed the *specificity* of the enzyme.

Energy needs to be supplied for most reactions to start. This is called the **activation energy**. Sometimes, the amount of energy needed is so large it prevents the reaction from happening under normal conditions. Enzymes help the molecules collide successfully, and therefore reduce the activation energy required. There are two hypotheses for how enzymes do this.

Lock and key hypothesis

An area within the tertiary structure of the enzyme has a shape that is complementary to the shape of a specific substrate molecule. This area is called the **active site**.

In the same way that only the right key will fit into a lock, only a specific substrate will 'fit' the active site of an enzyme. This is the lock and key hypothesis.

When the substrate is bound to the active site an **enzyme-substrate complex** is formed. The substrate or substrates then react and the product or products are formed in an **enzyme-product complex**. The product or products are then released, leaving the enzyme unchanged and able to take part in subsequent reactions.

The substrate is held in such a way by the enzyme that the right atom-groups are close enough to react. The R-groups within the active site of the enzyme will also interact with the substrate, forming temporary bonds. These put strain on the bonds within the substrate, which also helps the reaction along.

Induced-fit hypothesis

More recently, evidence from research into enzyme action suggests the active site of the enzyme actually changes shape slightly as the substrate enters. This is called the **induced-fit hypothesis** and is a modified version of the lock and key hypothesis. The initial interaction between the enzyme and substrate is relatively weak, but these weak interactions rapidly induce changes in the enzyme's tertiary structure that strengthen binding, putting strain on the substrate molecule. This can weaken a particular bond or bonds in the substrate, therefore lowering the activation energy for the reaction.

Intracellular enzymes

Enzymes have an essential role in both the structure and the function of cells and whole organisms. The synthesis of polymers from monomers, for example making polysaccharides from glucose, requires enzymes. Enzymes that act within cells are called intracellular enzymes.

Hydrogen peroxide is a toxic product of many metabolic pathways. The enzyme *catalase* ensures hydrogen peroxide is broken down to oxygen

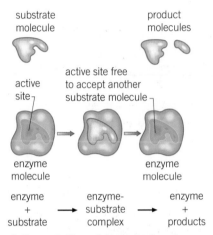

▲ Figure 1 *The substrate fits into the active site of the enzyme forming the enzyme-substrate complex where the reaction takes place and the products are released*

Study tip

The substrate should be described as having a complementary shape to its active site, not the same shape.

and water quickly, therefore preventing its accumulation. It is found in both plant and animal tissues.

Extracellular enzymes

All of the reactions happening within cells need substrates (raw materials) to make products needed by the organism. These raw materials need to be constantly supplied to cells to keep up with the demand. Nutrients (components necessary for survival and growth) present in the diet or environment of the organism supply these materials.

Nutrients are often in the form of polymers such as proteins and polysaccharides. These large molecules cannot enter cells directly through the cell-surface membrane. They need to be broken down into smaller components first.

Enzymes are released from cells to break down these large nutrient molecules into smaller molecules in the process of digestion. These enzymes are called extracellular enzymes. They work outside the cell that made them. In some organisms, for example fungi, they work outside the body.

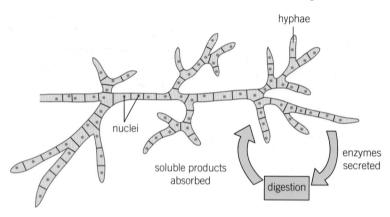

▲ **Figure 2** *Cells in fungal hyphae synthesise enzymes that are secreted outside the cell. The enzymes breakdown organic matter, so it can be absorbed and used for growth*

Both single-celled and multicellular organisms rely on extracellular enzymes to make use of polymers for nutrition.

Single-celled organisms, such as bacteria and yeast, release enzymes into their immediate environment. The enzymes break down larger molecules, such as proteins, and the smaller molecules produced, such as amino acids and glucose, are then absorbed by the cells.

Many multicellular organisms eat food to gain nutrients. Although the nutrients are taken into the digestive system the large molecules still have to be digested so smaller molecules can be absorbed into the bloodstream. From there they are transported around the body to be used as substrates in cellular reactions. Examples of extracellular enzymes involved in digestion in humans are amylase and trypsin.

Digestion of starch

The digestion of starch begins in the mouth and continues in the small intestine. Starch is digested in two steps, involving two different enzymes. Different enzymes are needed because each enzyme only catalyses one specific reaction.

1   Starch polymers are partially broken down into maltose, which is a disaccharide. The enzyme involved in this stage is called *amylase*. Amylase is produced by the salivary glands and the pancreas. It is released in saliva into the mouth, and in pancreatic juice into the small intestine.

Synoptic link

You learnt about starch in Topic 3.3, Carbohydrates.

2 Maltose is then broken down into glucose, which is a monosaccharide. The enzyme involved in this stage is called *maltase*. Maltase is present in the small intestine.

Glucose is small enough to be absorbed by the cells lining the digestive system and subsequently absorbed into the bloodstream.

Digestion of proteins

Trypsin is a **protease**, a type of enzyme that catalyses the digestion of proteins into smaller peptides, which can then be broken down further into amino acids by other proteases. Trypsin is produced in the pancreas and released with the pancreatic juice into the small intestine, where it acts on proteins. The amino acids that are produced by the action of proteases are absorbed by the cells lining the digestive system and then absorbed into the bloodstream.

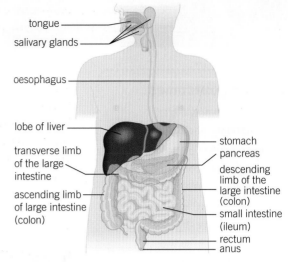

▲ Figure 3 *Human digestive system*

Summary questions

1 a State the type of biological molecule used to form enzymes. *(1 mark)*
  b Name the monomers that form this biological molecule. *(1 mark)*
  c Describe how the structure(s) of this biological molecule determines enzyme activity. *(4 marks)*

2 Explain how catabolism and anabolism are related to metabolism. *(3 marks)*

3 There are two theories explaining enzyme substrate interaction. The lock and key model and induced fit model of enzyme action.
  a Explain what is meant by the term model in the sentence above. *(2 marks)*
  b Explain how the following terms are relevant to each of the models. *complementary, flexibility, R group interactions, bond strain* *(4 marks)*

4 The transition state model adds more detail to the induced fit model, in the same way that the induced fit model was more detailed than the original lock and key model.

All reactions go through a transition state as the different chemical components interact with each other. The formation of this transition sate determines the activation energy of the reaction.

  a Explain the meaning of the term activation energy. *(2 marks)*
  b Explain the energy changes occurring resulting in the transition state model (the red line). *(4 marks)*
  c Discuss why the models of enzyme action have changed over time. *(4 marks)*

Synoptic link

You learnt about the structure of proteins in Topic 3.6, Structure of proteins.

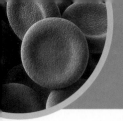

4.2 Factors affecting enzyme activity

Specification reference: 2.1.4

Learning outcomes

Demonstrate knowledge, understanding, and application of:

→ the effects of pH, temperature, enzyme concentration and substrate concentration on enzyme activity

→ practical investigations into factors effecting enzyme activity.

For enzymes to catalyse a reaction, they must come into contact with the substrate, and the enzyme must be the right shape (complementary) for the substrate. Enzymes are complex proteins and their structure can be affected by factors such as temperature and pH. These can cause changes in the shape of their active site. Enzymes are more likely to come into contact with the substrate if temperature and substrate concentration are increased.

Factors affecting enzyme action can be investigated by measuring the rate of the reactions they catalyse.

Temperature

Increasing the temperature of a reaction environment increases the kinetic energy of the particles. As temperature increases, the particles move faster and collide more frequently. In an enzyme-controlled reaction an increase in temperature will result in more frequent successful collisions between substrate and enzyme. This leads to an increase in the rate of reaction.

The **temperature coefficient**, Q_{10}, of a reaction (or process) is a measure of how much the rate of a reaction increases with a 10 °C rise in temperature. For enzyme-controlled reactions this is usually taken as two, which means that the rate of reaction doubles with a 10 °C temperature increase.

Denaturation from temperature

As enzymes are proteins their structure is affected by temperature. At higher temperatures the bonds holding the protein together vibrate more. As the temperature increases the vibrations increase until the bonds strain and then break. The breaking of these bonds results in a change in the precise tertiary structure of the protein. The enzyme has changed shape and is said to have been **denatured**.

When an enzyme is denatured the active site changes shape and is no longer complementary to the substrate. The substrate can no longer fit into the active sites and the enzyme will no longer function as a catalyst.

Optimum temperature

The optimum temperature is the temperature at which the enzyme has the highest rate of activity. The optimum temperature of enzymes can vary significantly. Many enzymes in the human body have optimum temperatures of around 40 °C, meanwhile thermophilic bacteria (found in hot springs) have enzymes with optimum temperatures of 70 °C, and psychrophilic organisms (that live in areas that are cold such as the antarctic and arctic regions) have enzymes with optimum temperatures below 5 °C.

Once the enzymes have denatured above the optimum temperature, the decrease in rate of reaction is rapid. There only needs to be a slight

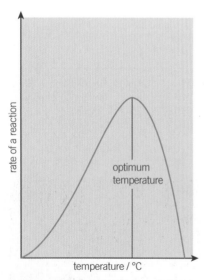

▲ **Figure 1** *The rate of reaction increases up to the optimum temperature and then decreases rapidly*

change in shape of an active site for it to no longer be complementary to its substrate. This happens to all of the enzyme molecules at about the same temperature so the loss of activity is relatively abrupt. At this point in an enzyme-controlled reaction the *temperature coefficient*, Q_{10} does not apply any more as the enzymes have denatured.

The decrease in the rate of reaction below the optimum temperature is less rapid. This is because the enzymes have not denatured, they are just less active.

Temperature extremes

The majority of living organisms have evolved to cope with living within a certain temperature range. Some organisms can also cope with extremes.

Examples of extremely cold environments are deep oceans, high altitudes, and polar regions. The enzymes controlling the metabolic activities of organisms living in these environments need to be adapted to the cold. Enzymes adapted to the cold tend to have more flexible structures, particularly at the active site, making them less stable than enzymes that work at higher temperatures. Smaller temperature changes will denature them.

Thermophiles are organisms adapted to living in very hot environments such as hot springs and deep sea hydrothermal vents. The enzymes present in these organisms are more stable than other enzymes due to the increased number of bonds, particularly hydrogen bonds and sulfur bridges, in their tertiary structures. The shapes of these enzymes, and their active sites, are more resistant to change as the temperature rises.

Enzymes in action

Siamese cats provide visual evidence of the effect of temperature on enzyme activity. Tyrosinase is an enzyme responsible for catalysing the production of melanin, the pigment responsible for dark coloured fur. Due to a mutation, Siamese cats produce a form of the enzyme tyrosinase that is denatured and therefore inactive at normal body temperature meaning that their fur is primarily white or cream coloured. The extremities of these cats – the tails, ears, and limbs – are at a slightly lower temperature, too low to denature mutant tyrosinase. This leads to the distinctive point coloration of these cats as melanin is produced in these areas.

1   Suggest why Siamese kittens are born completely white.

Study tip

Remember that enzymes are denatured not 'killed'.

A denatured protein is still a protein – it just has a different shape and can no longer function as it normally would.

pH

Proteins, and so enzymes, are also affected by changes in pH. Hydrogen bonds and ionic bonds between amino acid R-groups hold proteins in their precise three-dimensional shape. These bonds result from interactions between the polar and charged R-groups present on the amino acids forming the primary structure. A change in pH refers to a change in hydrogen ion concentration. More hydrogen ions are present in low pH (acid) environments and fewer hydrogen ions are present in high pH (alkaline) environments.

The active site will only be in the right shape at a certain hydrogen ion concentration. This is the **optimum pH** for any particular enzyme. When the pH changes from the optimum – becoming more acidic or alkaline – the structure of the enzyme, and therefore the active site, is altered. However, if the pH returns to the optimum then the protein will resume its normal shape and catalyse the reaction again. This is called renaturation.

When the pH changes more significantly (beyond a certain pH) the structure of the enzyme is irreversibly altered and the active site will no longer be complementary to the substrate. The enzyme is now said to be denatured and substrates can no longer bind to the active sites. This will reduce the rate of reaction.

Hydrogen ions interact with polar and charged R-groups. Changing the concentration of hydrogen ions therefore changes the degree of this interaction. The interaction of R-groups with hydrogen ions also affects the interaction of R-groups with each other.

The more hydrogen ions present (low pH), the less the R-groups are able to interact with each other. This leads to bonds breaking and the shape of the enzyme changing. The reverse is true when fewer hydrogen ions (high pH) are present. This means the shape of an enzyme will change as the pH changes and therefore it will only function within a narrow pH range. Table 1 shows the pH conditions under which the various enzymes of the human digestive system function. These are the optimum pHs for these enzymes.

> **Study tip**
>
> Different enzymes in different organisms also have a whole range of optimum pHs from very acid to very alkaline.

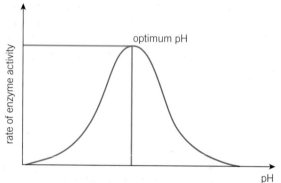

▲ Figure 2 *Effect of pH on the rate of an enzyme-controlled reaction. The rate of reaction decreases as the pH moves away from the optimum, becoming higher or lower. Note how this graph is symmetrical, unlike the graph showing rate of reaction against temperature. It shows reversible reduction in enzyme activity as the pH moves away from the optimum*

▼ Table 1 *The action of different enzymes at various points in the digestive system*

	Site of action	pH	Enzymes	Function
Saliva	mouth/throat	neutral (pH 7–8)	amylase	starch → maltose
Gastric juice	stomach	acidic (pH 1–2)	pepsin	proteins → polypeptides
Pancreatic juice	small intestine/ duodenum	slightly alkaline (pH 8)	trypsin	proteins → polypeptides
			lipase	triglycerides → glycerol + fatty acids
			amylase	starch → maltose
			maltase	maltose → glucose

Substrate and enzyme concentration

When the concentration of substrate is increased the number of substrate molecules, atoms, or ions in a particular area or volume increases. The increased number of substrate particles leads to a higher collision rate with the active sites of enzymes and the formation of more enzyme–substrate complexes. The rate of reaction therefore increases.

This is also true when the concentration of the enzyme increases. This will increase the number of available active sites in a particular area or volume, leading to the formation of enzyme-substrate complexes at a faster rate.

The rate of reaction increases up to its maximum (V_{max}). At this point all of the active sites are occupied by substrate particles and no more enzyme-substrate complexes can be formed until products are released from active sites. The only way to increase the rate of reaction would be to add more enzyme or increase the temperature.

If the concentration of the enzyme is increased more active sites are available so the reaction rate can rise towards a higher V_{max}. The concentration of substrate becomes the limiting factor again and increasing this will once again allow the reaction rate to rise until the new V_{max} is reached.

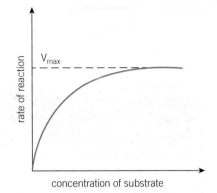

▲ **Figure 3** *Effect of substrate concentration on the rate of an enzyme-controlled reaction. The rate of reaction increases as the substrate concentration increases up to a maximum rate, V_{max}*

Investigations into the effects of different factors on enzyme activity

Investigating the effects of different factors on enzyme activity provides an insight into how enzymes work.

Catalase is an enzyme present in plant tissue and animal tissue, making it a good choice for use in investigations because it is readily available. Catalase catalyses the breakdown of hydrogen peroxide into water and oxygen. The volume of oxygen gas collected in a set length of time can be used as a measure of the rate of reaction.

In the second experiment the liver was boiled for five minutes before being placed in the hydrogen peroxide solution.

The graph below shows the results from the experiment.

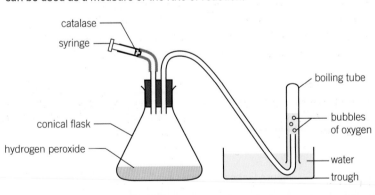

▲ **Figure 4** *Apparatus used to collect oxygen released when catalase reacts with hydrogen peroxide*

A student conducted a series of experiments to determine the effect of temperature on enzymes.

In the first experiment, using apparatus similar to that shown in the diagram above, liver tissue was put into hydrogen peroxide solution and the volume of oxygen released every five seconds was measured.

1 Explain why the student used liver tissue.
2 Show the chemical reaction that occurs resulting in the release of oxygen.
3 Explain the shape of the graph.
4 Describe and explain what has happened in the second experiment.

The student then investigated the effect of substrate concentration on enzyme activity.

Suggest a hypothesis for this experiment.

The results are shown below.

Percentage of Hydrogen peroxide	Volume of gas collected (cm³) after … (s)											
	5	10	15	20	25	30	35	40	45	50	55	60
Test 1												
100	43	62	68	74	76	77	77	77	77	77	77	77
30	12	21	24	26	27	27	27	27	27	27	27	27
10	10	14	16	18	18	18	18	18	18	18	18	18
Test 2												
100	42	62	70	65	76	77	77	77	77	77	77	77
30	13	21	25	27	28	28	28	28	28	28	28	28
10	8	15	17	19	19	19	19	19	19	19	19	19
Test 3												
100	43	63	68	74	75	76	77	77	77	77	77	77
30	12	20	24	26	27	27	27	27	27	27	27	27
10	8	14	17	18	18	18	18	18	18	18	18	18
Mean of three tests												
100	42.7	62.3	68.7	71.0	75.7	76.7	77.0	77.0	77.0	77.0	77.0	77.0
30	12.3	20.7	24.3	26.3	27.3	27.3	27.3	27.3	27.3	27.3	27.3	27.3
10	8.7	14.3	16.7	18.3	18.3	18.3	18.3	18.3	18.3	18.3	18.3	18.3

5 Describe how the student would have carried out the experiment at different temperatures.

6 a Name the independent variable and the dependent variable in this experiment.

 b List the controlled variables.

7 Explain the importance of keeping these variables (controlled) constant.

8 a Explain the term anomaly.

 b Identify any anomalies in the results above.

9 Explain why the student did the experiment three times and then calculated the means of the results.

10 Plot a graph of the results.

11 Explain the shape of the graph you have produced.

12 Write an evaluation for this experiment.

 ## Serial dilutions

A serial dilution is a repeated, stepwise dilution of a stock solution of known concentration. A serial dilution is usually done by factors of ten, to produce a range of concentrations. Serial dilutions are useful even if the concentration of the initial solution is unknown as they give us relative concentrations. Serial dilutions are used in many different ways – for example to investigate the effect of changing the concentration of an enzyme or a substrate in a reaction, and in determining the numbers of microorganisms in a culture.

Figure 5 shows how a serial dilution might be set up. Adding 1 ml of stock solution to 9 ml of distilled water gives 10 ml of dilute solution in which there is 1 ml stock/10 ml hence a 1/10 or 10 fold dilution. This step is repeated a number of times to give a serial dilution.

Catalase is an enzyme that catalyses the breakdown of hydrogen peroxide. To investigate the effect of different concentrations of catalase on the rate of breakdown of hydrogen peroxide, catalase-rich tissues such as liver or potato can be ground down to make a solution. The solution will contain catalase released from cells. Serial dilution of this solution will produce a range of solutions with different relative concentrations of catalase. The effect of the different concentrations on the rate of reaction can be investigated by adding equal volumes of a given concentration of hydrogen peroxide to each solution.

1 You are provided with a stock solution of enzyme X with a concentration of 20 mmol/dm^{-3}. Explain how you would prepare a range of five solutions with different concentrations of enzyme X using the stock solution. Show your working and state the concentrations produced.

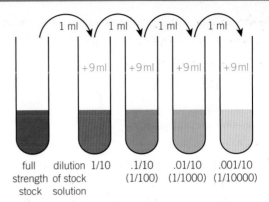

| full strength stock | dilution of stock solution | 1/10 | .1/10 (1/100) | .01/10 (1/1000) | .001/10 (1/10000) |

▲ **Figure 5** *This figure shows how a serial dilution might be set up. Adding 1 ml of stock solution to 9 ml of distilled water gives 10 ml of dilute solution in which there is 1 ml stock/10 ml hence a 1/10 or 10 fold dilution. This step is repeated a number of times to give a serial dilution.*

Summary questions

1 Explain the term 'denatured' with reference to enzymes. *(3 marks)*

2 The graph below shows the optimum pHs of three enzymes.

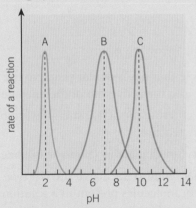

Pepsin is an enzyme that, along with stomach acid, digests proteins in the stomach.

State which of the curves, A, B or C, is likely to represent the activity of the enzyme pepsin over a range of pHs. Explain the reasons for your choice. *(2 marks)*

3 Bacteria that colonise hydrothermal vents, where temperatures are very high, have enzymes with very high optimum temperatures.

Suggest why these bacteria are unlikely to cause infections in humans. *(3 marks)*

4 Enzymes with very low optimum temperatures tend to have quite flexible structures. Using your knowledge of collision theory, explain why this flexibility is necessary. *(6 marks)*

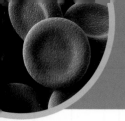

4.3 Enzyme inhibitors

Specification reference: 2.1.4

As you have seen, it is important that cellular conditions such as pH and temperature are kept within narrow limits so that enzyme activity is not delayed. This ensures that reactions can happen at a rate fast enough to sustain living processes, for example, respiration.

It is also important that reactions do not happen too fast as this could lead to the build-up of excess products. Living processes rarely involve just one reaction but are complex, multi-step reaction pathways. These pathways need to be closely regulated to meet the needs of living organisms without wasting resources.

Control of metabolic activity within cells

The different steps in reaction pathways are controlled by different enzymes. Controlling the activity of enzymes at crucial points in these pathways regulates the rate and quantity of product formation.

Enzymes can be activated with cofactors, (Topic 4.4 Cofactors, coenzymes, and prosthetic groups), or inactivated with **inhibitors**.

Inhibitors are molecules that prevent enzymes from carrying out their normal function of catalysis (or slow them down). There are two types of enzyme inhibition – competitive and non-competitive.

Competitive inhibition

Competitive inhibition works in the following way:

- A molecule or part of a molecule that has a similar shape to the substrate of an enzyme can fit into the active site of the enzyme.
- This blocks the substrate from entering the active site, preventing the enzyme from catalysing the reaction.
- The enzyme cannot carry out its function and is said to be inhibited.
- The non-substrate molecule that binds to the active site is a type of inhibitor. Substrate and inhibitor molecules present in a solution will compete with each other to bind to the active sites of the enzymes catalysing the reaction. This will reduce the number of substrate molecules binding to active sites in a given time and slows down the rate of reaction. For this reason such inhibitors are called **competitive inhibitors** and the degree of inhibition will depend on the relative concentrations of substrate, inhibitor, and enzyme.

Most competitive inhibitors only bind temporarily to the active site of the enzyme, so their effect is reversible. However there are some exceptions such as aspirin.

Effect on rates of reaction

A competitive inhibitor reduces the rate of reaction for a given concentration of substrate, but it does not change the V_{max} of the enzyme it inhibits. If the substrate concentration is increased enough

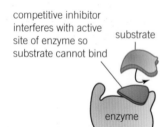

competitive inhibitor interferes with active site of enzyme so substrate cannot bind

substrate

enzyme

▲ **Figure 1** *Competitive inhibition. The shape of the inhibitor is similar in shape to the part of the substrate that binds to the active site. This means the inhibitor can temporarily bind to the active site and block the substrate from binding*

Learning outcomes

Demonstrate knowledge, understanding, and application of:

→ the effects of inhibitors on the rate of enzyme-controlled reactions.

Study tip

When describing a competitive inhibitor, do not say that it has the same shape as the substrate. It has a similar shape to part of or all of the substrate.

there will be so much more substrate than inhibitor that the original V_{max} can still be reached.

Examples of competitive inhibition

Statins are competitive inhibitors of an enzyme used in the synthesis of cholesterol. Statins are regularly prescribed to help people reduce blood cholesterol concentration. High blood cholesterol levels can result in heart disease.

Aspirin irreversibly inhibits the active site of COX enzymes, preventing the synthesis of prostaglandins and thromboxane, the chemicals responsible for producing pain and fever.

Non-competitive inhibition

Non-competitive inhibition works in the following way:

- The inhibitor binds to the enzyme at a location other than the active site. This alternative site is called an allosteric site.

- The binding of the inhibitor causes the tertiary structure of the enzyme to change, meaning the active site changes shape.

- This results in the active site no longer having a complementary shape to the substrate so it is unable to bind to the enzyme.

- The enzyme cannot carry out its function and it is said to be inhibited.

As the inhibitor does not compete with the substrate for the active site it is called a **non-competitive inhibitor**.

Effect on rates of reaction

Increasing the concentration of enzyme or substrate will not overcome the effect of a non-competitive inhibitor. Increasing the concentration of inhibitor, however, will decrease the rate of reaction further as more active sites become unavailable.

Examples of irreversible non-competitive inhibitors

As mentioned, the binding of the inhibitor may be reversible or non-reversible. Irreversible inhibitors cannot be removed from the part of the enzyme they are attached to. They are often very toxic, but not always. Organophosphates used as insecticides and herbicides irreversibly inhibit the enzyme acetyl cholinesterase, an enzyme necessary for nerve impulse transmission. This can lead to muscle cramps, paralysis, and even death if accidentally ingested.

Proton pump inhibitors (PPIs) are used to treat long-term indigestion. They irreversibly block an enzyme system responsible for secreting hydrogen ions into the stomach. This makes PPIs very effective in reducing the production of excess acid which, if left untreated, can lead to formation of stomach ulcers.

End-product inhibition

End-product inhibition is the term used for enzyme inhibition that occurs when the product of a reaction acts as an inhibitor to the enzyme that produces it. This serves as a *negative-feedback*

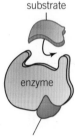

substrate

enzyme

non-competitive inhibitor changes shape of enzyme so it cannot bind to substrate

▲ Figure 2 *Non-competitive inhibition. The inhibitor binds to an allosteric site on the enzyme, changing the shape of the active site. This means the substrate can no longer bind the active site*

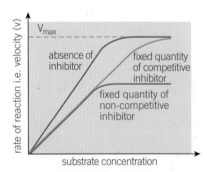

▲ Figure 3 *Comparison of competitive and non-competitive inhibition of an enzyme-controlled reaction. With a competitive inhibitor, as the substrate concentration is increased the effect of the inhibitor is almost overcome. With a non-competitive inhibitor, as the substrate concentration is increased the effect of the inhibitor is not overcome*

Synoptic link

You learnt about ATP in Topic 3.11, ATP.

control mechanism for the reaction. Excess products are not made and resources are not wasted. It is an example of non-competitive reversible inhibition.

Respiration is a metabolic pathway resulting in the production of ATP. Glucose is broken down in a number of steps. The first step involves the addition of two phosphate groups to the glucose molecule. The addition of the second phosphate group, which results in the initial breakdown of the glucose molecule, is catalysed by the enzyme phosphofructokinase (PFK). This enzyme is competitively inhibited by ATP. ATP therefore regulates its own production.

- When the levels of ATP are high, more ATP binds to the allosteric site on PFK, preventing the addition of the second phosphate group to glucose. Glucose is not broken down and ATP is not produced at the same rate.

- As ATP is used up, less binds to PFK and the enzyme is able to catalyse the addition of a second phosphate group to glucose. Respiration resumes, leading to the production of more ATP.

Summary questions

1 Explain why a non-competitive inhibitor does not need to have a similar shape to a substrate molecule. *(3 marks)*

2 Explain why increasing the concentration of substrate will never produce the V_{max} of a reaction after the addition of a non-competitive inhibitor. *(2 marks)*

3 End-product inhibition is likely to be competitive rather than non-competitive. Suggest reasons for this, and give an example of end-product inhibition. *(4 marks)*

4 Ethylene glycol present in antifreeze is poisonous when ingested. Ethylene glycol is oxidised using the same enzymes used to oxidise ethanol (alcohol). The products made during the breakdown of ethylene glycol, rather than ethylene glycol itself, are responsible for the toxic effects. Ethylene glycol is able to leave the body unchanged in urine.

   Suggest why ethanol is often used in emergency departments as an antidote to antifreeze poisoning. *(6 marks)*

4.4 Cofactors, coenzymes, and prosthetic groups

Specification reference: 2.1.4

The difference between cofactors and coenzymes

Some enzymes need a non-protein 'helper' component in order to carry out their function as biological catalysts. They may transfer atoms or groups from one reaction to another in a multi-step pathway or they may actually form part of the active site of an enzyme. These components are called **cofactors**, or if the cofactor is an organic molecule it is called a coenzyme.

Inorganic cofactors are obtained via the diet as minerals, including iron, calcium, chloride, and zinc ions. For example, the enzyme amylase, which catalyses the breakdown of starch (Topic 4.1, Enzyme action), contains a chloride ion that is necessary for the formation of a correctly shaped active site.

Many coenzymes are derived from vitamins, a class of organic molecule found in the diet. For example, vitamin B3 is used to synthesise NAD (nicotinamide adenine dinucleotide), a coenzyme responsible for the transfer of hydrogen atoms between molecules involved in respiration. NADP, which plays a similar role in photosynthesis, is also derived from vitamin B3.

Another example is vitamin B5, which is used to make coenzyme A. Coenzyme A is essential in the breakdown of fatty acids and carbohydrates in respiration.

Prosthetic groups

You have previously met prosthetic groups when you learnt about the globular protein, haemoglobin, in which the prosthetic group is an iron (Fe) ion. Prosthetic groups are cofactors – they are required by certain enzymes to carry out their catalytic function. While some cofactors are loosely or temporarily bound to the enzyme protein in order to activate them, prosthetic groups are tightly bound and form a permanent feature of the protein. For example zinc ions (Zn^{2+}) form an important part of the structure of carbonic anhydrase, an enzyme necessary for the metabolism of carbon dioxide.

Precursor activation

Many enzymes are produced in an inactive form, known as inactive precursor enzymes, particularly enzymes that can cause damage within the cells producing them or to tissues where they are released, or enzymes whose action needs to be controlled and only activated under certain conditions.

Precursor enzymes often need to undergo a change in shape (tertiary structure), particularly to the active site, to be activated. This can be achieved by the addition of a cofactor. Before the cofactor is added,

Learning outcomes

Demonstrate knowledge, understanding, and application of:

→ the need for coenzymes, cofactors, and prosthetic groups in some enzyme-controlled reactions.

→ the role of inactive precursors

▲ Figure 1 *A molecular model of the enzyme alpha amylase, which catalyses the breakdown of starch*

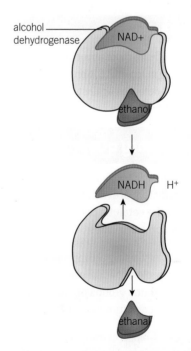

ethanal, H+, and NADH are released from the active site

▲ Figure 2 *Alcohol dehydrogenase needs NAD+ to accept hydrogen produced when ethanal is formed from ethanol*

the precursor protein is called an apoenzyme. When the cofactor is added and the enzyme is activated, it is called a holoenzyme.

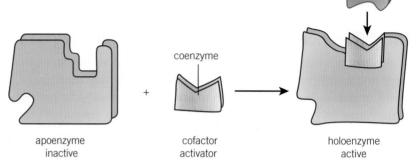

Synoptic link

You learnt about prosthetic groups in proteins in Topic 3.6, Structure of proteins.

▲ Figure 3 *Coenzymes, or cofactors, are often needed to produce the specific shape of the active site*

Sometimes the change in tertiary structure is brought about by the action of another enzyme, such as a protease, which cleaves certain bonds in the molecule. In some cases a change in conditions, such as pH or temperature, results in a change in tertiary structure and activates a precursor enzyme. These types of precursor enzymes are called zymogens or proenzymes.

When inactive pepsinogen is released into the stomach to digest proteins, the acid pH brings about the transformation into the active enzyme pepsin. This adaptation protects the body tissues against the digestive action of pepsin.

Synoptic link

You will learn more about blood clotting in Topic 12.5, Non-specific animal defences against pathogens.

Enzyme activation and the blood-clotting mechanism

Blood clotting, or coagulation, is an important biological response to tissue damage. The blood-clotting process only begins when platelets aggregate at the site of tissue damage. The aggregated platelets release clotting factors, including factor X.

Factor X is an important component in the blood-clotting mechanism. It is an enzyme that is dependent on the cofactor vitamin K for activation. Activated factor X catalyses the conversion of prothrombin into the enzyme thrombin by cleaving certain bonds in the molecule, thus altering its tertiary structure. Thrombin is a protease and catalyses the conversion of soluble fibrinogen to insoluble fibrin fibres. Fibrin molecules, together with platelets, form a blood clot.

This series of successive enzyme activations in blood clotting is called the coagulation cascade.

1   Explain the importance of enzyme activation in controlling blood clotting.

Summary questions

1   Describe two ways in which cofactors are necessary for the catalytic role of some enzymes.   (*2 marks*)

2   Explain, using an appropriate example of each, how prosthetic groups are different from coenzymes.   (*4 marks*)

3   Using blood clotting as your example, explain the different ways in which enzymes can be activated.   (*4 marks*)

Practice questions

1   Compounds containing some metal ions such as lead, mercury, copper or silver are poisonous. This is because ions of these metals are non-competitive inhibitors of several enzymes.

Other metal ions are required for enzymes to function. Magnesium ions, for example, are required by DNA polymerase.

Which of the following statements correctly describes the nature and function of magnesium ions?

Statement 1: minerals and coenzymes

Statement 2: vitamins and cofactors

Statement 3: minerals and cofactors

Statement 4: vitamins and coenzymes

(*1 mark*)

2   a   Describe what is happening at the points labelled on the graph below. You should use the terms active site, substrate and enzyme substrate complex in your answer.   (*5 marks*)

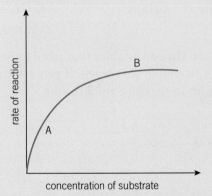

Amylase catalyses the hydrolysis of starch to maltose in the mouth. Amylase stops working as the food passes into the stomach. Maltase then catalyses the hydrolysis of maltose into glucose in the small intestine.

b   (i)   State which group of biological molecules amylase and maltase belong to.   (*1 mark*)

(ii)   Describe the meaning of the term hydrolysis.   (*2 marks*)

(iii)   Explain why maltase, not amylase, hydrolyses maltose completing the digestion of starch.   (*3 marks*)

(iv)   Explain why amylase stops working in the stomach.   (*4 marks*)

The acid stomach contents are neutralised by alkaline pancreatic juice in the small intestine.

c   Suggest why this is necessary to ensure digestion of maltose.   (*3 marks*)

d   Explain what is meant by the term Vmax with respect to enzymes.   (*3 marks*)

3   a   State the **type** of monomers that make up proteins.   (*1 mark*)

b   Malonate inhibits cellular respiration. Hans Krebs added malonate to minced pigeon muscle tissue when trying to work out the different steps in Krebs cycle, a series of reactions that occur in respiration.

$$
\begin{array}{c}
COO^- \\
| \\
CH_2 \\
| \\
CH_2 \\
| \\
COO^-
\end{array}
\quad
\xrightarrow[\text{dehydrogenase}]{\text{succinate}}
\quad
\begin{array}{c}
{}^-OOC {-} C {-} H \\
\| \\
H {-} C {-} COO^-
\end{array}
$$

succinate   fumarate

$$
\begin{array}{c}
COO^- \\
| \\
CH_2 \\
| \\
COO^-
\end{array}
\quad
\xrightarrow[\text{dehydrogenase}]{\text{succinate}}
\quad
\text{no reaction}
$$

malonate

In step 6 of Krebs cycle, succinate dehydrogenase catalyzes the oxidation of succinate to fumarate.

Explain why malonate inhibits this reaction.   (*4 marks*)

c Identify which curve on the graph below demonstrates competitive inhibition giving the reasons for your choice.

(1 marks)

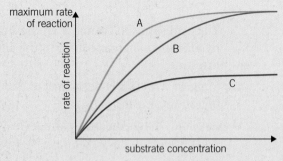

d (i) Explain why end product inhibitors have to be both reversible and competitive *(3 marks)*

(ii) Explain why end product enzyme inhibition is essential for the control of cellular activity. *(3 marks)*

4 Amylase is an enzyme that hydrolyses amylose to maltose. Maltose, like glucose, is a reducing sugar.

A student investigated the action of amylase on amylose. She mixed amylase with amylose and placed the mixture in a water bath.

Describe how she could measure the change in concentration of maltose (reducing sugar) as the reaction proceeds.

In you answer, you should ensure that the steps in the procedure are sequenced correctly *(7 marks)*

OCR F212 2011

5 Figure 4 shows the results that the student obtained from a practical procedure in which the rate of formation of maltose was measured in the presence and absence of chloride ions.

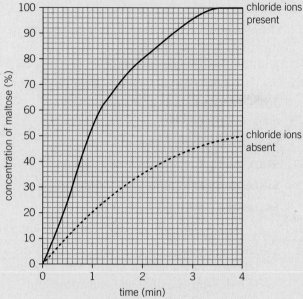

a Describe the effect of chloride ions on the rate of reaction. *(2 marks)*

OCR F212 2011

b State **three** variables that need to be controlled in this practical procedure in order to produce valid results.

(3 marks)

OCR F212 2011

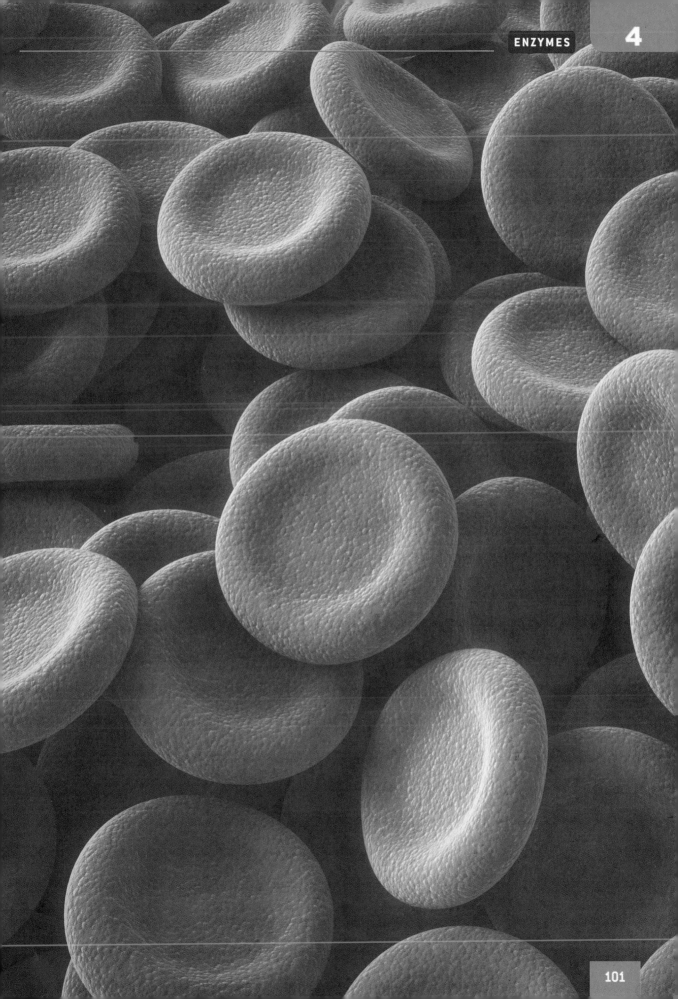

You learnt about the structure of cells in Chapter 2, Basic components of living systems. Membranes are the structures that separate the contents of cells from their environment. They also separate the different areas within cells (organelles) from each other and the cytosol. Some organelles are divided further by internal membranes.

The formation of separate membrane-bound areas in a cell is called **compartmentalisation**. Compartmentalisation is vital to a cell as metabolism includes many different and often incompatible reactions. Containing reactions in separate parts of the cell allows the specific conditions required for cellular reactions, such as chemical gradients, to be maintained, and protects vital cell components.

Membrane structure

All the membranes in a cell have the same basic structure. The cell surface membrane which separates the cell from its external environment is known as the **plasma membrane**.

Membranes are formed from a **phospholipid bilayer**. The hydrophilic phosphate heads of the phospholipids form both the inner and outer surface of a membrane, sandwiching the fatty acid tails of the phospholipids to form a hydrophobic core inside the membrane.

Cells normally exist in aqueous environments. The inside of cells and organelles are also usually aqueous environments. Phospholipid bilayers are perfectly suited as membranes because the outer surfaces of the hydrophilic phosphate heads can interact with water.

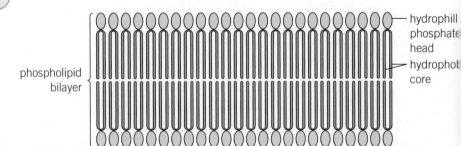

▲ Figure 1 *Phospholipids arranged in a double layer forming a hydrophobic core*

Cell membrane theory

Membranes were seen for the first time following the invention of electron microscopy, which allowed images to be taken with higher magnification and resolution. Images taken in the 1950s showed the membrane as two black parallel lines – supporting an earlier theory that membranes were composed of a lipid bilayer.

In 1972 American scientists Singer and Nicolson proposed a model, building upon an earlier lipid-bilayer model, in which proteins occupy various positions in the membrane. The model is known as the **fluid-mosaic model** because the phospholipids are free to move within the layer relative to each other (they are fluid), giving the membrane flexibility, and because the proteins embedded in the bilayer vary in shape, size, and position (in the same way as the tiles of a mosaic). This model forms the basis of our understanding of membranes today.

A closer look at cell membrane components

Plasma membranes contain various proteins and lipids – the type and number of which are particular to each cell type.

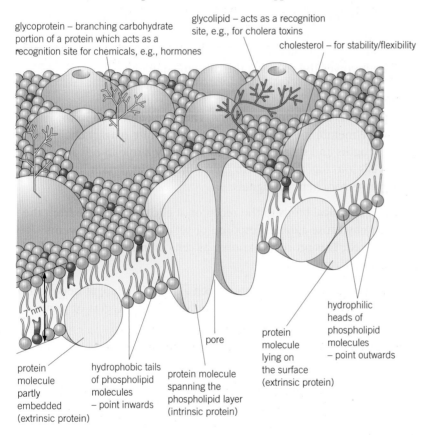

glycoprotein – branching carbohydrate portion of a protein which acts as a recognition site for chemicals, e.g., hormones

glycolipid – acts as a recognition site, e.g., for cholera toxins

cholesterol – for stability/flexibility

hydrophilic heads of phospholipid molecules – point outwards

protein molecule lying on the surface (extrinsic protein)

pore

protein molecule spanning the phospholipid layer (intrinsic protein)

hydrophobic tails of phospholipid molecules – point inwards

protein molecule partly embedded (extrinsic protein)

7 nm

▲ **Figure 2** *The fluid-mosaic model of a plasma membrane showing the many components, including cholesterol, glycolipids and glycoproteins, and membrane proteins*

The components of plasma membranes play an important role in the functions of the membrane and the cell or organelle they are part of.

Membrane proteins

Membrane proteins have important roles in the various functions of membranes. There are two types of proteins in the cell-surface membrane – intrinsic and extrinsic proteins.

Intrinsic proteins

Intrinsic proteins, or integral proteins, are transmembrane proteins that are embedded through both layers of a membrane. They have amino acids with hydrophobic R-groups on their external surfaces, which interact with the hydrophobic core of the membrane, keeping them in place.

Channel and carrier proteins are intrinsic proteins. They are both involved in transport across the membrane.

- **Channel proteins** provide a hydrophilic channel that allows the passive movement (Topic 5.3, Diffusion) of polar molecules and ions down a concentration gradient through membranes. They are held in position by interactions between the hydrophobic core of the membrane and the hydrophobic R-groups on the outside of the proteins.

- **Carrier proteins** have an important role in both passive transport (down a concentration gradient) and active transport (against a concentration gradient) into cells (Topic 5.4, Active transport). This often involves the shape of the protein changing.

Glycoproteins

Glycoproteins are intrinsic proteins. They are embedded in the cell-surface membrane with attached carbohydrate (sugar) chains of varying lengths and shapes. Glycoproteins play a role in cell adhesion (when cells join together to form tight junctions in certain tissues) and as **receptors** for chemical signals.

When the chemical binds to the receptor, it elicits a response from the cell. This may cause a direct response or set off a cascade of events inside the cell. This process is known as cell communication or **cell signalling**. Examples include:

- receptors for neurotransmitters such as acetylcholine at nerve cell synapses. The binding of the neurotransmitters triggers or prevents an impulse in the next neurone

- receptors for peptide hormones, including insulin and glucagon, which affect the uptake and storage of glucose by cells.

Some drugs act by binding to cell receptors. For example, β blockers are used to reduce the response of the heart to stress.

Glycolipids

Glycolipids are similar to glycoproteins. They are lipids with attached carbohydrate (sugar) chains. These molecules are called cell markers or antigens and can be recognised by the cells of the immune system as self (of the organism) or non-self (of cells belonging to another organism).

Extrinsic proteins

Extrinsic proteins or peripheral proteins are present in one side of the bilayer. They normally have hydrophilic R-groups on their outer surfaces and interact with the polar heads of the phospholipids or with intrinsic proteins. They can be present in either layer and some move between layers.

Cholesterol

Cholesterol is a lipid with a hydrophilic end and a hydrophobic end, like a phospholipid. It regulates the fluidity of membranes.

Cholesterol molecules are positioned between phospholipids in a membrane bilayer, with the hydrophilic end interacting with the heads and the hydrophobic end interacting with the tails, pulling them together. In this way cholesterol adds stability to membranes without making them too rigid. The cholesterol molecules prevent the membranes becoming too solid by stopping the phospholipid molecules from grouping too closely and crystallising.

Sites of chemical reactions

Like enzymes, proteins in the membranes forming organelles, or present within organelles, have to be in particular positions for chemical reactions to take place. For example, the electron carriers and the enzyme ATP synthase have to be in the correct positions within the cristae (inner membrane of mitochondrion) for the production of ATP in respiration. The enzymes of photosynthesis are found on the membrane stacks within the chloroplasts.

Synoptic link

You learnt about the structure of cholesterol in Topic 3.5, Lipids.

Summary questions

1   Define the term 'compartmentalisation'.   (2 marks)

2   Describe the difference between intrinsic and extrinsic
    proteins. State two examples of each.   (4 marks)

3   Alcohol, caffeine, and nicotine are all lipid-soluble molecules –
    they have an almost instant and widespread effect on the body.
    Explain why.   (2 marks)

4   Membranes, particularly those present within mitochondria,
    are often highly folded. Suggest what advantages this folding
    provides.   (6 marks)

Synoptic link

You learnt about the structure of mitochondria and chloroplasts in Topic 2.4, Eukaryotic cell structure, and Topic 2.5, The ultrastructure of plant cells. You will learn more about the production of ATP in respiration in Chapter 18, Respiration.

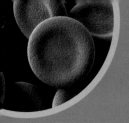

5.2 Factors affecting membrane structure

Specification reference: 2.1.5

Learning outcomes

Demonstrate knowledge, understanding, and application of:

→ factors affecting membrane structure and permeability

→ practical investigations into factors affecting membrane structure and permeability.

Membranes control the passage of different substances into and out of cells (and organelles). If membranes lose their structure, they lose control of this and cell processes will be disrupted. A number of factors affect membrane structure including temperature and the presence of solvents.

Temperature

Phospholipids in a cell membrane are constantly moving. When temperature is increased the phospholipids will have more kinetic energy and will move more. This makes a membrane more fluid and it begins to lose its structure. If temperature continues to increase the cell will eventually break down completely.

This loss of structure increases the permeability of the membrane, making it easier for particles to cross it.

Carrier and channel proteins in the membrane will be denatured at higher temperatures. These proteins are involved in transport across the membrane so as they denature, membrane permeability will be affected.

Solvents

Water, a polar solvent, is essential in the formation of the phospholipid bilayer. The non-polar tails of the phospholipids are orientated away from the water, forming a bilayer with a hydrophobic core. The charged phosphate heads interact with water, helping to keep the bilayer intact.

Many organic solvents are less polar than water for example alcohols, or they are non-polar like benzene. Organic solvents will dissolve membranes, disrupting cells. This is why alcohols are used in antiseptic wipes. The alcohols dissolve the membranes of bacteria in a wound, killing them and reducing the risk of infection.

Pure or very strong alcohol solutions are toxic as they destroy cells in the body. Less concentrated solutions of alcohols, such as alcoholic drinks, will not dissolve membranes but still cause damage. The non-polar alcohol molecules can enter the cell membrane and the presence of these molecules between the phospholipids disrupts the membrane.

When the membrane is disrupted it becomes more fluid and more permeable. Some cells need intact cell membranes for specific functions, for example, the transmission of nerve impulses by neurones (nerve cells). When neuronal membranes are disrupted, nerve impulses are no longer transmitted as normal.

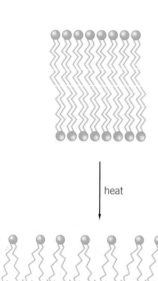

heat

▲ Figure 1 *The increase in kinetic energy of the phospholipids disrupts the structure of the membrane, creating gaps and making it more permeable*

Study tip

Unlike proteins, membranes are not denatured by high temperatures – when they lose their structure they should be described as disrupted or destroyed.

This also happens to neurones in the brain, explaining the changes seen in peoples' behaviour after consuming alcoholic drinks.

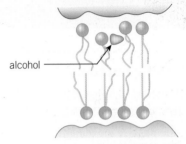

water surface (outside of cell)

alcohol

water surface (inside of cell)

▲ **Figure 2** *The presence of alcohol molecules between the phospholipids disrupts the structure of the membrane*

 Investigating membrane permeability

Beetroot cells contain betalain, a red pigment that gives them their distinctive colour, because of this they are useful for investigating the effects of temperature and organic solvents on membrane permeability. When beetroot cells membranes are disrupted the red pigment is released and the surrounding solution is coloured. The amount of pigment released into a solution is related to the disruption of the cell membranes.

To investigate the effect of temperature on the permeability of cell membranes a student carried out the following procedure. Five small pieces of beetroot of equal size were cut using a cork borer. The beetroot pieces were thoroughly washed in running water, they were then placed in 100 ml of distilled water in a water bath. The temperature of the water bath was increased in 10 °C intervals. Samples of the water containing the beetroot were taken five minutes after each temperature was reached. The absorbance of each sample was measured using a colorimeter with a blue filter. The experiment was done three times, each time with fresh beetroot pieces and a mean calculated for each temperature. Their results are shown in the graph below.

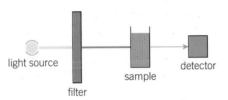

light source sample detector

filter

▲ **Figure 3** *Light first passes through a filter and then the sample. The intensity of light hitting the detector is recorded*

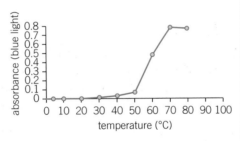

▲ **Figure 4**

1 Suggest why:
   a The beetroot pieces were washed in running water.
   b Samples of the water containing the beetroot were taken five minutes after each temperature was reached.
   c The experiment was repeated three times.
   d The absorbance of the samples was measure using a colorimeter with a blue filter
2 The absorbance of the solution can be calculated from the amount of light transmitted. Suggest how and explain why the absorbance would change as the amount of pigment increases.
3 Look at the graph and suggest at what point the membrane what disrupted.
4 Suggest how you would carry out an investigation to see the effects of organic solvents on membrane permeability.

Summary questions

1 Explain why solvents like water do not disrupt cell membranes. (*2 marks*)

2 Describe how the absorbance of light could be measured quantitatively as the concentration of released pigment increased with increasing temperature. Describe the graph you would plot to show these results. (*4 marks*)

3 Suggest how an excessive consumption of alcohol could lead to liver cell death and ultimately be fatal. (*6 marks*)

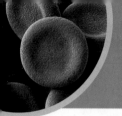

5.3 Diffusion

Specification reference: 2.1.5

Study tip

Diffusion only occurs between different concentrations of the *same* substance.

The exchange of substances between cells and their environment or between membrane-bound compartments within cells and the cell cytosol is defined as either active (requiring metabolic energy) or passive. All movement requires energy. Passive movement, however, utilises energy from the natural motion of particles, rather than energy from an another source. This topic will focus on **passive transport** methods.

Diffusion

Diffusion is the net, or overall, movement of particles (atoms, molecules or ions) from a region of higher concentration to a region of lower concentration. It is a passive process and it will continue until there is a concentration equilibrium between the two areas. Equilibrium means a balance or no difference in concentrations.

Diffusion happens because the particles in a gas or liquid have kinetic energy (they are moving). This movement is random and an unequal distribution of particles will eventually become an equal distribution. Equilibrium doesn't mean the particles stop moving, just that the movements are equal in both directions.

Particles move at high speeds and are constantly colliding, which slows down their overall movement. This means that over short distances diffusion is fast, but as diffusion distance increases the rate of diffusion slows down because more collisions have taken place.

For this reason cells are generally microscopic – the movement of particles within cells depends on diffusion and a large cell would lead to slow rates of diffusion. Reactions would not get the substrates they need quickly enough or ATP would be supplied too slowly to energy-requiring processes.

Factors affecting rate of diffusion

- temperature – the higher the temperature the higher the rate of diffusion. This is because the particles have more kinetic energy and move at higher speeds.

- concentration difference – the greater the difference in concentration between two regions the faster the rate of diffusion. because the overall movement from the higher concentration to lower concentration will be larger.

A concentration difference is said to be a concentration gradient, which goes from high to low concentration. Diffusion takes place *down* a concentration gradient. It takes a lot more energy to move substances *up* a concentration gradient.

So far diffusion in the absence of a barrier or membrane has been considered. This is **simple diffusion**.

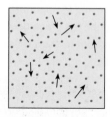

▲ Figure 1 *The random movement of particles means the initial unequal distribution eventually evens out*

Rate of diffusion and surface area

The rate of diffusion can be calculated in two ways – by distance travelled/time and volume filled/time. Distance travelled/time is not affected by changes in surface area, whilst volume/time varies depending on the surface area.

A student used different sized agar blocks to investigate how the rate of diffusion was affected by surface area.

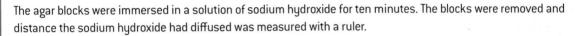

The agar used to make the blocks contained the indicator phenolphthalein with turns pink in the presence of an alkali.

The agar blocks were immersed in a solution of sodium hydroxide for ten minutes. The blocks were removed and distance the sodium hydroxide had diffused was measured with a ruler.

Results

Cube size (cm)	Surface area (cm²)	Volume (cm³)	Surface area / volume	Diffusion distance (cm)	Rate of diffusion using distance (cm / min)	Rate of diffusion using volume (cm³ / min)	Rate of diffusion using volume per 64 cm³ agar
4 × 4 × 4				0.3			
2 × 2 × 2				0.3			
1 × 1 × 1				0.3			

1 Copy and complete the table.
2 Explain what has happened in order for the diffusion distances to be measured.
3 Describe and explain the results.

Diffusion across membranes

Diffusion across membranes involves particles passing through the phospholipid bilayer. It can only happen if the membrane is permeable to the particles – non-polar molecules such as oxygen (O_2) diffuse through freely down a concentration gradient.

The hydrophobic interior of the membrane repels substances with a positive or negative charge (ions), so they cannot easily pass through. Polar molecules, such as water (H_2O) with partial positive and negative charges can diffuse through membranes, but only at a very slow rate. Small polar molecules pass through more easily than larger ones. Membranes are therefore described as **partially permeable**.

The rate at which molecules or ions diffuse across membranes is affected by:

- surface area – the larger the area of an exchange surface, the higher the rate of diffusion
- thickness of membrane – the thinner the exchange surface, the higher the rate of diffusion.

Synoptic link

You will learn about surface area to volume ratio in Topic 7.1, Specialised exchange surfaces.

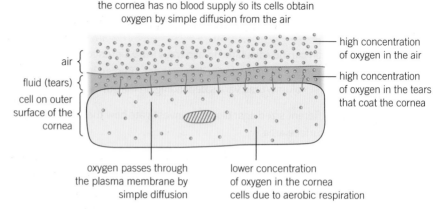

the cornea has no blood supply so its cells obtain
oxygen by simple diffusion from the air

air — high concentration of oxygen in the air

fluid (tears) — high concentration of oxygen in the tears that coat the cornea

cell on outer surface of the cornea

oxygen passes through the plasma membrane by simple diffusion

lower concentration of oxygen in the cornea cells due to aerobic respiration

▲ Figure 2 *Passive diffusion of oxygen into a cell of the cornea*

Facilitated diffusion

As you have learnt, the phospholipid bilayers of membranes are barriers to polar molecules and ions. However, membranes contain channel proteins through which polar molecules and ions can pass. Diffusion across a membrane through protein channels is called **facilitated diffusion**.

Membranes with protein channels are **selectively permeable** as most protein channels are specific to one molecule or ion.

Facilitated diffusion can also involve carrier proteins (Topic 5.4, Active transport), which change shape when a specific molecule binds. In facilitated diffusion the movement of the molecules is down a concentration gradient and does not require external energy.

The rate of facilitated diffusion is dependent on the temperature, concentration gradient, membrane surface area and thickness, but is also affected by the number of channel proteins present. The more protein channels, the higher the rates of diffusion overall.

 ## Investigations into the factors affecting diffusion rates in model cells

As you have already seen, cell membranes are highly complex structures involved in the active and passive transport of ions and molecules. The hydrophobic hydrocarbon core of the membrane is a barrier to ions and large polar molecules, but it allows the passage of non-polar molecules. Cells are too small and cell membranes too thin to use in practical investigations so dialysis tubing is used as a substitute membrane. This model enables us to investigate the effects of temperature and concentration on the rate of diffusion across membranes.

Dialysis tubing is partially permeable, with pores a similar size to those on a partially permeable membrane. This means that small molecules like water can pass through it, but larger molecules like starch cannot fit through the pores. The tubing is therefore a barrier to large molecules.

A model cell can be simulated by tying one end of a section of tubing, filling with a solution and then tying the other end. The 'cell' is then placed into another solution. The solutions could contain different sizes, or concentrations, of solute molecules.

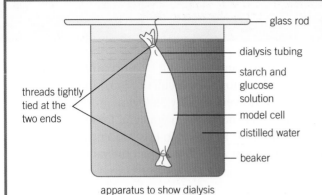

threads tightly tied at the two ends

glass rod
dialysis tubing
starch and glucose solution
model cell
distilled water
beaker

apparatus to show dialysis

▲ **Figure 3** *The apparatus above is used to demonstrate that glucose molecules are small enough to diffuse out of the 'cell' but the starch molecules are too large. After a set time the water is tested for the presence of starch and glucose*

The changes in concentration of solute molecules, both inside and outside the model cells, can be measured over time. Rates of diffusion across the tubing can then be calculated.

Glucose is a small molecule which can cross the tubing. Benedict's solution is used to test for the presence of glucose, and can also be used to estimate concentration.

Starch molecules are large and will not cross the tubing. Iodine is used to test for the presence of starch.

Water is a small molecule which will pass through the tubing while other solutes such as sucrose will not. Model cells can be placed in solutions with different solute concentrations. The rates of osmosis can be calculated using changes in volume or mass of the model cells over time.

Rates of diffusion at different temperatures can also be calculated using a water bath to change the temperature of the model cell. Other variables such as concentration must be then be kept constant.

1   Explain why Benedict's test is both quantitative and qualitative.
2   Explain what is meant by the term model cell.
3   a   Describe the differences between dialysis tubing and cell membranes with reference to transport across membranes.
    b   Explain why some ions can pass through dialysis tubing by diffusion but can only pass through cell membranes by facilitated diffusion.

Summary questions

1   Explain why the rate of diffusion increases as temperature increases.   (*2 marks*)

2   State two changes to the structure of a cell-surface membrane that would increase the rate at which polar molecules diffuse into a cell.   (*2 marks*)

3   Movement requires energy and yet the movement of molecules in diffusion is described as passive (not requiring energy). Explain this statement and state the source of the energy involved in diffusion.   (*2 marks*)

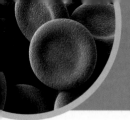

5.4 Active transport

Specification reference: 2.1.5

Learning outcomes

Demonstrate knowledge, understanding, and application of:

→ active transport of molecules across membranes

→ endocytosis and exocytosis as processes requiring ATP.

Synoptic link

You learnt about ATP as an energy supply in Topic 3.11, ATP.

Diffusion, by its nature, will ultimately result in concentration gradients being reduced until particles (atoms, molecules or ions) in the different regions reach equilibrium. However, many biological processes depend on the presence of a concentration gradient, for example, the transmission of nerve impulses. To maintain this concentration gradient, particles must be moved up it at a rate faster than the rate of diffusion. This is an energy-requiring process called **active transport.**

Active transport

Active transport is the movement of molecules or ions into or out of a cell from a region of lower concentration to a region of higher concentration. The process requires energy and carrier proteins. Energy is needed as the particles are being moved up a concentration gradient, in the opposite direction to diffusion. Metabolic energy is supplied by ATP.

Carrier proteins span the membranes and act as 'pumps'. The general process of active transport is described below – in this example transport is from outside to inside a cell (Figure 1).

1   The molecule or ion to be transported binds to receptors in the channel of the carrier protein on the outside of the cell.

2   On the inside of the cell ATP binds to the carrier protein and is hydrolysed into ADP and phosphate.

3   Binding of the phosphate molecule to the carrier protein causes the protein to change shape – opening up to the inside of the cell.

4   The molecule or ion is released to the inside of the cell.

5   The phosphate molecule is released from the carrier protein and recombines with ADP to form ATP.

6   The carrier protein returns to its original shape.

The process is selective – specific substances are transported by specific carrier proteins.

Bulk transport

Bulk transport is another form of active transport. Large molecules such as enzymes, hormones, and whole cells like bacteria are too large to move through channel or carrier proteins, so they are moved into and out of cell by bulk transport.

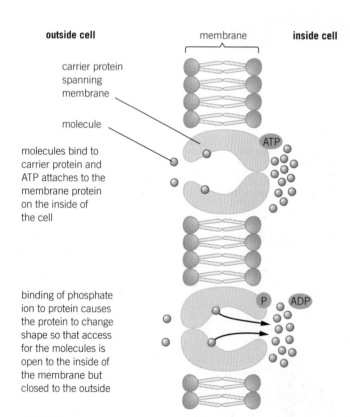

outside cell   membrane   inside cell

carrier protein spanning membrane

molecule

molecules bind to carrier protein and ATP attaches to the membrane protein on the inside of the cell

binding of phosphate ion to protein causes the protein to change shape so that access for the molecules is open to the inside of the membrane but closed to the outside

▲ Figure 1 Active transport. The shape of the carrier protein changes to move a particle from one side of the membrane to the other

- **Endocytosis** is the bulk transport of material *into* cells. There are two types of endocytosis, **phagocytosis** for solids and **pinocytosis** for liquids – the process is the same for both. The cell-surface membrane first invaginates (bends inwards) when it comes into contact with the material to be transported. The membrane enfolds the material until eventually the membrane fuses, forming a vesicle. The vesicle pinches off and moves into the cytoplasm to transfer the material for further processing within the cell. For example, vesicles containing bacteria are moved towards lysosomes, where the bacteria are digested by enzymes.

- **Exocytosis** is the reverse of endocytosis. Vesicles, usually formed by the Golgi apparatus, move towards and fuse with the cell surface membrane. The contents of the vesicle are then released *outside* of the cell.

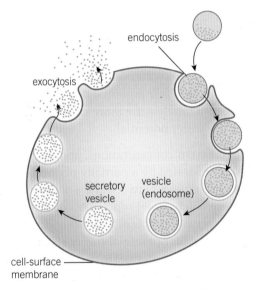

▲ **Figure 2** *Exocystosis is the reverse of endocytosis*

Energy in the form of ATP is required for movement of vesicles along the cytoskeleton, changing the shape of cells to engulf materials, and the fusion of cell membranes as vesicles form or as they meet the cell-surface membrane.

Summary questions

1 Explain why facilitated diffusion is not a form of active transport.
(2 marks)

2 Cells that carry out active transport usually have more mitochondria than cells that do not. Explain why. *(2 marks)*

3 Plant roots take up mineral ions from the soil. The concentration of mineral ions in the soil water is very low. Suggest why active transport is very important in root hair cells. *(4 marks)*

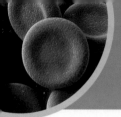

5.5 Osmosis

Study tip

Remember that all water potential values are negative. Pure water has a water potential of zero.

Osmosis is a particular type of diffusion – specifically the diffusion of water across a partially permeable membrane. As with all types of diffusion it is a passive process and energy is not required.

Water potential

A solute is a substance dissolved in a solvent (for example water) forming a solution.

The amount of solute in a certain volume of aqueous solution is the concentration. **Water potential** is the pressure exerted by water molecules as they collide with a membrane or container. It is measured in units of pressure pascals (Pa) or kilopascals (kPa). The symbol for water potential is the Greek letter psi Ψ.

Pure water is defined as having a water potential of 0 kPa (at standard temperature and atmospheric pressure – 25 °C and 100 kPa). This is the highest possible value for water potential, as the presence of a solute in water lowers the water potential below zero. All solutions have negative water potentials – the more concentrated the solution the more negative the water potential.

When solutions of different concentrations, and therefore different water potentials, are separated by a partially permeable membrane, the water molecules can move between the solutions but the solutes usually cannot. There will be a net movement of water from the solution with the higher water potential (less concentrated) to the solution with the lower water potential (more concentrated). This will continue until the water potential is equal on both sides of the membrane (equilibrium).

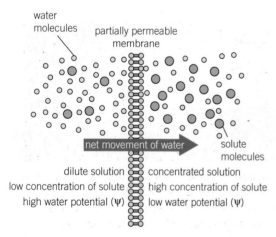

▲ **Figure 1** *Due to the greater number of water molecules on the left-hand side of the partially permeable membrane, diffusion occurs until the number of water molecules is equal on both sides of the membrane. This movement is called osmosis*

Effects of osmosis on plant and animal cells

The diffusion of water into a solution leads to an increase in volume of this solution. If the solution is in a closed system, such as a cell, this results in an increase in pressure. This pressure is called **hydrostatic pressure** and has the same units as water potential, kPa. At the cellular level this pressure is relatively large and potentially damaging.

Animal cells

If an animal cell is placed in a solution with a higher water potential than that of the cytoplasm, water will move into the cell by osmosis, increasing the hydrostatic pressure inside the cell. All cells have thin cell-surface membranes (around 7 nm) and no cell walls. The cell-surface membrane cannot stretch much and cannot withstand the increased pressure. It will break and the cell will burst, an event called **cytolysis**.

If an animal cell is placed in a solution that has a lower water potential than the cytoplasm it will lose water to the solution by osmosis down the water potential gradient. This will cause a reduction in the volume of the cell and the cell-surface membrane to 'pucker', referred to as crenation (Figure 2).

To prevent either cytolysis or crenation, multicellular animals usually have control mechanisms to make sure their cells are continuously surrounded by aqueous solutions with an equal water potential (isotonic). In blood the aqueous solution is blood plasma.

▼ Table 1 *Osmosis in a red blood cell*

Water potential (Ψ) of external solution compared to cell solution	Higher (less negative)	Equal	Lower (more negative)
Net movement of water	Enters cell	Water constantly enters and leaves, but at equal rates	Leaves cell
State of cell	Swells and bursts	No change	Shrinks
	contents, including haemoglobin, are released / remains of cell surface membrane	normal red blood cell	haemoglobin is more concentrated, giving cell a darker appearance / cell shrunken and shrivelled

Plant cells

Like animal cells, plant cells contain a variety of solutes, mainly dissolved in a large vacuole. However, unlike animals, plants are unable to control the water potential of the fluid around them, for example, the roots are usually surrounded by almost pure water.

Plants cells have strong cellulose walls surrounding the cell-surface membrane. When water enters by osmosis the increased hydrostatic pressure pushes the membrane against the rigid cell walls. This pressure against the cell wall is called **turgor**. As the turgor pressure increases it resists the entry of further water and the cell is said to be turgid.

When plant cells are placed in a solution with a lower water potential than their own, water is lost from the cells by osmosis. This leads to a reduction in the volume of the cytoplasm, which eventually pulls the cell-surface membrane away from the cell wall – the cell is said to be plasmolysed.

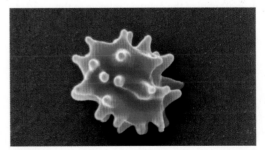

▲ Figure 2 *Scanning electron micrograph of a red blood cell that has been placed in a solution of lower water potential than the cytoplasm and become crenated by osmosis (× 5000 magnification)*

▼ Table 2 *Osmosis in a plant cell*

Water potential (Ψ) of external solution compared to cell solution	Higher (less negative)	Equal	Lower (more negative)
Net movement of water	Enters cell	Water constantly enters and leaves, but at equal rates	Leaves cell
Condition of protoplast	Swells and becomes turgid	No change	Plasmolysis, contents shrink
	protoplast pushed against cell wall nucleus cellulose cell wall protoplast	protoplast beginning to pull away from the cell wall	protoplast completely pulled away from the cell wall space filled with external solution of lower water potential

Osmosis investigations

The effect of solutions with different water potentials can be observed in both plant and animal cells.

Plant cells

Pieces of potato or onion can be placed into sugar or salt solutions with different concentrations, and therefore different water potentials. Water will move into or out of cells depending on the water potential of the solution relative to the water potential of the plant tissue. As the plant tissue gains or loses water it will increase or decrease in mass and size, and vice versa.

A student used potato cores and their knowledge of osmosis to investigate the water potential of potato cells.

The following results were obtained.

Sugar concentration mol dm^{-3}	Original mass (g)	Final mass (g)	Difference in mass (g)	% mass change	Mean % mass change
0.0	3.0	4.0			
	3.0	4.1			
	3.3	4.2			
0.1	3.0	3.5			
	3.2	3.6			
	2.9	3.3			
0.3	3.0	3.0			
	2.9	3.0			
	3.2	3.2			
0.5	3.2	2.8			
	3.0	2.6			
	3.1	2.7			
0.7	3.1	2.2			
	3.3	2.4			
	3.0	2.0			

1 Copy and complete the table to show final masses, percentage mass changes and mean percentage changes.
2 Plot a graph of the results.
3 Describe and explain the shape of the graph.
4 Write a short evaluation using the information given and suggest any improvements to the investigation.

Animal cells

Eggs can be used to demonstrate osmosis in animal cells. A chicken's egg is not exactly a single cell, but with the shell removed a single membrane-bound structure remains and it will behave in the same way as a cell when placed in solutions of varying water potentials.

To investigate osmosis, eggs without their shells are placed in different concentrations of sugar syrup. Over time, osmosis takes place and there will have been a net movement of water into or out of the eggs, depending on the concentration of the syrup they were in. (Note that if the egg is hard boiled for easier handling that this will damage the membrane.)

Summary questions

1 Copy the diagram below and use arrows to show the net movement of water. *(2 marks)*

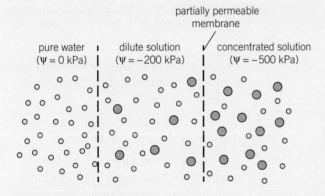

2 Explain why it is not possible to have a positive water potential. *(2 marks)*

3 At which point on the graph in Figure 3 is the water potential of the solution equal to the water potential of the cells? *(1 mark)*

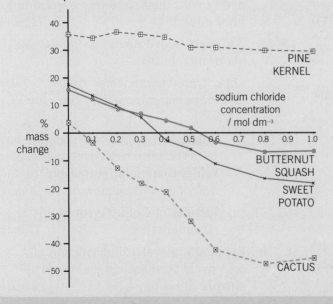

◀ **Figure 3** *Mass changes in plant tissues bathed in salt solutions*

4 Explain why it is important to keep the concentrations of electrolytes (solutes) in body tissues at the correct level to ensure proper hydration. *(3 marks)*

5 Look at Figure 3 again. State which plant tissue in these data has the highest solute concentration (lowest water potential). Suggest a reason for this. *(4 marks)*

Practice questions

1 Solution A has a more negative water potential than solution B.

Which of the following statements is/are correct?

Statement 1: Solution B has a higher water potential than solution A

Statement 2: There would be a net movement of water from solution A to solution B

Statement 3: There would be a net movement of water from solution A to distilled water

A 1, 2 and 3 are correct

B Only 1 and 2 are correct

C Only 2 and 3 are correct

D Only 1 is correct   (*1 mark*)

2 Membrane proteins are essential in order for cells to interact with their environment and other cells. About a third of the genes in a human genome code for membrane proteins.

a (i) Explain what is meant by the term simple protein.   (*2 marks*)

(ii) List the different levels of protein structure.   (*3 marks*)

b Outline the roles of proteins in a cell surface membrane.   (*5 marks*)

3 A plasma membrane has a very complex structure which is affected by changes in the cells environment. Changes in pH or temperature affect the permeability of plasma membranes.

a Describe how you would use a colorimeter to investigate how cell membrane permeability changes with temperature.   (*6 marks*)

b Describe what you would do to ensure your results were:

(i) *valid*   (*2 marks*)

(ii) *reliable*   (*2 marks*)

c Suggest why alcohol is used in antiseptic wipes.   (*3 marks*)

4 The graph below shows the permeability of a plasma membrane to different organic solvents.

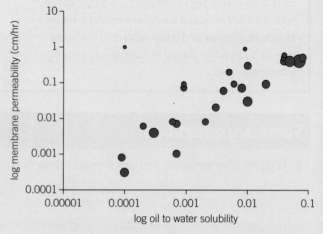

a Describe the relationship between lipid solubility and membrane permeability of organic molecules.   (*2 marks*)

b The size of the circles is proportional to the size of the organic molecule.

It is known that with two organic molecules of the same lipid solubility, the one with greater molecular weight, or size, will cross the membrane more slowly.

State whether the data in the graph above agrees with this fact. Suggest a reason for your answer.   (*3 marks*)

5 a A student stated the following answer about membranes

The cell membrane is a thin semi-permeable membrane that surrounds the cytoplasm of a cell

(i) Suggest why the student should have used the term 'selectively permeable' rather than 'semi-permeable' to describe a cell membrane.   (*2 marks*)

(ii) Identify one other error in this statement   (*1 mark*)

b Explain how water molecules cross selectively permeable membranes by **simple** diffusion.   (*2 marks*)

6   A classic experiment investigated the effect of temperature on the rate of sugar transport in a potted plant.

Aphid mouthparts were used to take samples of sugar solution from the transport tissue in the stem. The sugary solution dripped from the mouthparts. The number of drips per minute was counted.

The procedure was repeated at different temperatures.

Table 1 shows the results obtained

Temperature (°C)	Number of drips per minute
5	3
10	6
20	14
30	26
40	19
50	0

Suggest brief explanations for these results.

(3 marks)

OCR F211 2012

7   The bilayer is the fundamental structure of all cell membranes. The bilayer is composed of two lipid layers which provide an effective barrier to aqueous environments. This allows for compartmentalisation and the formation of cells and organelles.

Discuss the roles of the different lipid components of plasma membranes.   (6 marks)

6 CELL DIVISION
6.1 The cell cycle
Specification reference: 2.1.6

The **cell cycle** is a highly ordered sequence of events that takes place in a cell, resulting in division of the cell, and the formation of two genetically identical daughter cells.

Phases of the cell cycle

In eukaryotic cells the cell cycle has two main phases – interphase and mitotic (division) phase.

Interphase

Cells do not divide continuously – long periods of growth and normal working separate divisions. These periods are called **interphase** and a cell spends the majority of its time in this phase.

Interphase is sometimes referred to as the resting phase as cells are not actively dividing. However, this is not an accurate description – interphase is actually a very active phase of the cell cycle, when the cell is carrying out all of its major functions such as producing enzymes or hormones, while also actively preparing for cell division.

During interphase:

- DNA is replicated and checked for errors in the nucleus
- protein synthesis occurs in the cytoplasm
- mitochondria grow and divide, increasing in number in the cytoplasm
- chloroplasts grow and divide in plant and algal cell cytoplasm, increasing in number
- the normal metabolic processes of cells occur (some, including cell respiration, also occur throughout cell division).

The three stages of interphase, as shown in Figure 1 are:

- G_1 – the first growth phase: proteins from which organelles are synthesised are produced and organelles replicate. The cell increases in size.
- S – synthesis phase: DNA is replicated in the nucleus.
- G_2 – the second growth phase: the cell continues to increase in size, energy stores are increased and the duplicated DNA is checked for errors.

Mitotic phase

The mitotic phase is the period of cell division. Cell division involves two stages:

G2 the cell checks the duplicated chromosomes for error, making any repairs that are needed.

mitosis

cytokinesis

interphase

S each of the chromosomes is duplicated

G1 cellular contents, apart from the chromosomes are duplicated.

G0

▲ Figure 1 *The cell cycle showing the stages of interphase*

Synoptic link

You learnt about DNA replication in Topic 3.9, DNA replication and the genetic code. You learnt about protein synthesis in Topic 2.4, Eukaryotic cell structure and Topic 3.10, Protein synthesis.

Study tip

Interphase is not a stage in cell division (mitosis) but the stage between cell divisions. Interphase is the longest phase of the cell cycle.

- **Mitosis** – the nucleus divides.
- **Cytokinesis** – the cytoplasm divides and two cells are produced.

The processes that take place during mitosis and cytokinesis (division of the cell into two separate cells) are discussed in more detail in Topic 6.2, Mitosis.

G_0

G_0 is the name given to the phase when the cell leaves the cycle, either temporarily or permanently. There are a number of reasons for this including:

- Differentiation – A cell that becomes specialised to carry out a particular function (differentiated) is no longer able to divide. It will carry out this function indefinitely and not enter the cell cycle again (you will learn more about cell specialisation and differentiation in Topic 6.4, The organisation and specialisation of cells).

- The DNA of a cell may be damaged, in which case it is no longer viable. A damaged cell can no longer divide and enters a period of permanent cell arrest (G_0). The majority of normal cells only divide a limited number of times and eventually become senescent.

- As you age, the number of these cells in your body increases. Growing numbers of senescent cells have been linked with many age related diseases, such as cancer and arthritis.

A few types of cells that enter G_0 can be stimulated to go back into the cell cycle and start dividing again, for example lymphocytes (white blood cells) in an immune response.

Control of the cell cycle

It is vital to ensure a cell only divides when it has grown to the right size, the replicated DNA is error-free (or is repaired) and the chromosomes are in their correct positions during mitosis. This is to ensure the fidelity of cell division – that two identical daughter cells are created from the parent cell.

Checkpoints are the control mechanisms of the cell cycle. They monitor and verify whether the processes at each phase of the cell cycle have been accurately completed before the cell is allowed to progress into the next phase.

Checkpoints occur at various stages of the cell cycle:

- G_1 checkpoint – This checkpoint is at the end of the G_1 phase, before entry into S phase. If the cell satisfies the requirements of this checkpoint (Figure 2) it is triggered to begin DNA replication. If not, it enters a resting state (G_0).

- G_2 checkpoint – This checkpoint is at the end of G_2 phase, before the start of

> **Synoptic link**
>
> You will learn about the primary immune response in Topic 12.6, The specific immune system.

> **Synoptic link**
>
> You learnt about the cytoskeleton of cells, including spindle fibres, in Topic 2.4, Eukaryotic cell structure.

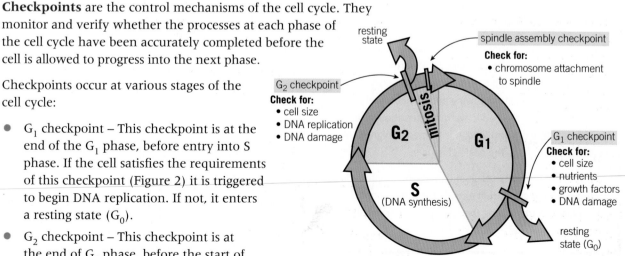

▲ **Figure 2** *Checkpoints controlling the cell cycle*

the mitotic phase. In order for this checkpoint to be passed, the cell has to check a number of factors (Figure 2), including whether the DNA has been replicated without error. If this checkpoint is passed, the cell initiates the molecular processes that signal the beginning of mitosis (Topic 6.2, Mitosis).

- Spindle assembly checkpoint (also called **metaphase** checkpoint): This checkpoint is at the point in mitosis where all the chromosomes should be attached to spindles and have aligned (metaphase – Topic 6.2, Mitosis). Mitosis cannot proceed until this checkpoint is passed.

Cell-cycle regulation and cancer

The passing of a cell-cycle checkpoint is brought about by kinases. These are a class of enzyme that catalyse the addition of a phosphate group to a protein (phosphorylation). Phosphorylation changes the tertiary structure of checkpoint proteins, activating them at certain points in the cell cycle.

Kinases involved in cell-cycle regulation are activated by binding to a variety of checkpoint proteins called cyclins. Binding of the correct cyclin to the appropriate kinase forms a cyclin-dependent kinase (CDK) complex. These complexes are activated by enzymes.

CDK complexes catalyse the activation of key cell-cycle proteins by phosphorylation. This ensures a cell progresses through the different phases of its cycle at the appropriate times. Different enzymes break down cyclins when they are not needed, signalling a cell to move into the next phase of the cycle.

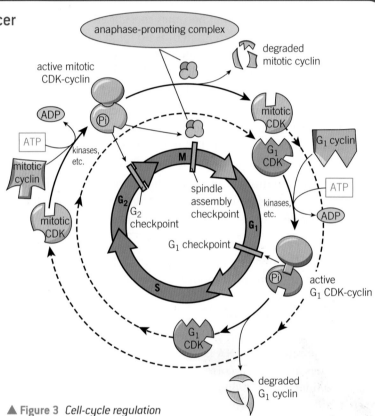

▲ **Figure 3** *Cell-cycle regulation*

Cancer is a group of many different diseases caused by the uncontrolled division of cells. An abnormal mass of cells is called a tumour. Tumours can be benign, meaning that they stop growing and do not travel to other locations in the body. If a tumour continues to grow unchecked and uncontrolled, it is termed malignant. A malignant tumour is the basis of cancer.

Tumours are often the result of damage or spontaneous mutation of the genes that encode the proteins that are involved in regulating the cell cycle, including the checkpoint proteins.

For example, if overexpression of a cyclin gene results from mutation, the abnormally large quantity of cyclins produced would disrupt the regulation of the cell cycle, resulting in uncontrolled cell division, tumour formation, and possibly leading to cancer.

Cyclin-dependent kinases can be used as a possible target for chemical inhibitors in the treatment of cancer. If the activity of CDKs can be reduced it may reduce or stop cell division and therefore cancer formation.

1 Compare the roles of cyclins and enzyme inhibitors. Describe any similarities and/or differences between their roles.

Summary questions

1   Mitosis and cytokinesis are processes involved in the production of new cells. Explain the difference between mitosis and cytokinesis.   (*2 marks*)

2   Explain, with reference to the structure and function of proteins, the importance of G_2 checkpoint.   (*3 marks*)

3   Telomeres are repetitive sequences of DNA at the ends of chromosomes. They protect the genes at the end of chromosomes and stop the ends of chromosomes fusing. DNA is not replicated all the way to the end so every time DNA replication occurs the telomeres shorten. This limits the number of times a cell can replicate, it is called the Hayflick limit.

   a   Suggest a disadvantage of indefinite cell division.   (*3 marks*)

   b   Telomerases are enzymes that result in the elongation of telomeres. They are not usually present in differentiated cells. Describe what the presence of telomerases could cause.   (*2 marks*)

4   A typical human cell contains 3×10^9 base pairs of DNA divided into 46 chromosomes.

   DNA replication in eukaryotic cells takes place at the rate of about 50 base pairs per minute.

   a   Suggest why the length of DNA is usually given by the number of base pairs rather than number of nucleotides.   (*1 mark*)

   b   Calculate the time it would take to replicate a section of DNA of this length, assuming replication started at one end and didn't stop until reaching the other end.   (*3 marks*)

   c   The DNA of a eukaryotic cell is usually replicated in eight hours. Explain how this is possible.   (*2 marks*)

   d   Suggest why it takes a much shorter time to replicate the genome of a prokaryotic cell.   (*1 mark*)

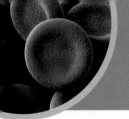
The importance of mitosis

Mitosis is the term usually used to describe the entire process of cell division in eukaryotic cells. It actually refers to nuclear division (division of the nucleus), an essential stage in cell division. Mitosis ensures that both daughter cells produced when a parent cell divides are genetically identical (except in the rare events where mutations occur). Each new cell will have an exact copy of the DNA present in the parent cell and the same number of chromosomes.

Mitosis is necessary when all the daughter cells have to be identical. This is the case during growth, replacement and repair of tissues in multicellular organisms such as animals, plants, and fungi. Mitosis is also necessary for **asexual reproduction**, which is the production of genetically identical offspring from one parent in multicellular organisms including plants, fungi, and some animals, and also in eukaryotic single-celled organisms such as *Ameoba* species. Prokaryotic organisms, including bacteria, do not have a nucleus and they reproduce asexually by a different process known as binary fission.

Chromosomes

Before mitosis can occur, all of the DNA in the nucleus is replicated during interphase (Topic 6.1, The cell cycle). Each DNA molecule (chromosome) is converted into two identical DNA molecules, called **chromatids**.

The two chromatids are joined together at a region called the **centromere**. It is necessary to keep the chromatids together during mitosis so that they can be precisely manoeuvred and segregated equally, one each into the two new daughter cells (Figures 1 and 2).

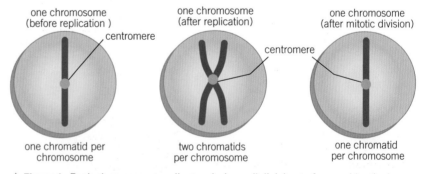

▲ Figure 1 *Each chromosome replicates during cell division to form an identical copy, or chromatid. The chromatids of a pair are joined at the centromere*

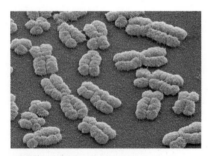

▲ Figure 2 *Scanning electron micrograph of replicated human chromosomes consisting of chromatid pairs joined at the centromere (× 3000 magnification)*

The stages of mitosis

There are four stages of mitosis – **prophase**, **metaphase**, **anaphase**, and **telophase**. We describe them separately but in fact they flow seamlessly from one to another. Each of these phases can be viewed and indentified using a light microscope. Dividing cells can be easily obtained from growing root tips of plants. The root tips can be treated with a chemical to allow the cells to be separated – then they can be squashed to form a single layer of cells on a microscope slide. Stains that bind DNA are used to make the chromosomes clearly visible.

The description of the four stages with example micrographs and labelled diagrams will help you to identify each phase in your own cell sections:

Prophase

1 During prophase, chromatin fibres (complex made up of various proteins, RNA and DNA) begin to coil and condense to form chromosomes that will take up stain to become visible under the light microscope. The nucleolus, a distinct area of the nucleus responsible for RNA synthesis, disappears. The nuclear membrane begins to break down.

2 Protein microtubules form spindle-shaped structures linking the poles of the cell. The fibres forming the spindle are necessary to move the chromosomes into the correct positions before division.

3 In animal cells and some plant cells, two centrioles migrate to opposite poles of the cell. The centrioles are cylindrical bundles of proteins that help in the formation of the spindle.

4 The spindle fibres attach to specific areas on the centromeres and start to move the chromosomes to the centre of the cell.

5 By the end of prophase the nuclear envelope has disappeared.

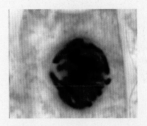

▲ **Figure 3** *Light micrograph of an onion (*Allium *spp.) root tip cell during* **prophase**. *The chromosomes are beginning to condense into defined units, which will allow organised and equal segregation into daughter cells*

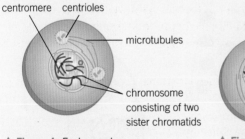

▲ **Figure 4** *Early prophase*

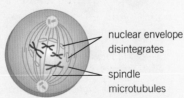

▲ **Figure 5** *Late prophase*

Metaphase

During metaphase the chromosomes are moved by the spindle fibres to form a plane in the centre of the cell, called the **metaphase plate**, and then held in position.

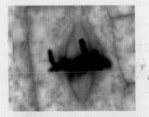

▲ **Figure 6** *Light micrograph of an onion root tip cell during* **metaphase**. *The chromosomes are organized in a plane across the centre of the cell*

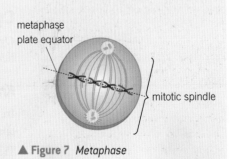

▲ **Figure 7** *Metaphase*

Anaphase

The centromeres holding together the pairs of chromatids in each chromosome divide during anaphase. The chromatids are separated – pulled to opposite poles of the cell by the shortening spindle fibres.

The characteristic 'V' shape of the chromatids moving towards the poles is a result of them being dragged by their centromeres through the liquid cytosol.

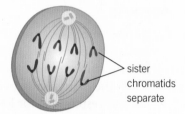

▲ Figure 9 *Anaphase*

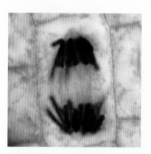

▲ Figure 8 *Light micrograph of an onion root tip cell during anaphase. The sister chromatids are moved to separate poles by shortening spindle fibres*

Telophase

In telophase the chromatids have reached the poles and are now called chromosomes. The two new sets of chromosomes assemble at each pole and the nuclear envelope reforms around them. The chromosomes start to uncoil and the nucleolus is formed. Cell division – or cytokinesis, begins.

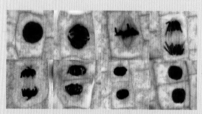

▲ Figure 10 *Light micrograph of onion root tip cells during* **telophase***. Sister chromatids have been moved to separate poles by spindle fibres and cell division begins*

▲ Figure 11 *Telophase*

Cytokinesis

Cytokinesis, the actual division of the cell into two separate cells, begins during telophase.

Animal cells

In animal cells a cleavage furrow forms around the middle of the cell. The cell-surface membrane is pulled inwards by the cytoskeleton until it is close enough to fuse around the middle, forming two cells (Figure 13).

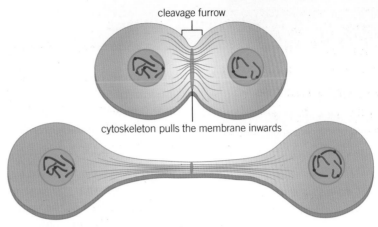

▲ Figure 13 *Cytokinesis in an animal cell*

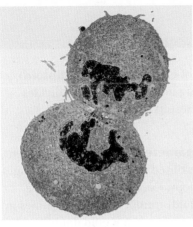

▲ Figure 12 *Cytokinesis in a human embryonic kidney cell. Coloured transmission electron micrograph × 800 magnification*

Plant cells

Plant cells have cell walls so it is not possible for a cleavage furrow to be formed. Vesicles from the Golgi apparatus begin to assemble in the same place as where the metaphase plate was formed. The vesicles fuse with each other and the cell surface membrane, dividing the cell into two (Figure 15).

New sections of cell wall then form along the new sections of membrane (if the dividing cell wall were formed before the daughter cells separated they would immediately undergo osmotic lysis from the surrounding water).

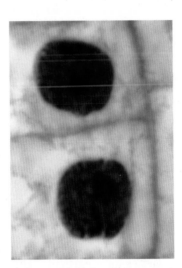

▲ Figure 14 *Light micrograph of cytokinesis in an onion root tip cell*

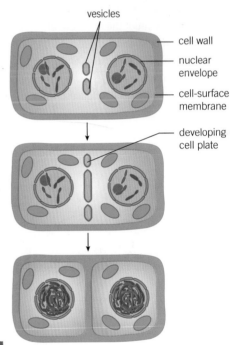

▲ Figure 15 *Cytokinesis in plant cells*

Summary questions

1 Explain why we normally see chromosomes as a double structure containing two chromatids. *(2 marks)*

2 Explain why it is essential that DNA replication results in two exact copies of the genetic material. *(2 marks)*

3 Describe the differences between cytokinesis in animal and plant cells and give reasons for these differences. *(4 marks)*

4 How many chromatids would be present in a human cell at prophase and at G_1 during interphase? *(2 marks)*

5 Explain why plant root tips are a good source of cells to examine for mitosis. *(4 marks)*

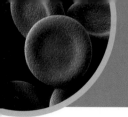

6.3 Meiosis

Specification reference: 2.1.6

Normal cells have two chromosomes of each type (termed **diploid**) – one inherited from each parent. During mitosis the nucleus divides once following DNA replication. This results in two genetically identical diploid daughter cells.

In sexual reproduction two sex cells (**gametes**), one from each parent, fuse to produce a fertilised egg. The fertilised egg (**zygote**) is the origin of all the cells that the organism develops. Gametes must therefore only contain half of the standard (diploid) number of chromosomes in a cell or the chromosome number of an organism would double with every round of reproduction.

Gametes are formed by another form of cell division known as **meiosis**. Unlike in mitosis, the nucleus divides twice to produce four daughter cells – the gametes. Each gamete contains half of the chromosome number of the parent cell – it is **haploid**. Meiosis is therefore known as **reduction division**.

Homologous chromosomes

As you will remember, each characteristic of an organism is coded for by two copies of each gene, one from each parent. Each nucleus of the organism's cells contains two full sets of genes, a pair of genes for each characteristic. Therefore each nucleus contains matching sets of chromosomes, called **homologous chromosomes**, and is termed diploid. Each chromosome in a homologous pair has the same genes at the same loci.

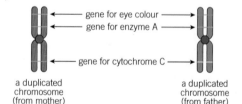

gene for eye colour
gene for enzyme A
gene for cytochrome C

a duplicated chromosome (from mother)   a duplicated chromosome (from father)

▲ **Figure 1** *A homologous pair of chromosomes. The same genes and all the alleles of that gene will appear at the same position (locus) on the chromosome*

Alleles

Genes for a particular characteristic may vary, leading to differences in the characteristic, for example blue eyes and brown eyes. The genes are still the same type as they both code for eye colour but the colour is different, meaning they are different versions of the same gene. Different versions of the same gene are called **alleles** (also known as gene variants). The different alleles of a gene will all have the same locus (position on a particular chromosome).

As homologous chromosomes have the same genes in the same positions, they will be the same length and size when they are visible in prophase. The centromeres will also be in the same positions.

The stages of meiosis

As discussed at the start of this topic, meiosis involves two divisions:

● **Meiosis I** – the first division is the reduction division when the pairs of homologous chromosomes are separated into two cells. Each intermediate cell will only contain one full set of genes instead of two, so the cells are haploid.

- **Meiosis II** – the second division is similar to mitosis, and the pairs of chromatids present in each daughter cell are separated, forming two more cells. Four haploid daughter cells are produced in total.

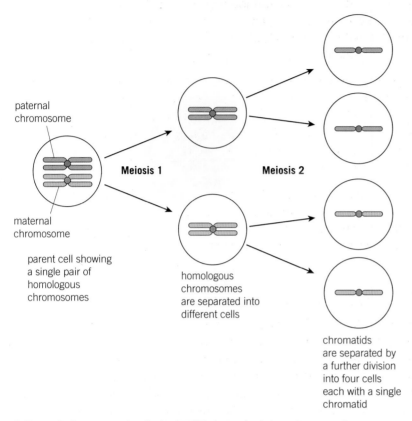

paternal chromosome

maternal chromosome

parent cell showing a single pair of homologous chromosomes

Meiosis 1

homologous chromosomes are separated into different cells

Meiosis 2

chromatids are separated by a further division into four cells each with a single chromatid

▲ Figure 2 *Summary of meiosis simplified to a single homologous pair*

Meiosis I

Prophase 1

During prophase 1, chromosomes condense, the nuclear envelope disintegrates, the nucleolus disappears and spindle formation begins, as in prophase of mitosis.

The difference in **prophase 1** is that the homologous chromosomes pair up, forming **bivalents**. Chromosomes are large molecules of DNA and moving them through the liquid cytoplasm as they are brought together results in the chromatids entangling. This is called **crossing over** (Figure 3).

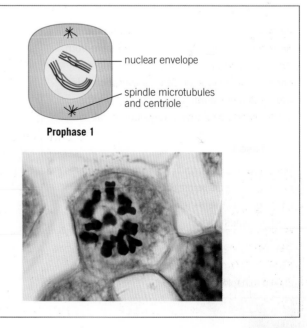

nuclear envelope

spindle microtubules and centriole

Prophase 1

Metaphase 1

Metaphase 1 is the same as metaphase in mitosis except that the homologous pairs of chromosomes assemble along the metaphase plate instead of the individual chromosomes.

The orientation of each homologous pair on the metaphase plate is random and independent of any other homologous pair. The maternal or paternal chromosomes can end up facing either pole. This is called **independent assortment**, and can result in many different combinations of alleles facing the poles (Figure 5). Independent assortment of chromosomes in metaphase 1 results in genetic variation.

bivalents aligned on the equator

Metaphase 1

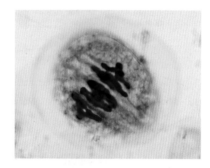

Anaphase 1

Anaphase 1 is different from anaphase of mitosis as the homologous chromosomes are pulled to the opposite poles and the chromatids stay joined to each other.

Sections of DNA on 'sister' chromatids, which became entangled during crossing over, now break off and rejoin – sometimes resulting in an exchange of DNA. The points at which the chromatids break and rejoin are called **chiasmata**.

When exchange occurs this forms **recombinant** chromatids, with genes being exchanged between chromatids. The genes being exchanged may be different alleles of the same gene, meaning the combination of alleles on the recombinant chromatids will be different from the allele combination on either the original chromatids (Figure 4). **Genetic variation** arises from this new combinations of alleles – the sister chromatids are no longer identical.

homologous chromosomes being pulled to opposite poles

Anaphase 1

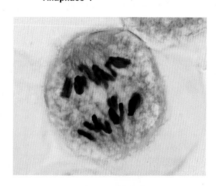

Telophase 1

Telophase 1 is essentially the same as telophase in mitosis. The chromosomes assemble at each pole and the nuclear membrane reforms. Chromosomes uncoil.

The cell undergoes cytokinesis and divides into two cells. The reduction of chromosome number from diploid to haploid is complete.

cell will divide across the equator

Telophase 1

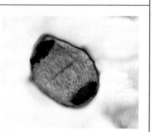

Meiosis II

Prophase 2 In prophase 2 the chromosomes, which still consist of two chromatids, condense and become visible again. The nuclear envelope breaks down and spindle formation begins.	 **Prophase II**	
Metaphase 2 Metaphase 2 differs from metaphase 1, as the individual chromosomes assemble on the metaphase plate, as in metaphase in mitosis. Due to crossing over, the chromatids are no longer identical so there is **independent assortment** again and more **genetic variation** produced in metaphase II.	 **Metaphase II**	
Anaphase 2 Unlike anaphase 1, **anaphase 2** results in the chromatids of the individual chromosomes being pulled to opposite poles after division of the centromeres – the same as in anaphase of mitosis.	 **Anaphase II**	
Telophase 2 The chromatids assemble at the poles at **telophase 2** as in telophase of mitosis. The chromosomes uncoil and form chromatin again. The nuclear envelope reforms and the nucleolus becomes visible. Cytokinesis results in division of the cells forming four daughter cells in total. The cells will be *haploid* due to the *reduction division*. They will also be genetically different from each other, and from the parent cell, due to the processes of *crossing over* and *independent assortment*.	 **Telophase II**	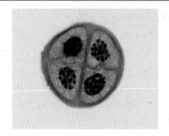

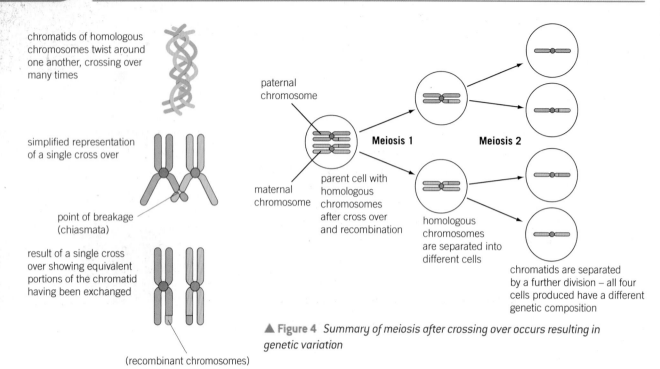

chromatids of homologous chromosomes twist around one another, crossing over many times

simplified representation of a single cross over

point of breakage (chiasmata)

result of a single cross over showing equivalent portions of the chromatid having been exchanged

(recombinant chromosomes)

▲ **Figure 3** *Crossing over*

paternal chromosome

maternal chromosome

Meiosis 1

Meiosis 2

parent cell with homologous chromosomes after cross over and recombination

homologous chromosomes are separated into different cells

chromatids are separated by a further division – all four cells produced have a different genetic composition

▲ **Figure 4** *Summary of meiosis after crossing over occurs resulting in genetic variation*

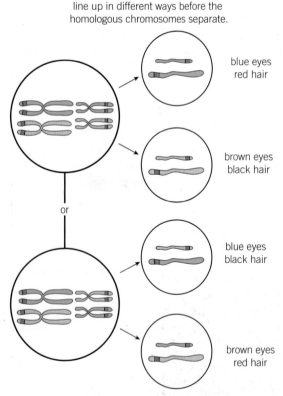

During meiosis I, chromosomes can line up in different ways before the homologous chromosomes separate.

blue eyes
red hair

brown eyes
black hair

or

blue eyes
black hair

brown eyes
red hair

▲ **Figure 5** *The pole that maternal or paternal homologous chromosomes face is due to random independent assortment and can result in many different combinations of alleles on either side of the metaphase plate*

Summary questions

1 a State which division in meiosis is a reduction division. *(1 mark)*
   b Explain why a reduction division is necessary in the production of gametes. *(2 marks)*

2 Explain the meaning of the term homologous chromosomes. *(2 marks)*

3 Outline how you could observe meiosis in a plant cell. *(4 marks)*

4 Explain how crossing over and independent assortment lead to genetic variation. *(4 marks)*

5 a Copy the diagram below and label the alleles on the recombinant chromosomes. *(3 marks)*

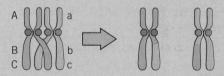

   b Suggest the importance of the creation of different allele combinations in populations. *(4 marks)*

6.4 The organisation and specialisation of cells

Specification reference: 2.1.6

As you have learnt, the basic unit of life is a cell. But many organisms are multicellular – they are made up of not one but hundreds, thousands or millions of cells. Although these cells within a single organism have common features such as membranes, organelles, and nuclei, they are not all identical. Different cells within an organism are **specialised** for different roles and organised into efficient biological structures, each with a particular function.

Levels of organisation in multicellular organisms

The organisation of a multicellular organism can be summarised as:

specialised cells → tissues → organs → organ systems → whole organism

Specialised cells

The cells within a multicellular organism are **differentiated**, meaning they are specialised to carry out very specific functions. You will explore cell differentiation further in the next topic.

Some examples of specialised cells are given in the following tables.

▼ Table 1 *Specialised animal cells. Dimensions given are according to cells in a human*

Erythrocytes or red blood cells have a flattened biconcave shape, which increases their surface area to volume ratio. This is essential to their role of transporting oxygen around the body. In mammals these cells do not have nuclei or many other organelles, which increases the space available for haemoglobin, the molecule that carries oxygen. They are also flexible so that they are able to squeeze through narrow capillaries.	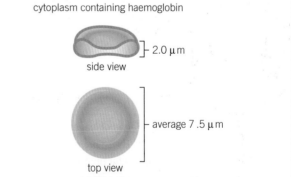 cytoplasm containing haemoglobin — 2.0 μm — side view — average 7.5 μm — top view
Neutrophils (a type of white blood cell) play an essential role in the immune system. They have a characteristic multi-lobed nucleus, which makes it easier for them to squeeze through small gaps to get to the site of infections. The granular cytoplasm contains many lysosomes that contain enzymes used to attack pathogens.	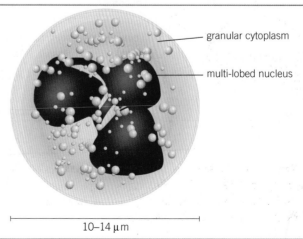 granular cytoplasm — multi-lobed nucleus — 10–14 μm

Sperm cells are male gametes. Their function is to deliver genetic information to the female gamete, the ovum (or egg). Sperm have a tail or flagellum, so they are capable of movement and contain many mitochondria to supply the energy needed to swim. The acrosome on the head of the sperm contains digestive enzymes, which are released to digest the protective layers around the ovum and allow the sperm to penetrate, leading to fertilisation.

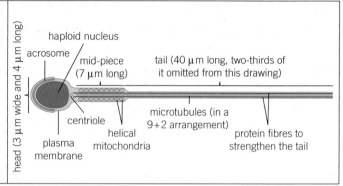

▼ Table 2 *Specialised plant cells*

Palisade cells present in the mesophyll contain chloroplasts to absorb large amounts of light for photosynthesis. The cells are rectangular box shapes, which can be closely packed to form a continuous layer. They have thin cell walls, increasing the rate of diffusion of carbon dioxide. They have a large vacuole to maintain turgor pressure. Chloroplasts can move within the cytoplasm in order to absorb more light.	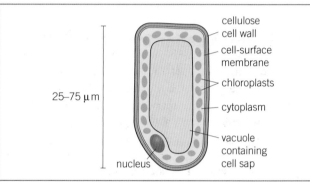
Root hair cells, present at the surfaces of roots near the growing tips, have long extensions called root hairs, which increase the surface area of the cell. This maximises the uptake of water and minerals from the soil.	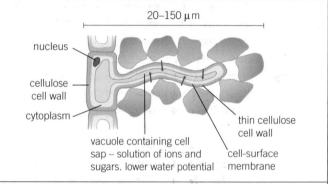
Pairs of *guard cells* on the surfaces of leaves form small openings called stomata. These are necessary for carbon dioxide to enter plants for photosynthesis. When guard cells lose water and become less swollen as a result of osmotic forces, they change shape and the stoma closes to prevent further water loss from the plant. The cell wall of a guard cell is thicker on one side so the cell does not change shape symmetrically as its volume changes.	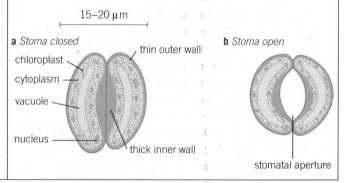

Tissues

A **tissue** is made up of a collection of differentiated cells that have a specialised function or functions. As a result, each tissue is adapted for a particular function within the organism.

There are four main categories of tissues in animals:

- nervous tissue, adapted to support the transmission of electrical impulses
- epithelial tissue, adapted to cover body surfaces, internal and external
- muscle tissue, adapted to contract
- connective tissue, adapted either to hold other tissues together or as a transport medium.

Some examples of specialised tissues in animals are given in Table 3.

Synoptic link

You will learn about water transport in plants in Topic 9.2, Water transport in multicellular plants.

▼ **Table 3** *Specialised animal tissues*

Squamous epithelium, made up of specialised squamous epithelial cells, is sometimes known as pavement epithelium due to its flat appearance. It is very thin due to the squat or flat cells that make it up and also because it is only one cell thick. It is present when rapid diffusion across a surface is essential. It forms the lining of the lungs and allows rapid diffusion of oxygen into the blood.	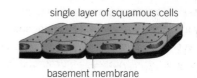 single layer of squamous cells / basement membrane
Ciliated epithelium is made up of ciliated epithelial cells. The cells have 'hair-like' structures called cilia on one surface that move in a rhythmic manner. Ciliated epithelium lines the trachea, for example, causing mucus to be swept away from the lungs. Goblet cells are also present, releasing mucus to trap any unwanted particles present in the air. This prevents the particles, which may be bacteria, from reaching the alveoli once inside the lungs.	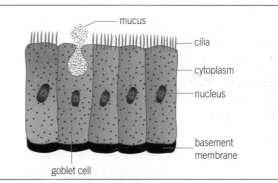 mucus / cilia / cytoplasm / nucleus / basement membrane / goblet cell
Cartilage is a connective tissue found in the outer ear, nose and at the ends of (and between) bones. It contains fibres of the proteins elastin and collagen. Cartilage is a firm, flexible connective tissue composed of chondrocyte cells embedded in an extracellular matrix. Cartilage, among other things, prevents the ends of bones from rubbing together and causing damage. Many fish have whole skeletons made of cartilage, not bone.	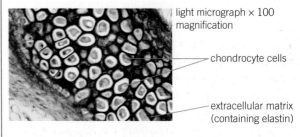 light micrograph × 100 magnification / chondrocyte cells / extracellular matrix (containing elastin)
Muscle is a tissue that needs to be able to shorten in length (contract) in order to move bones, which in turn move the different parts of the body. There are different types of muscle fibres. Skeletal muscle fibres (muscles which are attached to bone) contain myofibrils (dark pink bands on the micrograph) which contain contractile proteins. The skeletal muscle micrograph shown here has several individual muscle fibres (pink) separated by connective tissue (thin white strips).	light micrograph × 100 magnification (longitudinal section) / muscle fibres / connective tissue

Synoptic link

You will learn more about the structure and function of the xylem and phloem in Topic 9.1, Transport systems in dicotolyedonous plants.

There are a number of different tissues in plants, including:

- epidermis tissue, adapted to cover plant surfaces

- vascular tissue, adapted for transport of water and nutrients.

Some examples of specialised tissues in plants are given in the Table 4.

▼ Table 4 *Specialised plant tissues*

The *epidermis* is a single layer of closely packed cells covering the surfaces of plants. It is usually covered by a waxy, waterproof cuticle to reduce the loss of water. Stomata, formed by a pair of guard cells that can open and close (Table 2) are present in the epidermis. They allow carbon dioxide in and out, and water vapour and oxygen in and out.	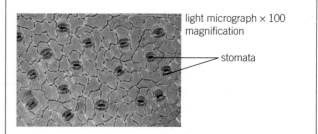 light micrograph × 100 magnification — stomata
Xylem tissue is a type of vascular tissue responsible for transport of water and minerals throughout plants. The tissue is composed of vessel elements, which are elongated dead cells. The walls of these cells are strengthened with a waterproof material called lignin (pink rings in the micrograph), which provides structural support for plants.	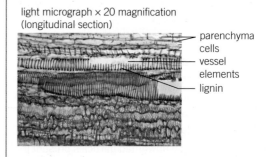 light micrograph × 20 magnification (longitudinal section) — parenchyma cells — vessel elements — lignin
Phloem tissue is another type of vascular tissue in plants, responsible for the transport of organic nutrients, particularly sucrose, from leaves and stems where it is made by photosynthesis to all parts of the plant where it is needed. It is composed of columns of sieve tube cells separated by perforated walls called sieve plates.	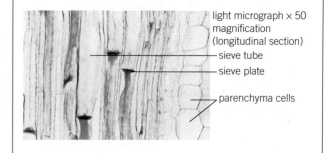 light micrograph × 50 magnification (longitudinal section) — sieve tube — sieve plate — parenchyma cells

Synoptic link

You will learn about the structure of the heart in Topic 8.5 The heart.

Organs

An **organ** is a collection of tissues that are adapted to perform a particular function in an organism. For example, the mammalian heart is an organ that is adapted for pumping blood around the body. It is made up of muscle tissue and connective tissue. The leaf is a plant organ that is adapted for photosynthesis. it contains epidermis tissues and vascular tissue (Figure 1).

Organ systems

Large multicellular organisms have coordinated **organ systems**. Each organ system is composed of a number of organs working together to carry out a major function in the body. Animal examples include:

- the digestive system, which takes in food, breaks down the large insoluble molecules into small soluble ones, absorbs the nutrients into the blood, retains water needed by the body and removes any undigested material from the body

- the cardiovascular system, which moves blood around the body to provide an effective transport system for the substances it carries

- the gaseous exchange system, which brings air into the body so oxygen can be extracted for respiration, and carbon dioxide can be expelled.

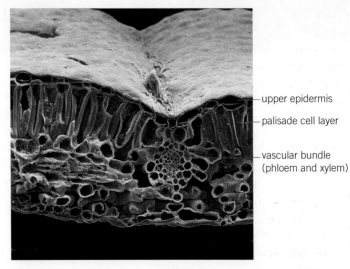

upper epidermis

palisade cell layer

vascular bundle (phloem and xylem)

▲ Figure 1 *A scanning electron micrograph of a leaf cross-section showing the arrangement of different tissues in this organ. × 150 magnification*

Summary questions

1 State two examples of epithelial tissue and describe how each is adapted for its function. *(6 marks)*

2 Describe two specialised cells, one that is not usually part of a tissue and one that is usually found as part of a tissue. In each case explain how they are adapted to their functions. *(4 marks)*

3 The cardiac muscle that makes up most of the heart is a tissue, but the heart itself is an organ. Explain the difference. *(4 marks)*

4 Using the digestive system as an example, explain the relationship between organs in an organ system. *(6 marks)*

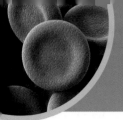

6.5 Stem cells

Learning outcomes

Demonstrate knowledge, understanding, and application of:

→ the features and differentiation of stem cells

→ the production of erythrocytes and neutrophils

→ the production of xylem vessels and phloem sieve tubes

→ the potential uses of stem cells in research and medicine.

As you explored in the previous topic, the different cells in a multicellular organism are specialised for different functions. The process of a cell becoming specialised is called **differentiation**. Despite being differentiated in structure and function, all body cells within an organism have the same DNA (except those like erythrocytes and sieve tube elements which don't have a nucleus). Differentiation involves the expression of some genes but not others in the cell's genome.

Stem cells

All cells in plants and animals begin as **undifferentiated** cells and originate from mitosis or meiosis. They are not adapted to any particular function (they are unspecialised) and they have the potential to differentiate to become any one of the range of specialised cell types in the organism. These undifferentiated cells are called **stem cells**.

Stem cells are able to undergo cell division again and again, and are the source of new cells necessary for growth, development, and tissue repair. Once stem cells have become specialised they lose the ability to divide, entering the G_0 phase of the cell cycle.

The activity of stem cells has to be strictly controlled. If they do not divide fast enough then tissues are not efficiently replaced, leading to ageing. However, if there is uncontrolled division then they form masses of cells called tumours, which can lead to the development of cancer.

Stem cell potency

A stem cell's ability to differentiate into different cell types is called **potency**. The greater the number of cell types it can differentiate into, the greater its potency. Stem cells differ depending on the type of cell they can turn into:

- **Totipotent** – these stem cells can differentiate into any type of cell. A fertilised egg, or zygote and the 8 or 16 cells from its first few mitotic divisions are totipotent cells, which are destined eventually to produce a whole organism. They can also differentiate into extra-embryonic tissues like the amnion and umbilicus.

- **Pluripotent** – these stem cells can form all tissue types but not whole organisms. They are present in early embryos and are the origin of the different types of tissue within an organism.

- **Multipotent** – these stem cells can only form a range of cells within a certain type of tissue. Haematopoetic stem cells in bone marrow are multipotent because this gives rise to the various types of blood cell.

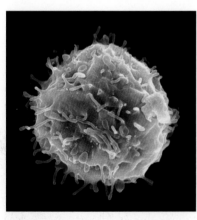

▲ Figure 1 *The stem cell in this coloured scanning electron micrograph is destined to become a blood cell. Some differentiated red blood cells can also be seen, × 100 magnification*

Differentiation

Multicellular organisms like animals and plants have evolved from unicellular (single-celled) organisms because groups of cells with different functions working together as one unit can make use of resources more efficiently than single cells operating on their own.

In multicellular organisms cells have to specialise to take on different roles in tissues and organs. They may be required to form barriers such as skin or be motile such as sperm cells. Cells have adapted to different roles in an organism and so have many shapes (and sizes) and often contain different organelles.

Erythrocytes (red blood cells) and neutrophils (white blood cells) are both present in blood (Topic 6.4, Table 1). They look very different because they have different functions. When cells differentiate they become adapted to their specific role. What form this adaptation takes is dependent on the function of the tissue, organ and organ system to which the cell belongs.

All blood cells are derived from stem cells in the **bone marrow**.

Replacement of red and white blood cells

Mammalian erythrocytes are essential for the transport of oxygen around the body. They are adapted to maximise their oxygen-carrying capacity by having only a few organelles so there is more room for haemoglobin.

Due to the lack of nucleus and organelles they only have a short lifespan of around 120 days. They therefore need to be replaced constantly. The stem cell colonies in the bone marrow produce approximately three billion erythrocytes per kilogram of body mass per day to keep up with the demand.

Neutrophils have an essential role in the immune system. They live for only about 6 hours and the colonies of stem cells in bone marrow produce in the region of 1.6 billion per kg per hour. This figure will increase during infection.

Sources of animal stem cells

- Embryonic stem cells – these cells are present at a very early stage of embryo development and are totipotent. After about seven days a mass of cells, called a blastocyst, has formed and the cells are now in a pluripotent state. They remain in this state in the fetus until birth.
- Tissue (adult) stem cells – these cells are present throughout life from birth. They are found in specific areas such as bone marrow. They are multipotent, although there is growing evidence that they can be artificially triggered to become pluripotent. Stem cells can also be harvested from the umbilical cords of newborn babies. The advantages of this source are the plentiful supply of umbilical cords and that invasive surgery is not needed. These stem cells can be stored in case they are ever

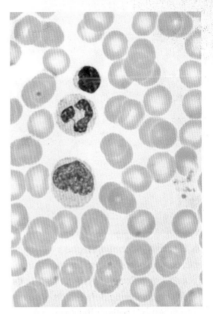

▲ Figure 2 *Light micrograph of human blood – red blood cells (erythrocytes) are pink and the purple cells are white blood cells (from left to right a monocyte, a neutrophil and a lymphocyte), × 500 magnification*

▲ Figure 3 *Scanning electron micrograph of human embryonic stem cell, × 500 magnification*

needed by the individual in the future, and tissues cultured from such stem cells would not be rejected in a transplant to the umbilicus' owner.

Sources of plant stem cells

Stem cells are present in **meristematic tissue** (**meristems**) in plants. This tissue is found wherever growth is occurring in plants, for example at the tips of roots and shoots (termed apical meristems).

Meristematic tissue is also located sandwiched between the phloem and xylem tissues and this is called the vascular cambium. Cells originating from this region differentiate into the different cells present in xylem and phloem tissues (Topic 6.4, The organisation and specialisation of cells). In this way the vascular tissue grows as the plant grows. The pluripotent nature of stem cells in the meristems continues throughout the life of the plant.

Uses of stem cells

Stem cells transplanted into specific areas have the potential to treat certain diseases, such as:

- heart disease – muscle tissue in the heart is damaged as a result of a heart attack, normally irreparably – this has been tried experimentally with some success already

- type 1 diabetes – with insulin-dependent diabetes the body's own immune system destroys the insulin-producing cells in the pancreas; patients have to inject insulin for life – this has been tried experimentally with some success already

- Parkinson's disease – the symptoms (shaking and rigidity) are caused by the death of dopamine-producing cells in the brain; drugs currently only delay the progress of the disease

- Alzheimer's disease – brain cells are destroyed as a result of the build up of abnormal proteins; drugs currently only alleviate the symptoms

- macular degeneration – this condition is responsible for causing blindness in the elderly and diabetics; scientists are currently researching the use of stem cells in its treatment and early results are very encouraging

- birth defects – scientists have already successfully reversed previously untreatable birth defects in model organisms such as mice.

- spinal injuries – scientists have restored some movement to the hind limbs of rats with damaged spinal cords using stem cell implants.

Stem cells are already used in such diverse areas as:

- the treatment of burns – stem cells grown on biodegradable meshes can produce new skin for burn patients, this is quicker than the normal process of taking a graft from another part of the body

- drug trials – potential new drugs can be tested on cultures of stem cells before being tested on animals and humans

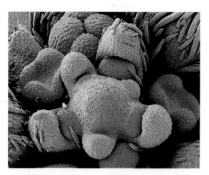

▲ Figure 4 *Scanning electron micrograph of a shoot apical meristem (purple) on a flowering plant. Floral buds (red) are appearing between developing young leaves (green). The coloration seen here is computer generated, × 300 magnification*

- developmental biology – with their ability to divide indefinitely and differentiate into almost any cell within an organism, stem cells have become an important area of study in developmental biology. This is the study of the changes that occur as multicellular organisms grow and develop from a single cell, such as a fertilised egg – and why things sometimes go wrong.

Ethics

Stem cells have been used in medicine for many years in the form of bone marrow transplants. More recently, the use of embryonic stem cells in therapies and research has lead to controversy and debates regarding the ethics of such use.

The embryos used originally were donated from those left over after fertility treatment. More recently the law in the UK has changed so that embryos can be specifically created in the laboratory as a source of stem cells.

The removal of stem cells from embryos normally results in the destruction of the embryos, although techniques are being developed that will allow stem cells to be removed without damage to embryos.

There are not only religious objections to the use of embryos in this way but moral objections too – many people believe that life begins at conception and the destruction of embryos is, therefore, murder. There is a lack of consensus as to when the embryo itself has rights, and also who owns the genetic material that is being used for research.

This controversy is holding back progress that could lead to the successful treatment of many incurable diseases. The use of umbilical cord stem cells overcomes these issues to a large extent, but these cells are merely multipotent, not pluripotent like embryonic stem cells, thus restricting their usefulness. Adult tissue stem cells can also be used but they do not divide as well as umbilical stem cells and are more likely to have acquired mutations. Developments are being made towards artificially transforming tissue stem cells into pluripotent cells. Induced pluripotent stem cells (iPSCs) are adult stem cells that have been genetically modified to act like embryonic stem cells and so are pluripotent.

The use of plant stem cells does not raise the same ethical issues as animal cells.

Synoptic link

You will learn about T cells and B cells in immunity in Topic 12.6, The specific immune system.

 ### Gene therapy using stem cells

Children born with the rare genetic condition Severe Combined Immunodeficiency (SCID) are extremely vulnerable to all infections and without treatment are unlikely to live for more than a year. They produce no T cells, and without T cells the B cells do not function either (T cells and B cells are types of white blood cell).

Normally SCID is treated with a bone marrow transplant, which depends on finding a matching donor. The transplanted stem cells divide and differentiate into the different types of white blood cells needed for a healthy immune system.

More recently experimental gene therapy has been used to treat SCID. The aim is that stem cells from the patient's own bone marrow are removed and genetically modified so that they function normally to produce the white blood cells needed. These are then put back into the patient and the condition should be corrected.

This treatment was initially successful in a small number of children, but in some of the children another gene was damaged in the process and they went on to develop leukaemia. However, gene therapy is still seen as having the most potential for treating SCID in the future.

1. Explain how this condition can be cured using a bone marrow transplant.
2. Explain the risks of a bone marrow transplant from a donor, and how the gene therapy described removes this risk.
3. Suggest why some patients receiving this gene therapy developed cancer.

Plant stem cells and medicines

Plant stem cells have a huge potential role to play in medicine. Many drugs used in medicines are derived from plants. Plant stem cells can be cultured, leading to an unlimited, and cheap, supply of plant-based drugs.

Paclitaxel is a common drug used in the treatment of breast and lung cancer. It cannot be chemically synthesised and must be obtained from the bark of yew trees (*Taxus brevifolia*). The trees have to be mature, which means the supply is limited and the extraction process difficult and expensive. An alternative way of producing the drug was developed using a related plant but it is still a difficult and expensive process. Recently stem cells from the yew tree have been used to produce paclitaxel cheaply and in sustainable quantities.

1. Suggest an environmental benefit of the use of plant stem cells in medicine production.

Summary questions

1. Describe the difference between pluripotent and multipotent with regard to stem cells, and state a source for both cell types in animals. *(4 marks)*

2. State where you would find meristematic tissue in a plant and explain the importance of the position of meristematic tissue to a plant. *(5 marks)*

3. Evaluate the advantages and disadvantages of using embryonic stem cells in medical research. *(4 marks)*

4. Alzheimer's is a progressive disease resulting in the loss of neurones in the brain. An area known as the cerebral cortex is primarily affected. There is a reduction in the quantity of the neurotransmitter acetylcholine released, which is necessary for memory and learning. This leads to the symptoms of dementia such as memory loss, mood swings, and confusion which get progressively worse over time. At the present time there is no cure and the few drugs available only temporarily relieve the symptoms.

   Parkinson's disease is due to neurones in a part of the brain called the substantia nigra dying. This leads to a reduction in the amount of the neurotransmitter dopamine released. The lack of dopamine leads to loss of the fine control of movement causing shaking, slowness, and rigidity. There are drugs available which treat the symptoms effectively but as the disease progresses the increasing doses needed result in more and more side effects.

   a. Explain why the use of stem cells is called regenerative medicine. *(2 marks)*

   b. Evaluate the use of stem cells in the potential treatment of the two diseases described above. *(5 marks)*

Practice questions

1 The cell cycle is a highly ordered sequence of events that takes place in a cell, resulting in division of the cell.

Which of the following statements is/are correct with respect to the cell cycle?

Statement 1: S – DNA is replicated in the nucleus.

Statement 2: G_1 – the cell continues to increase in size, energy stores are increased and the DNA is checked for errors.

Statement 3: G_2 – proteins from which organelles are synthesised are produced and organelles replicate.

A 1, 2 and 3 are correct

B Only 1 and 2 are correct

C Only 2 and 3 are correct

D Only 1 is correct (*1 mark*)

2 Fertilised eggs are transported along a structure called the fallopian tube to the uterus. The fallopian tubes are lined with ciliated epithelium.

a Define the term 'differentiation'. (*2 marks*)

b Describe how ciliated epithelial cells are adapted to their function (*3 marks*)

c Explain the difference between ciliated cells and ciliated epithelium (*4 marks*)

d Suggest how the ciliated epithelium is involved in the transport of fertilised eggs. (*2 marks*)

e List the letters in the correct order to show correctly increasing levels of organisation in a living organism.

A is a collection of tissues that are adapted to perform a particular function in an organism

B is made up of a collection of differentiated cells that have a specialised function

C is made up of not one but hundreds, thousands or millions of cells

D is the smallest structural and functional unit of **B**

E is composed of more than one **A** working together to carry out a major function in the organism (*1 mark*)

3 a The figure shows some drawings of a cell during different stages of mitosis.

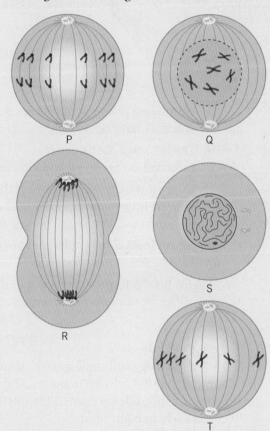

Place stages P, Q, R, S and T in the correct sequence

The first stage has been identified for you.

S.. (*4 marks*)

b Mitosis is part of the cell cycle. The figure shows a diagram of the cell cycle

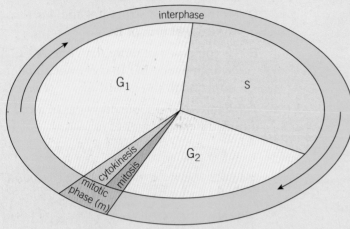

(i) Name one process that occurs during stages G_1 and G_2 *(1 mark)*

(ii) During stage S, the genetic information is copied and checked.

Suggest what might happen if the genetic information is not checked. *(2 marks)*

c During **meiosis** a cell undergoes two divisions.

Suggest how cells produced by meiosis may differ from those produced by mitosis. *(2 marks)*

OCR F211 2009

4 a A fertilised egg cell undergoes mitosis producing two cells in 36 hours. The cells continue to undergo mitosis completing each cycle in eight hours.

Calculate how many cells there will be present three days after fertilisation. *(2 marks)*

b (i) Explain why the nuclei of most human cells contain 46 pairs of chromosomes. *(2 marks)*

(ii) Name the type of cell which contain half the number of chromosomes present in a diploid cell. *(1 mark)*

c Copy and complete the table below. *(4 marks)*

	Meiosis	Mitosis
Homologous chromosomes	pair up	
Daughter cells n/2n		
Number of cell divisions		
Crossing over		✗

The figure below shows drawings of the six chromosomes inside an animal cell viewed during later prophase of mitosis.

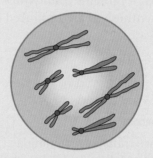

d (i) Identify **one pair** of homologous chromosomes in the diagram by drawing around each chromosome in the pair **on the diagram**. *(1 mark)*

The nucleus of a sperm cell is produced by **meiosis.**

(ii) Draw a diagram in the space below to represent the chromosomes that are present in the nucleus of a sperm cell from **the same animal**. *(2 marks)*

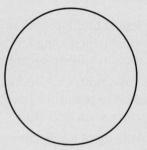

OCR F211 2010

e Explain why prokaryotic cells do not undergo meiosis. *(3 marks)*

5 The diagram below shows the relative times that cells can spend in each stage of the cell cycle.

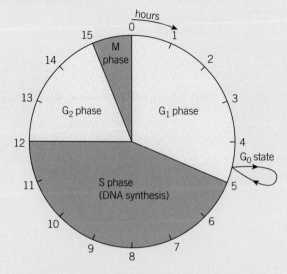

a Calculate the percentage of the overall time that the cell spends in interphase.
(2 marks)

b Outline the importance of strict regulation of the cell cycle. *(3 marks)*

6 Stem cells are being investigated as a possible cure for certain types of diabetes. In type 1 diabetes the immune system recognises the beta cells in the pancreas, which produce insulin, as foreign and destroys them. Stem cells, which can be obtained from different sources, can be used to replace the beta cells that have been destroyed.

Some of these stem cells are particular useful as they do not trigger a response by the immune system.

a State what is meant by the term stem cell. *(2 marks)*

b List three sources of stem cells. *(3 marks)*

c Explain the meaning of stem cell potency. *(5 marks)*

d Describe three examples of the use of stem cells to reverse disease. *(6 marks)*

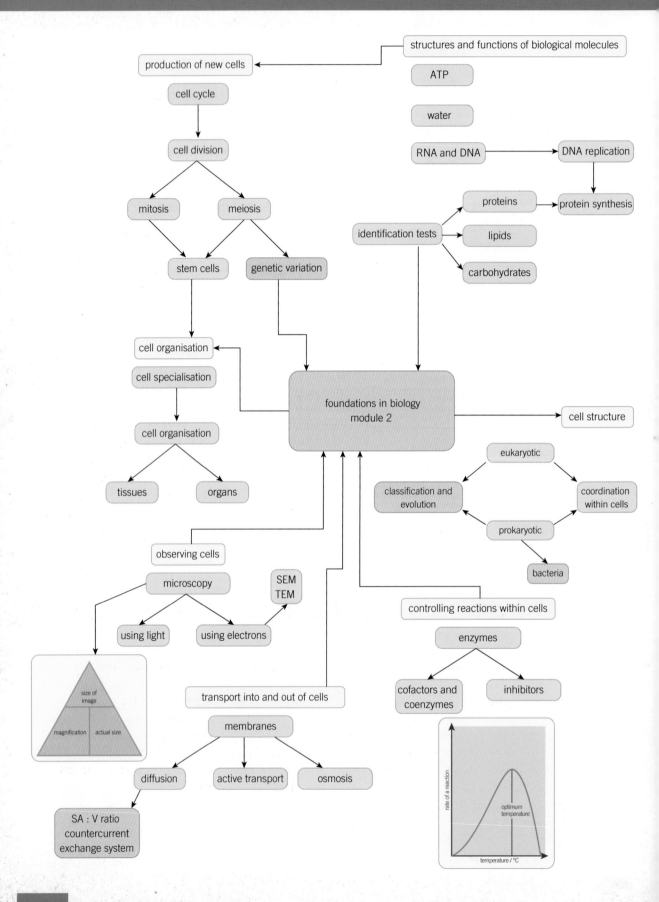

Application

A study published in the New England Journal of Medicine in 2013 showed that a Mediterranean diet, high in olive oil, nuts, fruit and vegetables and low in red meat and pastries gave a substantial reduction in the risk of death from heart attacks, strokes and cardiovascular disease. 7447 people took part in the study. All were between 55 and 80 and judged to be at relatively high risk of cardiovascular disease. The mean participation time was 4.8 years. One group had a Mediterranean diet with extra olive oil, one group had a Mediterranean diet with added nuts and the control group had a low fat diet.

One set of results is shown below:

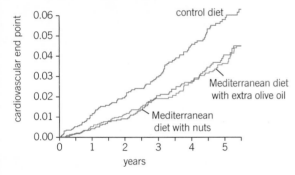

▲ **Figure 1** *Graph to show the effect of diet on the cardiovascular end point, which is the occurrence of myocardial infarctions (heart attacks), stroke and deaths from cardiovascular disease.*

1 In this study one of the major components controlled in the diets of the people taking part was their lipid consumption.
   a What are lipids?
   b Where would you expect to find lipids in the ultrastructure of a cell?
   c What has been the role of the electron microscope in helping us understand the role of lipids in the ultrastructure of the cell?

2 In the study, people on the Mediterranean diet had to eat lots of olive oil or nuts. These both contain monounsaturated fatty acids. What are these, and how do they differ from other fatty acids?

3 a Summarise the data from Figure 1 to describe the effect of the three different diets on the risk of cardiovascular disease.
   b Suggest two additional pieces of evidence you might want to see before accepting the findings of this study.
   c In 2014 a meta-analysis study comparing the results from 72 studies showed no statistically relevant link between saturated fats and heart disease, and no protective effect of cetain types of polyunsaturated fats. Write a news report on this large study, explaining the potential impact of the findings and the importance of the study to the general public.

Extension

1 If a patient has a suspected myocardial infarction there are a number of biochemical markers which occur in the blood which can help doctors make a diagnosis of what is happening inside the body. Research each of these key markers and write a brief report, describing the biochemistry of the molecule and explaining why they are an indicator that a patient has had a heart attack:
   a Cardiac troponins I and T
   b Creatine kinase
   c myoglobin
   d natriuretic peptides (peptides)

2 Obesity is closely linked with many health issues, including cardiovascular health. Prepare a poster presentation looking at the biochemistry and cell biology of obesity – show the biochemistry of lipids and carbohydrates, summarise how they are linked to obesity and investigate fat cells, including their microscopic appearance and adaptations to their functions in the body.

MODULE 3
Exchange and transport

Chapters in this module

7 Exchange surfaces and breathing

8 Transport in animals

9 Transport in plants

Introduction

All living organisms need to move materials into and out of their cells. Commonly this includes getting nutrients and oxygen in and carbon dioxide and other waste products out. Simple diffusion works well for single celled organisms but larger, multi-cellular organisms need to transport materials from the outside word into their bodies before they can pass into the individual cells. Exchange surfaces and transport systems are vital for these exchanges to take place in plants and animals alike.

Exchange surfaces explores the need for specialised exchange surfaces, and what makes an effective one, before moving on to specific examples. You will learn about the mammalian gaseous exchange system, how air is moved in and out of the system and the interrelationships between the volume of the lungs and the rate of breathing. You will compare the gas exchange surfaces with those of insects and fish, adapted for very different bodies and lifestyles.

Transport in animals explains why, as animals become larger and more active, transport systems become essential to supply nutrients and oxygen to and waste products from individual cells. The key roles of the blood, the blood vessels and the heart in this transport are fully explored including the electrical control of the heart beat and how this can be recorded using an ECG.

Transport in plants describes the key transport systems in plants. You will learn how both the supply of nutrients from the soil and the movement of the products of photosynthesis around the plant depend on the flow of water through the vascular system made up of the xylem and the phloem.

Knowledge and understanding checklist

From your Key Stage 4 study you should be able to answer the following questions. Work through each point, using your Key Stage 4 notes and other resources. There is also support available on Kerboodle.

☐ Explain how substances are transported into and out of cells though diffusion, osmosis, and active transport.

☐ Explain the need for exchange surfaces and a transport system in terms of surface area: volume ratio.

☐ Describe some of the substances transported into and out of a range of organisms in terms of the requirements of those organisms to include oxygen, carbon dioxide, water, dissolved food molecules, mineral ions, and urea.

☐ Describe the human circulatory system, including the relationship with the gaseous exchange system, and explain how the structure of the heart and the blood vessels are adapted to their functions.

☐ Explain how red blood cells, white blood cells, platelets and plasma are adapted to their functions in the blood.

☐ Explain how the structure of the xylem and the phloem are adapted to their functions in the plant.

☐ Explain how water and mineral ions are taken up by plants, relating the structure of the root hair cells to their function.

☐ Describe the processes of transpiration and translocation, including the structure and function of the stomata.

☐ Explain the effect of a variety of environmental factors on the rate of water uptake by a plant, to include light intensity, air movement, and temperature.

Maths skills checklist

In this module, you will need to use the following maths skills.

☐ **Recognise and make use of appropriate units in calculations.** You will need to be able to do this in all your calculations of volume, breathing rates, etc.

☐ **Recognise and use expressions in decimal and standard form.** You will need this to analyse and interpret primary and secondary data relating to lung volumes and breathing rates.

☐ **Use ratios, fractions, and percentages.** You will need this to calculate surface area:volume ratios.

☐ **Estimate results.** You will need this to understand surface area: volume ratios.

MyMaths.co.uk
Bringing Maths Alive

EXCHANGE SURFACES AND BREATHING
7.1 Specialised exchange surfaces
Specification reference: 3.1.1

▲ **Figure 1** *Getting oxygen into the muscle cells of a dolphin isn't as easy as getting it into an* Amoeba

Imagine a microscopic single-celled organism such as *Amoeba* drifting in the ocean currents. Now think of a dolphin, swimming at high speeds, hunting fish or playing with other dolphins. They both need glucose and oxygen for cellular respiration and produce waste carbon dioxide, which must be removed. However, the quantities involved and the distances the substances need to travel, are very different.

The need for specialised exchange surfaces

In microscopic organisms such as *Amoeba* all of the oxygen needed by the organism, and the waste carbon dioxide produced, can be exchanged with the external environment by diffusion through the cell surface. The distances the substances have to travel are very small.

There are two main reasons why diffusion alone is enough to supply the needs of single-celled organisms:

- The metabolic activity of a single-celled organism is usually low, so the oxygen demands and carbon dioxide production of the cell are relatively low.
- The surface area to volume (SA:V) ratio of the organism is large (see below).

As organisms get larger they can be made up of millions or even billions of cells arranged in tissues, organs, and organ systems. Their metabolic activity is usually much higher than most single-celled organisms. Think of the dolphin in Figure 1. The amount of energy used in moving through the water means the oxygen demands of the muscle cells deep in the body will be high and they will produce lots of carbon dioxide. The distance between the cells where the oxygen is needed and the supply of oxygen is too far for effective diffusion to take place. What's more, the bigger the organism, the smaller the SA:V ratio. So gases can't be exchanged fast enough or in large enough amounts for the organism to survive.

Surface area : volume ratio – modelling an organism

The SA:V ratio is important in many areas of biology. A sphere is a useful shape for modelling cells or organisms. A series of simple calculations shows clearly how the SA:V ratio changes as the organism gets bigger, and why size matters so much.

Worked example: SA:V ratios

Compare the SA:V ratios of organisms with a radius of 1, 3 and 10 arbitrary units (au) and explain how this affects their ability to exchange materials with the environment.

The surface area of a sphere is calculated using the formula $4\pi r^2$

The volume of a sphere is calculated using the equation $\frac{4}{3}\pi r^3$

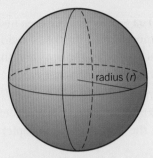

radius (*r*)

▲ **Figure 2** *The radius of a sphere*

To do these calculations you use 3.14 as the value of π.

Example 1: Radius = 1 au

Surface area is $4\pi r^2$ so: $4 \times 3.14 \times 1 \times 1 = 12.6$ (1dp)

Volume is $\frac{4}{3}\pi r^3$ so: $\frac{(4 \times 3.14 \times 1 \times 1 \times 1)}{3} = 4.2$ (1dp)

SA:V ratio is $\frac{12.6}{4.2} = 3:1$

Example 2: Radius = 3 au

Surface area is $4\pi r^2$ so: $4 \times 3.14 \times 3 \times 3 = 113.0$ (1dp)

Volume is $\frac{4}{3}\pi r^3$ so: $\frac{(4 \times 3.14 \times 3 \times 3 \times 3)}{3} = 113.0$ (1dp)

SA:V ratio is $\frac{113.0}{113.0} = 1:1$

Example 3: Radius = 10 au

Surface area is $4\pi r^2$ so: $4 \times 3.14 \times 10 \times 10 = 1256.0$ (1dp)

Volume is $\frac{4}{3}\pi r^3$ so: $\frac{(4 \times 3.14 \times 10 \times 10 \times 10)}{3} = 4186.7$ (1dp)

SA:V ratio is $\frac{1256}{4186} = 1:3$ (to the nearest whole number)

The bigger the organism, the smaller the surface area to volume ratio becomes, and the distances that substances need to travel from the outside to reach the cells at the centre of the body get longer. This makes it harder and ultimately impossible to absorb enough oxygen through the available surface area to meet the needs of the body.

Synoptic link

You will apply the same principles of SA:V ratio when you consider how nutrients are supplied to the cells of multicellular organisms in Chapter 8, Transport in animals and Chapter 9, Transport in plants.

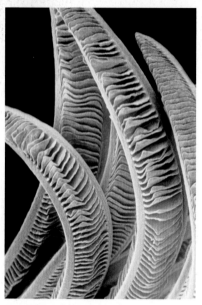

▲ **Figure 3** *This scanning electron micrograph shows the large surface area of the gills of a fish, × 200 magnification*

Synoptic link

You will learn more about specialised exchange surfaces in many other topics. More detail on the increased surface area of root hair cells in plants and the villi in the small intestine can be found in Topic 9.2, Water transport in multicellular animals and Topic 8.3, Blood, tissue fluid, and lymph respectively.

Specialised exchange surfaces

Large, multicellular organisms have evolved specialised systems for the exchange of the substances they need and the substances they must remove.

All effective **exchange surfaces** have certain features in common. You will be looking at many of them in detail in this chapter. Here is a summary of the characteristic features of effective exchange surfaces, along with some examples:

- **Increased surface area** – provides the area needed for exchange and overcomes the limitations of the SA:V ratio of larger organisms. Examples include root hair cells in plants and the villi in the small intestine of mammals.

- **Thin layers** – these mean the distances that substances have to diffuse are short, making the process fast and efficient. Examples include the alveoli in the lungs (see next topic) and the villi of the small intestine.

- **Good blood supply** – the steeper the concentration gradient, the faster diffusion takes place. Having a good blood supply ensures substances are constantly delivered to and removed from the exchange surface. This maintains a steep concentration gradient for diffusion. For example the alveoli of the lungs, the gills of a fish and the villi of the small intestine.

- **Ventilation to maintain diffusion gradient** – for gases, a ventilation system also helps maintain concentration gradients and makes the process more efficient, for example the alveoli and the gills of a fish where ventilation means a flow of water carrying dissolved gases (see Topic 7.4, Ventilation and gas exchange in other organisms).

Summary questions

1 Explain why single-celled organisms do not need specialised exchange surfaces. *(4 marks)*

2 Describe the main features of any efficient exchange surface and explain how the structures relate to their functions. *(6 marks)*

3 One roughly spherical organism has a radius of 2 au. Another has a radius of 6 au. Compare the SA:V ratios of the organisms and use these to explain why the larger organisms need specialised exchange surfaces. *(6 marks)*

7.2 The mammalian gaseous exchange system

Specification reference: 3.1.1

Animals that live on the land face a continual conflict between the need for gaseous exchange and the need for water. Gaseous exchange surfaces are moist, so oxygen dissolves in the water before diffusing into the body tissues. As a result the conditions needed to take in oxygen successfully are also ideal for the evaporation of water. Mammals have evolved complex systems that allow them to exchange gases efficiently but minimise the amount of water lost from the body. You are going to look at the human **gaseous exchange system** as an example of the specialised systems common to all mammals.

Learning outcomes

Demonstrate knowledge, understanding, and application of:

→ the structures and functions of the components of mammalian gaseous exchange system

→ the mechanism of ventilation in mammals.

The human gaseous exchange system

Mammals are relatively big – they have a small SA:V ratio and a very large volume of cells. They also have a high metabolic rate because they are active and maintain their body temperature independent of the environment. As a result they need lots of oxygen for cellular respiration and they produce carbon dioxide, which needs to be removed. This exchange of gases takes place in the lungs.

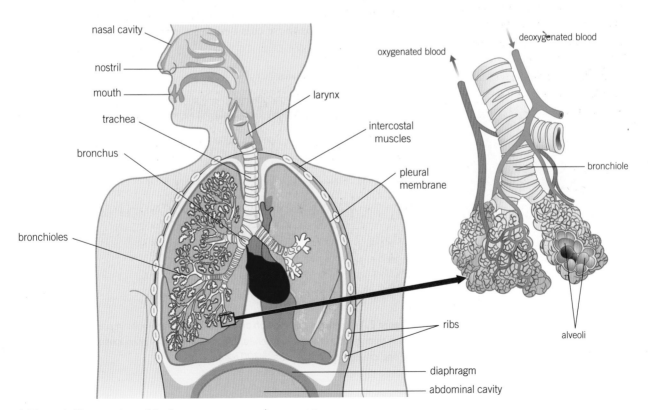

▲ Figure 1 *The structure of the human gaseous exchange system*

Key structures

Nasal cavity

The nasal cavity has a number of important features:

- a large surface area with a good blood supply, which warms the air to body temperature
- a hairy lining, which secretes mucus to trap dust and bacteria, protecting delicate lung tissue from irritation and infection
- moist surfaces, which increase the humidity of the incoming air, reducing evaporation from the exchange surfaces.

After passing through the nasal cavity, the air entering the lungs is a similar temperature and humidity to the air already there.

Trachea

The **trachea** is the main airway carrying clean, warm, moist air from the nose down into the chest. It is a wide tube supported by incomplete rings of strong, flexible **cartilage**, which stop the trachea from collapsing. The rings are incomplete so that food can move easily down the oesophagus behind the trachea.

The trachea and its branches are lined with a **ciliated epithelium**, with **goblet cells** between and below the epithelial cells (Figure 2). Goblet cells secrete mucus onto the lining of the trachea, to trap dust and microorganisms that have escaped the nose lining. The cilia beat and move the mucus, along with any trapped dirt and microorganisms, away from the lungs. Most of it goes into the throat and is swallowed and digested. One of the effects of cigarette smoke is that it stops these cilia beating.

> **Synoptic link**
>
> You learnt about cilia in Topic 2.4, Eukaryotic cell structure.

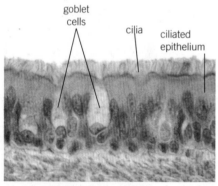
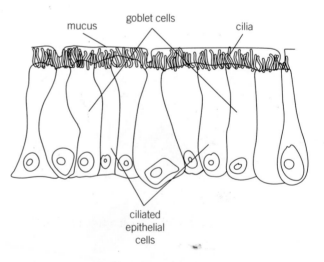

▲ **Figure 2** Left: *A light micrograph and. Right: a diagrammatic representation of a section through the lining of a trachea. The lining consists of mucus-secreting goblet cells and ciliated epithelial cells, × 300 magnification*

Bronchus

In the chest cavity the trachea divides to form the left bronchus (plural bronchi), leading to the left lung, and the right bronchus leading to the right lung. They are similar in structure to the trachea, with the same supporting rings of cartilage, but they are smaller.

Bronchioles

In the lungs the bronchi divide to form many small bronchioles. The smaller bronchioles (diameter 1 mm or less) have no cartilage rings. The walls of the bronchioles contain smooth muscle. When the smooth muscle contracts, the bronchioles constrict (close up). When it relaxes, the bronchioles dilate (open up). This changes the amount of air reaching the lungs. Bronchioles are lined with a thin layer of flattened epithelium, making some gaseous exchange possible.

Alveoli

The alveoli (singular alveolus) are tiny air sacs, which are the main gas exchange surfaces of the body. Alveoli are unique to mammalian lungs. Each alveolus has a diameter of around 200–300 μm and consists of a layer of thin, flattened epithelial cells, along with some collagen and elastic fibres (composed of elastin). These elastic tissues allow the alveoli to stretch as air is drawn in. When they return to their resting size, they help squeeze the air out. This is known as the **elastic recoil** of the lungs.

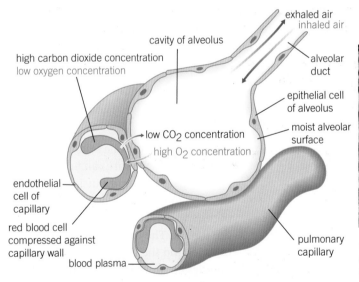

exhaled air
inhaled air
cavity of alveolus
high carbon dioxide concentration
low oxygen concentration
alveolar duct
epithelial cell of alveolus
moist alveolar surface
low CO_2 concentration
high O_2 concentration
endothelial cell of capillary
red blood cell compressed against capillary wall
blood plasma
pulmonary capillary

▲ **Figure 3** *Gaseous exchange in an alveolus*

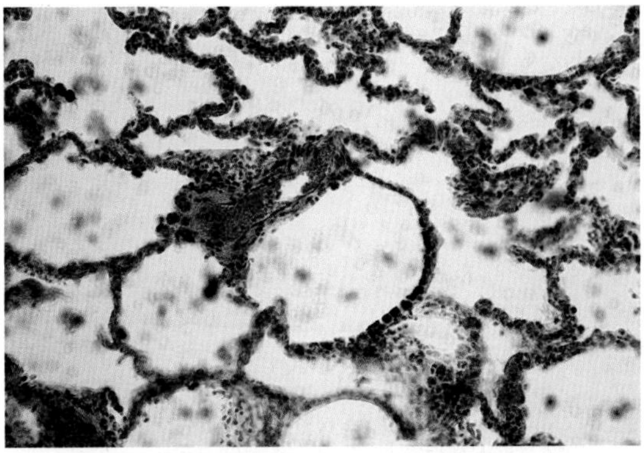

▲ **Figure 4** *The single cell walls of the alveoli and the capillaries (containing red blood cells) that make up the structure of the lung can be seen clearly in this light micrograph, × 60 magnification*

The main adaptations of the alveoli for effective gaseous exchange include:

- Large surface area – there are 300–500 million alveoli per adult lung. The alveolar surface area for gaseous exchange in the two lungs combined is around 50–75 m^2. The average floor area of a 4-bedroom

Synoptic link

You learnt about specialised epithelial cells in Topic 6.4, The organisation and specialisation of cells.

Synoptic link

You learnt about collagen and elastin in Topic 3.6, Structure of proteins.

Synoptic link

You learnt about cell membranes and how they are freely permeable to gases such as oxygen and carbon dioxide in Topic 5.3, Diffusion.

house in the UK is only 67 m². If the lungs were simple, balloon-like structures, the surface area would not be big enough for the amount of oxygen needed to diffuse into the body. This demonstrates again the importance of the SA:V ratio (Topic 7.1, Specialised exchange surfaces).

● Thin layers – both the alveoli and the capillaries that surround them have walls that are only a single epithelial cell thick, so the diffusion distances between the air in the alveolus and the blood in the capillaries are very short.

● Good blood supply – the millions of alveoli in each lung are supplied by a network of around 280 million capillaries. The constant flow of blood through these capillaries brings carbon dioxide and carries off oxygen, maintaining a steep concentration gradient for both carbon dioxide and oxygen between the air in the alveoli and the blood in the capillaries.

● Good ventilation – breathing moves air in and out of the alveoli, helping maintain steep diffusion gradients for oxygen and carbon dioxide between the blood and the air in the lungs.

The inner surface of the alveoli is covered in a thin layer of a solution of water, salts and **lung surfactant**. It is this surfactant that makes it possible for the alveoli to remain inflated. Oxygen dissolves in the water before diffusing into the blood, but water can also evaporate into the air in the alveoli. Several of the adaptations of the human gas exchange system are to reduce this loss of water.

Ventilating the lungs

Air is moved in and out of the lungs as a result of pressure changes in the thorax (chest cavity) brought about by the breathing movements. This movement of air is called ventilation.

The rib cage provides a semi-rigid case within which pressure can be lowered with respect to the air outside it. The diaphragm is a broad, domed sheet of muscle, which forms the floor of the thorax. The external intercostal muscles and the internal intercostal muscles are found between the ribs. The thorax is lined by the pleural membranes, which surround the lungs. The space between them, the pleural cavity, is usually filled with a thin layer of lubricating fluid so the membranes slide easily over each other as you breathe.

Inspiration

Inspiration (taking air in or inhalation) is an energy-using process.

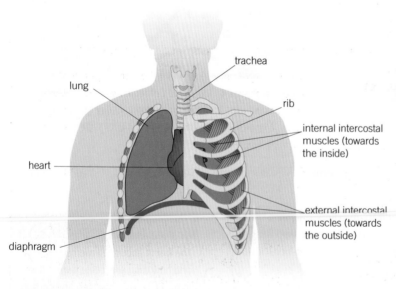

▲ **Figure 5** *The arrangement of the rib cage, diaphragm, and intercostal muscles*

The dome-shaped diaphragm contracts, flattening, and lowering. The external intercostal muscles contract, moving the ribs upwards and outwards. The volume of the thorax increases so the pressure in the thorax is reduced. It is now lower than the pressure of the atmospheric air, so air is drawn through the nasal passages, trachea, bronchi, and bronchioles into the lungs. This equalises the pressures inside and outside the chest.

Expiration

Normal expiration (breathing out or exhalation) is a passive process.

The muscles of the diaphragm relax so it moves up into its resting domed shape. The external intercostal muscles relax so the ribs move down and inwards under gravity. The elastic fibres in the alveoli of the lungs return to their normal length. The effect of all these changes is to decrease the volume of the thorax. Now the pressure inside the thorax is greater than the pressure of the atmospheric air, so air moves out of the lungs until the pressure inside and out is equal again.

You can exhale forcibly using energy. The internal intercostal muscles contract, pulling the ribs down hard and fast, and the abdominal muscles contract forcing the diaphragm up to increase the pressure in the lungs rapidly.

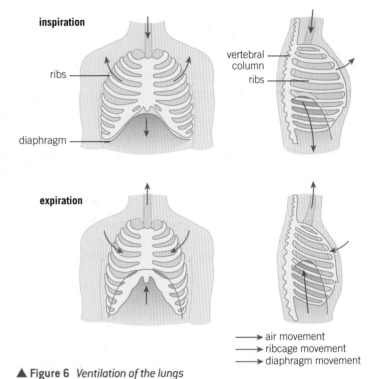

▲ **Figure 6** *Ventilation of the lungs*

Attacking asthma

5.4 million people in the UK are currently being treated for asthma. They have airways that are sensitive to everyday triggers including house dust mites, cigarette smoke, pollen, and stress.

During an asthma attack, the cells lining the bronchioles release histamines, chemicals that make the epithelial cells become inflamed and swollen. Histamines stimulate the goblet cells to make excess mucus, and the smooth muscle in the bronchiole walls to contract. As a result, the airways narrow and fill with mucus, making it difficult to breathe.

Asthma medicines have been developed to reduce the symptoms and even prevent attacks. The drugs are delivered straight into the breathing system using an inhaler.

There are two main ways of treating asthma:

Relievers give immediate relief from the symptoms. They are chemicals similar to the hormone adrenaline. They attach to active sites on the surface membranes of smooth muscle cells in the bronchioles, making them relax and dilating the airways.

Preventers are often steroids, which are taken every day to reduce the sensitivity of the lining of the airways.

1   Explain how reliever medicines would overcome the symptoms of an asthma attack.
2   Explain how steroids could reduce the likelihood of an asthma attack.

The first breath

The first breath a newborn baby takes needs a force 15–20 times greater than any normal inhalation to inflate the lungs. The lungs are enormously stretched as the air flows in, and the elastic tissue never returns to its original length. This intake of breath is only possible because of special chemicals called lung surfactants containing phospholipids and both hydrophilic and hydrophobic proteins. The surfactant stops the alveoli collapsing and sticking together as the baby exhales. Without it, the second breath would be as difficult as the first, and continued breathing impossible.

Babies born at full term have alveoli coated in lung surfactant all ready for breathing. However, the cells of the alveoli do not produce enough surfactant for the lungs to work properly until around the 30th week of pregnancy. This is one reason why premature babies can struggle to breathe and may die. In recent years artificial lung surfactants have been produced. A tiny amount sprayed into the lungs of a premature baby coats the alveoli just like the natural surfactant, making breathing easier, helping to prevent lung damage and enabling many more babies to survive.

1   Discuss why the first breath taken is so much harder than any subsequent breaths and outline how artificial lung surfactants have improved survival rates for premature babies.

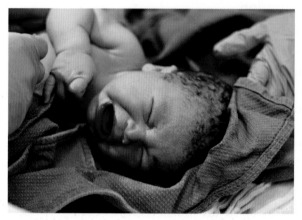

▲ **Figure 7** *The first breath a newborn takes is the hardest – and without lung surfactant it would be impossible*

Synoptic link

You learnt about phospholipids and proteins in Chapter 2, Basic components of living systems.

Summary questions

1   Explain how the structures below are adapted to make gaseous exchange as effective as possible.
   a   nose   (3 marks)
   b   trachea   (3 marks)
   c   bronchioles   (3 marks)

2   a   Explain how the alveoli are adapted for gaseous exchange.   (4 marks)
   b   In some diseases, the structure of the alveoli breaks down to give much bigger air sacs. Explain (with example calculations) how this reduces their effectiveness for gaseous exchange.   (5 marks)

3   Smokers get more infections of the breathing system than non-smokers. Suggest why.   (6 marks)

7.3 Measuring the process

Specification reference: 3.1.1

The amount of gaseous exchange that needs to take place in your lungs will vary a lot depending on your size and level of activity. The gaseous exchange system has to be able to respond to the differing demands of your body.

Measuring the capacity of the lungs

The volume of air that is drawn in and out of the lungs can be measured in a variety of different ways:

- A peak flow meter (Figure 1) is a simple device that measures the rate at which air can be expelled from the lungs. People who have asthma often use these to monitor how well their lungs are working.

- Vitalographs are more sophisticated versions of the peak flow meter. The patient being tested breathes out as quickly as they can through a mouthpiece, and the instrument produces a graph of the amount of air they breathe out and how quickly it is breathed out. This volume of air is called the forced expiratory volume in 1 second.

- A spirometer is commonly used to measure different aspects of the lung volume, or to investigate breathing patterns. There are many different forms of the spirometer but they all use the principle shown in Figure 2.

Components of the lung volume

There are several different aspects of the lung volume that can be measured.

- **Tidal volume** is the volume of air that moves into and out of the lungs with each resting breath. It is around 500 cm^3 in most adults at rest, which uses about 15% of the vital capacity of the lungs.

- **Vital capacity** is the volume of air that can be breathed in when the strongest possible exhalation is followed by the deepest possible intake of breath.

- **Inspiratory reserve volume** is the maximum volume of air you can breathe in over and above a normal inhalation.

- **Expiratory reserve volume** is the extra amount of air you can force out of your lungs over and above the normal tidal volume of air you breathe out.

- **Residual volume** is the volume of air that is left in your lungs when you have exhaled as hard as possible. This cannot be measured directly.

> ### Learning outcomes
> Demonstrate knowledge, understanding, and application of:
> → the relationship between vital capacity, tidal volume, breathing rate and oxygen uptake.

▲ **Figure 1** *Peak flow meters give a useful quick measure of how much air can be moved out of (and therefore into) the lungs*

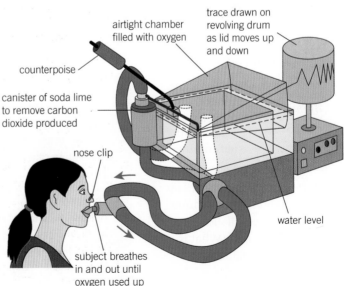

▲ **Figure 2** *Spirometers can be used to measured the volumes of gas breathed in and out under different conditions*

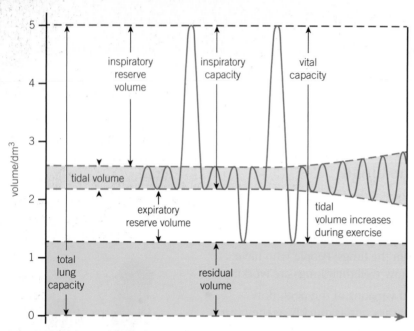

▲ **Figure 3** *These volumes are averages taken from many individuals of different sexes, sizes, and levels of fitness*

- **Total lung capacity** is the sum of the vital capacity and the residual volume.

Recordings from a spirometer show the different volumes of air moved in and out of the lungs.

Breathing rhythms

The pattern and volume of breathing changes as the demands of the body change. The **breathing rate** is the number of breaths taken per minute. The ventilation rate is the total volume of air inhaled in one minute.

ventilation rate = tidal volume × breathing rate (per minute)

When the oxygen demands of the body increase, for example during exercise, the tidal volume of air moved in and out of the lungs with each breath can increase from 15% to as much as 50% of the vital capacity. The breathing rate can also increase. In this way the ventilation of the lungs and so the oxygen uptake during gaseous exchange can be increased to meet the demands of the tissues.

🖩 Worked example: Breathing calculations

- The normal tidal volume of a male is 500 cm^3. His ventilation rate is 6 dm^3 per minute. What is his resting breathing rate?

ventilation rate = tidal volume × breathing rate (per minute)

6 dm^3 = 6000 cm^3

6000 = 500 × breathing rate

Breathing rate = $\dfrac{6000}{500}$

= 12 breaths per minute

- During physical exertion, his breathing rate goes up to 20 breaths per minute and the ventilation rate to 15 dm^3. What is the new tidal volume?

15 dm^3 = 15 000 cm^3

15 000 = tidal volume × 20

tidal volume = $\dfrac{15\,000}{20}$

= 750 cm^3

Summary questions

1 Describe how you could investigate breathing rates in a school laboratory. *(3 marks)*

2 Describe the relationships between tidal volume, breathing rate, and oxygen uptake. *(4 marks)*

3 A dog is under stress during a visit to the vet and pants but the ventilation rate remains steady. Suggest a possible explanation for these observations. *(2 marks)*

4 The normal breathing rate of a healthy 50 year old woman is 18 breaths per minute and her tidal volume is 500 cm^3.
  a During strenuous exercise her ventilation rate goes up to 45 000 cm^3 per minute and she is breathing 30 times a minute. What is her tidal volume during exercise and what increase is this over her normal tidal volume? *(6 marks)*
  b She develops a chest infection and her breathing rate increases to 25 breaths per minute, but her tidal volume falls to 300. By what percentage does her ventilation rate fall compared wirh her normal resting rate as a result of the infection? *(6 marks)*

7.4 Ventilation and gas exchange in other organisms

Specification reference: 3.1.1

Internal gas-exchange systems such as the lungs in mammals are not the only way for multicellular organisms to get the oxygen needed by the cells. The gaseous exchange systems of insects are effective but very different, yet they have many key features in common with mammalian systems.

Gaseous exchange systems in insects

Many insects are very active during parts of their life cycles. They are mainly land-dwelling animals with relatively high oxygen requirements. However, they have a tough **exoskeleton** through which little or no gaseous exchange can take place. They do not usually have blood pigments that can carry oxygen. They need a different way of exchanging gases. The gaseous exchange system of insects has evolved to deliver the oxygen directly to the cells and to remove the carbon dioxide in the same way (Figure 1).

How does gas exchange take place in insects?

Along the thorax and abdomen of most insects are small openings known as **spiracles**. Air enters and leaves the system through the spiracles, but water is also lost. Just like mammals, insects need to maximise the efficiency of gaseous exchange, but minimise the loss of water. In many insects the spiracles can be opened or closed by sphincters. The spiracle sphincters are kept closed as much as possible to minimise water loss.

When an insect is inactive and oxygen demands are very low, the spiracles will all be closed most of the time. When the oxygen demand is raised or the carbon dioxide levels build up, more of the spiracles open.

Leading away from the spiracles are the **tracheae**. These are the largest tubes of the insect respiratory system, up to 1 mm in diameter, and they carry air into the body. They run both into and along the

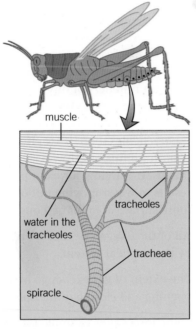

▲ **Figure 1** *The gaseous exchange system in an insect*

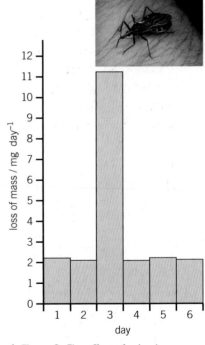

▲ **Figure 2** *The effect of spiracle opening on water loss in the blood-sucking bug* Rhodnius *(top). The insect was kept in dry air for 6 days. It was not fed, to keep it relatively inactive. On day 3 carbon dioxide levels were raised, which resulted in the spiracles opening. The air was returned to normal on day 4*

▲ Figure 3 *Careful dissection of an insect such as this immature desert locust (*<u>Schistocerca</u> <u>gregaria</u>*) can show the network of tracheae which spread out from the spiracles. Here the locust dissection has been submerged in water to help show the tracheae which are filled with air and appear a pearly white. Very fine tracheae can be seen floating up into the water*

body of the insect. The tubes are lined by spirals of chitin, which keep them open if they are bent or pressed. Chitin is the material that makes up the cuticle. It is relatively impermeable to gases and so little gaseous exchange takes place in the trachea.

The tracheae branch to form narrower tubes until they divide into the tracheoles, minute tubes of diameter 0.6–0.8 μm. Each tracheole is a single, greatly elongated cell with no chitin lining so they are freely permeable to gases. Because of their very small size they spread throughout the tissues of the insect, running between individual cells. This is where most of the gaseous exchange takes place between the air and the respiring cells.

In most insects, for most of the time, air moves along the tracheae and tracheoles by diffusion alone, reaching all the tissues. The vast numbers of tiny tracheoles give a very large surface area for gaseous exchange. Oxygen dissolves in moisture on the walls of the tracheoles and diffuses into the surrounding cells. Towards the end of the tracheoles there is **tracheal fluid**, which limits the penetration of air for diffusion. However, when oxygen demands build up – when the insect is flying, for example – a lactic acid build up in the tissues results in water moving out of the tracheoles by osmosis. This exposes more surface area for gaseous exchange.

All of the oxygen needed by the cells of an insect is supplied to them by the tracheal system.

The extent of gas exchange in most insects is controlled by the opening and closing of the spiracles.

Some insects, for example larger beetles, locusts and grasshoppers, bees, wasps and flies, have very high energy demands. To supply the extra oxygen they need, these insects have alternative methods of increasing the level of gaseous exchange. These include:

● mechanical ventilation of the tracheal system – air is actively pumped into the system by muscular pumping movements of the thorax and/or the abdomen. These movements change the volume of the body and this changes the pressure in the tracheae and tracheoles. Air is drawn into the tracheae and tracheoles, or forced out, as the pressure changes

● collapsible enlarged tracheae or air sacs, which act as air reservoirs – these are used to increase the amount of air moved through the gas exchange system. They are usually inflated and deflated by the ventilating movements of the thorax and abdomen.

Discontinuous gas exchange cycles in insects

Discontinuous gas exchange cycles (DGC) have been found to be relatively common in many species of insects. In DCG spiracles have three states – closed, open, and fluttering.

- When the spiracles are closed no gases move in or out of the insect. Oxygen moves into the cells by diffusion from the tracheae and carbon dioxide diffuses into the body fluids of the insect where it is held in a process called buffering.

- When the spiracles flutter, they open and close rapidly. This moves fresh air into the tracheae to renew the supply of oxygen, while minimising water loss.

- When carbon dioxide levels build up really high in the body fluids of the insect, the spiracles open widely and carbon dioxide diffuses out rapidly. There may also be pumping movements of the thorax and abdomen when the spiracles are open to maximise gaseous exchange.

Originally scientists thought discontinuous gas exchange was an adaptation for water conservation in insects. Now the evidence suggests this is not the case and there are a number of conflicting theories about the adaptive advantages of discontinuous gas exchange for insects, which include helping gaseous exchange in insects that spend at least part of their lives in enclosed spaces such as burrows, or reducing the entry of fungal spores, which can parasitise an insect. This is still an area of very active research and argument.

1 Suggest one way in which each of the given theories on the value of DGC as an adaptation might be investigated.

Respiratory systems in bony fish

Animals that get their oxygen from water do not need to try and prevent water loss from their gaseous exchange surfaces as land animals do, but there are other difficulties to overcome.

Water is 1000 times denser than air. It is 100 times more viscous (thick) and has a much lower oxygen content. To cope with the viscosity of water and the slow rate of oxygen diffusion, fish have evolved very specialised respiratory systems that are different from those of land-dwelling animals. It would use up far too much energy to move dense, viscous water in and out of lung-like respiratory organs. Moving water in one direction only is much simpler and more economical in energy terms.

Gills

Bony fish such as trout and cod are relatively big, active animals that live almost exclusively in water. Because they are very active, their cells have a high oxygen demand. Their SA:V ratio means that diffusion would not be enough to supply their inner cells with the oxygen they need, and their scaly outer covering does not allow gaseous exchange. However, bony fish have evolved a ventilatory system adapted to take oxygen from the water and get rid of carbon dioxide into the water. They maintain a flow of water in one direction over the **gills**, which are their organs of gaseous exchange. Gills have the large surface area, good blood supply, and thin layers needed for successful gaseous exchange. In bony fish they are contained in a gill cavity and covered by a protective **operculum** (a bony flap), which is also active in maintaining a flow of water over the gills (Figure 4).

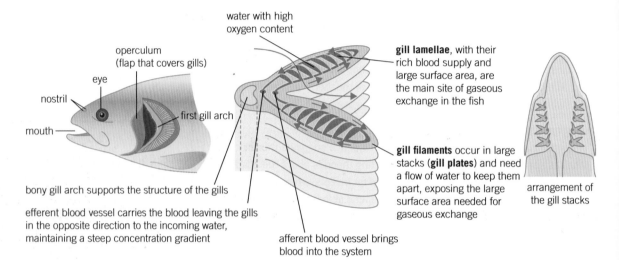

water with high oxygen content

operculum (flap that covers gills)

eye

nostril

mouth

first gill arch

gill lamellae, with their rich blood supply and large surface area, are the main site of gaseous exchange in the fish

gill filaments occur in large stacks (**gill plates**) and need a flow of water to keep them apart, exposing the large surface area needed for gaseous exchange

arrangement of the gill stacks

bony gill arch supports the structure of the gills

efferent blood vessel carries the blood leaving the gills in the opposite direction to the incoming water, maintaining a steep concentration gradient

afferent blood vessel brings blood into the system

▲ **Figure 4** *The gills are the site of gaseous exchange in bony fish and have a number of adaptations for their role in supplying oxygen and removing carbon dioxide*

The gills make up the gaseous exchange surface of the fish. They have many features in common with both mammalian and insect gaseous exchange surfaces. However, they also have particular challenges. To allow efficient gas exchange at all times, fish need to maintain a continuous flow of water over the gills, even when they are not moving. They also need to carry out gaseous exchange as effectively as possible in water, a medium where diffusion is slower than in air.

Water flow over the gills

When fish are swimming they can keep a current of water flowing over their gills simply by opening their mouth and operculum. However, when the fish stops moving, the flow of water also stops. The more primitive cartilaginous fish such as the sharks and rays often rely on continual movement to ventilate the gills. This is known as ram ventilation – they just ram the water past the gills. However, most bony fish do not rely on movement-generated water flow over the gills. They have evolved a sophisticated system involving the operculum, which allows them to move water over their gills all the time (Figure 5).

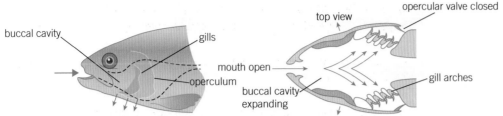

The mouth is opened and the floor of the buccal cavity (mouth) is lowered. This increases the volume of the buccal cavity. As a result the pressure in the cavity drops and water moves into the buccal cavity. At the same time the opercular valve is shut and the opercular cavity containing the gills expands. This lowers the pressure in the opercular cavity containing the gills. The floor of the buccal cavity starts to move up, increasing the pressure there so water moves from the buccal cavity over the gills.

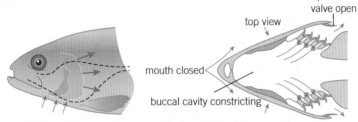

The mouth closes, the operculum opens and the sides of the opercular cavity move inwards. All of these actions increase the pressure in the opercular cavity and force water over the gills and out of the operculum. The floor of the buccal cavity is steadily moved up, maintaining a flow of water over the gills.

▲ **Figure 5** *Although this diagram shows the ventilation of the gills in stages, this is actually a continuous process, which ensures that water is constantly flowing over the gills*

Effective gaseous exchange in water

Gills have a large surface area for diffusion, a rich blood supply to maintain steep concentration gradients for diffusion, and thin layers so that diffusing substances have only short distances to travel. Gills have two extra adaptations that help to ensure the most effective possible gaseous exchange occurs in the water:

- The tips of adjacent gill filaments overlap. This increases the resistance to the flow of water over the gill surfaces and slows down the movement of the water. As a result there is more time for gaseous exchange to take place.

- The water moving over the gills and the blood in the gill filaments flow in different directions. A steep concentration gradient is needed for fast, efficient diffusion to take place. Because the blood and water flow in opposite directions, a countercurrent exchange system is set up. This adaptation ensures that steeper concentration gradients are maintained than if blood and water flowed in the same direction (known as a parallel system). As a result, more gaseous exchange can take place. The bony fish, with their countercurrent systems, remove about 80% of the oxygen from the water flowing over them. The cartilaginous fish have parallel systems and can only extract about 50% of the oxygen from the water flowing over them (Figure 6).

▲ **Figure 6** *Careful dissection of a trout (genera* Onchorhynchus*). The operculum (top) helps to protect the gills. Removal of the operculum shows thin feathery gills which have a large surface area (middle). Here you can see that the water flows in through the mouth over the gills and out (bottom)*

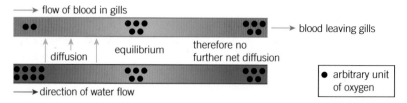

parallel system: blood in the gills and water flowing over the gills travel in the same direction, which gives an initial steep oxygen concentration gradient between blood and water. Diffusion takes place until the oxygen concentration of the blood and water are in equilibrium, then no net movement of oxygen into the blood occurs.

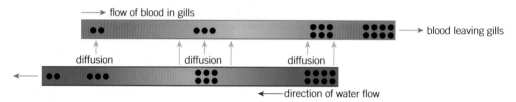

countercurrent system: blood and water flow in opposite directions so an oxygen concentration gradient between the water and the blood is maintained all along the gill. Oxygen continues to diffuse down the concentration gradient so a much higher level of oxygen saturation of the blood is achieved.

▲ **Figure 7** *The advantages of a countercurrent exchange system are clear to see. The system also enables bony fish to remove more carbon dioxide from the blood than a parallel system*

Dissecting, examining, and drawing gaseous exchange systems

Dissecting an animal gives you a unique insight into the complexity of a multicellular living organism. The process of evolution over millions of years has resulted in internal systems of great efficiency and elegance.

When you dissect an animal it can be confusing. You will be looking at one particular body system but they do not exist in isolation. You cannot see the gas exchange system, for example, without getting through the body wall.

You will need specialist equipment including boards and pins with which to display your dissection. The tools needed for successful dissections include sharp scissors and scalpels along with tweezers and mounted needles to lift and tease out tissues.

When you carry out a dissection the aim is to be as precise and clean in your work as possible. You should observe and display the relevant features of an organism to the best of your ability, and then record what you have seen in a clear and well-labelled diagram. It may be useful to take a photograph of your dissection and so preserve what it actually looked like alongside your labelled diagram. In Figure 3 and Figure 6 you can see examples of dissections of the gas exchange system of an insect and the gas exchange system of a bony fish.

1 Drawings from dissections are always done in pencil – suggest why.
2 Make careful labelled drawings of the dissections shown in Figure 3 and Figure 6.

The histology of exchange surfaces

Some of the key features of exchange surfaces – the large surface area, the short diffusion distances, and the proximity of the blood supply often cannot be seen when you look at the whole organ system.

However using prepared slides with the light microscope can give you an insight into the detailed adaptations of these surfaces for their role in gaseous exchange.

Example 1 – The network of tiny air sacs with walls that are a single cell thick that make up the surface area of the alveoli can clearly be seen in light micrographs of lung tissue , × 40 magnification.

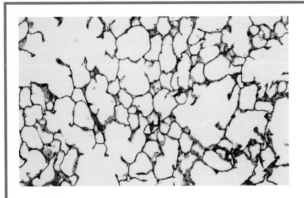

Example 3 – The structure of the gills in a bony fish show clearly in a light micrograph, showing the delicate structures and large surface area as well as the major blood vessels , × 278 magnification.

Example 2 – The delicate rings of chitin which support the trachea in insects, holding them open so that air can diffuse through them to the tracheoles and the tissues themselves are revealed in this light micrograph , × 37 magnification.

1 Make carefully labelled drawings of each of these micrographs, using the skills you learned in Chapter 2.
2 What are the advantages of using the light microscope to look at these tissues?

Summary questions

1 Suggest why a fish will die when it is left out of water for too long. *(2 marks)*

2 Compile a table to compare the gaseous exchange systems of a human, an insect and a bony fish. *(5 marks)*

3 Explain how insects that have particularly high energy requirements can increase the amount of gaseous exchange taking place in their bodies. *(6 marks)*

4 Explain how the structure of the gas exchange system of a bony fish maximises the amount of oxygen that can be taken from the water. *(6 marks)*

Study tip

Make sure you are clear about the key features of a successful gaseous exchange surface and can recognise those common features in the gaseous exchange systems of a variety of organisms.

Practice questions

1 There are several different aspects of the lung volume that can be measured.

   Which of the following statements is/are correct with respect to lung volumes?

   Statement 1: tidal volume is the volume of air which moves into and out of the lungs with each resting breath

   Statement 2: vital capacity is the maximum amount of air you can breathe in over and above a normal inhalation

   Statement 3: inspiratory reserve volume is the volume of air which can be breathed out by the strongest possible exhalation followed by the deepest possible intake of breath

   A   1, 2 and 3 are correct

   B   Only 1 and 2 are correct

   C   Only 2 and 3 are correct

   D   Only 1 is correct   *(1 mark)*

2 a State what a peak flow meter is used to measure   *(2 marks)*

   b The chart below shows the range of normal peak flow for both men and women of different ages.

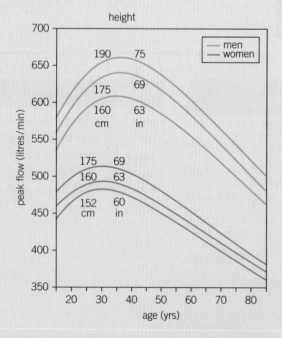

Using the graph, describe how lung function varies with both age and gender.
               (4 marks)

   c Draw another line on the graph for a man who had untreated asthma from birth. Explain the reasons for the position of your line and state why it would be unlikely to see someone produce this line.
               (3 marks)

   d Explain why inhalation is an active process but normal exhalation is a passive process.   *(3 marks)*

3 Fick's law states that:

$$\text{Rate of diffusion} = \frac{\text{Area of diffusion surface} \times \text{Difference in concentration}}{\text{Thickness of surface over which diffusion takes place}}$$

   a Explain how the human respiratory system is designed to obey this law.
               (3 marks)

   b The gaseous exchange surfaces in fish and mammals are designed to perform the same function: the uptake of oxygen and removal of carbon dioxide.

   Discuss how they are each adapted to carry out this function in different environments; fish in water and mammals in air.   *(3 marks)*

4 Insects have an exoskeleton.

   a State the main component of this exoskeleton.   *(1 mark)*

   b The component named in a is a long-chain polymer of a N-acetylglucosamine, a derivative of glucose.

   Explain what is meant by the term polymer, with reference to this component.   *(2 marks)*

   c Although relatively small, insects have a high oxygen demand. The exoskeleton is impermeable to oxygen.

   Describe how insects are adapted to ensure their tissues receive oxygen at a sufficient rate.   *(4 marks)*

   d Describe one similarity and one difference between the trachea of a mammal and the trachea of an insect.   *(2 marks)*

5 The diagram below shows the lung volumes at different phases of the respiratory cycle.

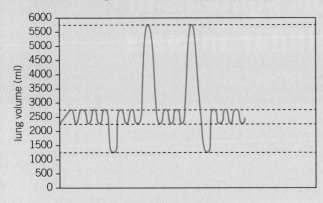

a Calculate, using the diagram, the inspiratory reserve volume and tidal volume of the lungs. *(2 marks)*

b Copy the diagram and indicate the residual volume and expiratory reserve volume. *(2 marks)*

c The diagram below compares the lung volumes of a woman before and during pregnancy.

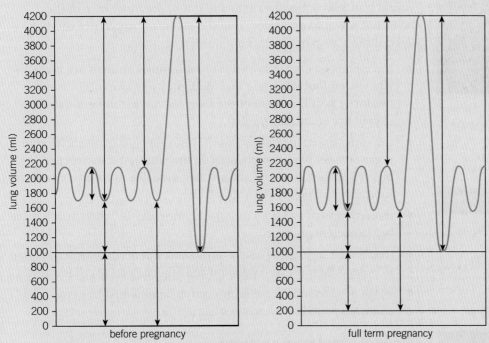

Describe the changes in lung volumes that occur during pregnancy and explain the reasons for these changes. *(6 marks)*

8 TRANSPORT IN ANIMALS
8.1 Transport systems in multicellular animals
Specification reference: 3.1.2

▲ **Figure 1** *When a multicellular animal takes in food, the nutrients have to be transported to all the cells in the body*

Synoptic link

You learnt about surface area to volume ratio in Topic 7.1, Specialised exchange surfaces.

Synoptic link

It will be useful now to recap your understanding of the various mechanisms for the movement of molecules across membranes in Topics 4.3–4.5 Diffusion, Active transport, and Osmosis.

In the previous chapter you looked at why multicellular organisms need exchange surfaces. They also need **transport systems** to supply oxygen and nutrients to the sites where they are needed and to remove waste products from the individual cells.

The need for specialised transport systems in animals

In single-celled organisms, processes such as diffusion, osmosis, active transport, endocytosis and exocytosis can supply everything the cell needs to import or export. These processes are also important in multicellular organisms, transporting substances within and between individual cells. However, as organisms get bigger, the distances between the cells and the outside of the body get greater. Diffusion would transport substances into and out of the inner core of the body, but it would be so slow that the organism would not survive. Specialised transport systems are needed because:

* the metabolic demands of most multicellular animals are high (they need lots of oxygen and food, they produce lots of waste products) so diffusion over the long distances is not enough to supply the quantities needed

* the surface area to volume (SA : V) ratio gets smaller as multicellular organisms get bigger so not only do the diffusion distances get bigger but the amount of surface area available to absorb or remove substances becomes relatively smaller

* molecules such as hormones or enzymes may be made in one place but needed in another

* food will be digested in one organ system, but needs to be transported to every cell for use in respiration and other aspects of cell metabolism

* waste products of metabolism need to be removed from the cells and transported to excretory organs.

Types of circulatory systems

Most large, multicellular animals have specialised **circulatory systems** (transport systems) which carry gases such as oxygen and carbon dioxide, nutrients, waste products and hormones around the body. Most circulatory systems have features in common:

* They have a liquid transport medium that circulates around the system (blood).

- They have vessels that carry the transport medium.
- They have a pumping mechanism to move the fluid around the system.

When substances are transported in a mass of fluid with a mechanism for moving the fluid around the body it is known as a **mass transport system**. Large, multicellular animals usually have either an open circulatory system or a closed circulatory system.

Open circulatory systems
In an **open circulatory system** there are very few vessels to contain the transport medium. It is pumped straight from the heart into the body cavity of the animal. This open body cavity is called the haemocoel. In the haemocoel the transport medium is under low pressure. It comes into direct contact with the tissues and the cells. This is where exchange takes place between the transport medium and the cells. The transport medium returns to the heart through an open-ended vessel (Figure 2).

These open-ended circulatory systems are found mainly in invertebrate animals, including most insects and some molluscs. Remember that in insects, gas exchange takes place in the tracheal system. Insect blood is called **haemolymph**. It doesn't carry oxygen or carbon dioxide. It transports food and nitrogenous waste products and the cells involved in defence against disease. The body cavity is split by a membrane and the heart extends along the length of the thorax and the abdomen of the insect. The haemolymph circulates but steep diffusion gradients cannot be maintained for efficient diffusion. The amount of haemolymph flowing to a particular tissue cannot be varied to meet changing demands.

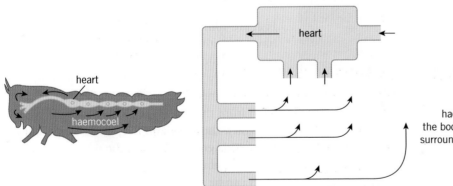

▲ Figure 2 *The open circulatory system of a locust supplies the cells with food and removes the nitrogenous waste products*

Closed circulatory systems
In a **closed circulatory system**, the blood is enclosed in blood vessels and does not come directly into contact with the cells of the body. The heart pumps the blood around the body under pressure and relatively quickly, and the blood returns directly to the heart. Substances leave and enter the blood by diffusion through the walls of the blood vessels.

Synoptic link

You learnt about mechanisms of ventilation and gaseous exchange in insects in Topic 7.4, Ventilation and gas exchange in other organisms.

The amount of blood flowing to a particular tissue can be adjusted by widening or narrowing blood vessels. Most closed circulatory systems contain a blood pigment that carries the respiratory gases.

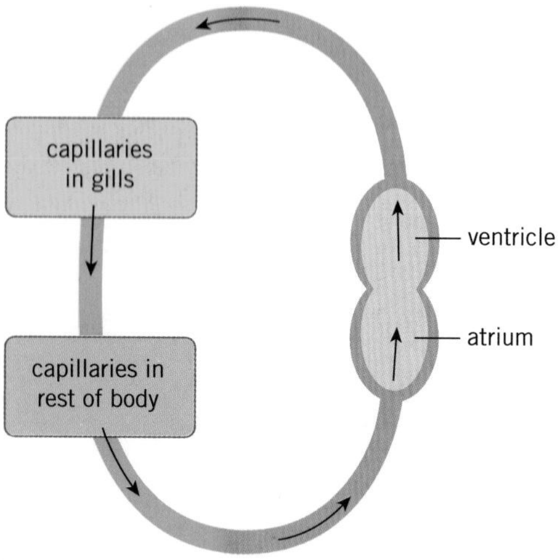

Closed circulatory systems are found in many different animal phyla, including echinoderms (animals such as sea urchins and starfish), cephalopod molluscs including the octopods and squid, annelid worms including the common earthworm, and all of the vertebrate groups, including the mammals.

Single closed circulatory systems

Single closed circulatory systems are found in a number of groups including fish and annelid worms. In **single circulatory systems** (Figure 3) the blood flows through the heart and is pumped out to travel all around the body before returning to the heart. In other words, the blood travels only once through the heart for each complete circulation of the body.

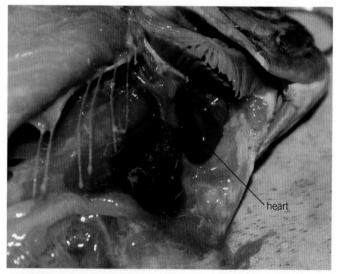

▲ Figure 3 *The single closed circulatory system of a fish*

In a single closed circulation, the blood passes through two sets of capillaries (microscopic blood vessels) before it returns to the heart. In the first, it exchanges oxygen and carbon dioxide. In the second set of capillaries, in the different organ systems, substances are exchanged between the blood and the cells. As a result of passing through these two sets of very narrow vessels, the blood pressure in the system drops considerably so the blood returns to the heart quite slowly. This limits the efficiency of the exchange processes so the activity levels of animals with single closed circulations tends to be relatively low.

Fish are something of an exception. They have a relatively efficient single circulatory system, which means they can be very active. They have a countercurrent gaseous exchange mechanism in their gills that allows them to take a lot of oxygen from the water. Their body weight is supported by the water in which they live and they do not maintain their own body temperature. This greatly reduces the metabolic demands on their bodies and, combined with their efficient gaseous exchange, explains how fish can be so active with a single closed circulatory system.

▲ Figure 4 *The single closed circulatory system of a fish. Dissection of this trout shows the close proximity of the fish heart to the gills. Fish hearts have one atrium and one ventricle*

Double closed circulatory systems

Birds and most mammals are very active land animals that maintain their own body temperature. This way of life is made possible in part by their double closed circulatory system (Figure 5). This is the most efficient system for transporting substances around the body. It involves two separate circulations:

Synoptic link

You learnt about the mechanisms of ventilation and gaseous exchange in fish in Topic 7.4, Ventilation and gas exchange in other organisms.

- Blood is pumped from the heart to the lungs to pick up oxygen and unload carbon dioxide, and then returns to the heart.
- Blood flows through the heart and is pumped out to travel all around the body before returning to the heart again.

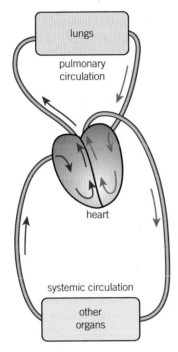

▲ **Figure 5** *The double closed circulatory system of a mammal*

So in a **double circulatory system**, the blood travels twice through the heart for each circuit of the body. Each circuit – to the lungs and to the body – only passes through one capillary network, which means a relatively high pressure and fast flow of blood can be maintained.

Summary questions

1 Describe the function of a circulatory system.   *(3 marks)*

2 Explain why circulatory systems are found in multicellular organisms but not unicellular organisms.   *(2 marks)*

3 Compare open and closed circulatory systems.   *(6 marks)*

4 Land predators such as foxes have a double closed circulatory system. Aquatic predators such as pike are effective with a single closed circulatory system. Explain why these two types of predator have different circulatory systems.   *(6 marks)*

Study tip

Make sure you are clear about the differences between open and closed circulatory systems, and between single and double closed circulatory systems.

Be aware of the advantages and disadvantages of the different systems.

8.2 Blood vessels

Specification reference: 3.1.2

Circulation in humans is typical of a mammalian circulatory system. It is estimated that if all the blood vessels of an average adult human were laid end to end they would stretch to 100 000 miles – that is the equivalent of about four times around the circumference of the Earth.

There are several different types of blood vessels in the body and their structural composition is closely related to their function. Some examples of different components utilised in some blood vessels are:

- Elastic fibres – these are composed of elastin and can stretch and recoil, providing vessel walls with flexibility.
- Smooth muscle – contracts or relaxes, which changes the size of the lumen (the channel within the blood vessel).
- Collagen – provides structural support to maintain the shape and volume of the vessel.

The sections below detail which components are found in each of the blood vessel types.

Arteries and arterioles

The arteries carry blood away from the heart to the tissues of the body. They carry **oxygenated blood** *except* in the pulmonary artery, which carries deoxygenated blood from the heart to the lungs, and (during pregnancy) the umbilical artery, which carries deoxygenated blood from the fetus to the placenta. The blood in the arteries is under higher pressure than the blood in the veins.

Artery walls contain elastic fibres, smooth muscle and collagen (Figure 1). The elastic fibres enable them to withstand the force of the blood pumped out of the heart and stretch (within limits maintained by collagen) to take the larger blood volume. In between the contractions of the heart, the elastic fibres recoil and return to their original length. This helps to even out the surges of blood pumped from the heart to give a continuous flow. However, you can still feel a pulse (surge of blood) when the heart contracts, which the elastic fibres cannot completely eliminate. The lining of an artery (endothelium) is smooth so the blood flows easily over it.

Arterioles link the arteries and the capillaries. They have more smooth muscle and less elastin in their walls than arteries, as they have little pulse surge, but can constrict or dilate to control the flow of blood into individual organs. When the smooth muscle in the arteriole contracts it constricts the vessel and prevents blood flowing into a

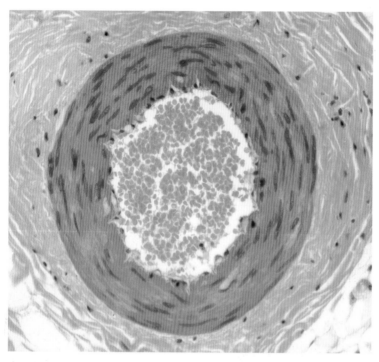

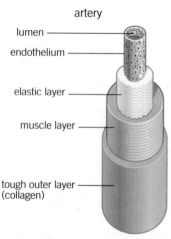

artery
lumen
endothelium
elastic layer
muscle layer
tough outer layer (collagen)

▲ **Figure 1** *The structure of the artery wall is closely related to its functions, and changes as the arteries get smaller, × 250 magnification*

capillary bed. This is *vasoconstriction*. When the smooth muscle in the wall of an arteriole relaxes, blood flows through into the capillary bed. This is *vasodilation*.

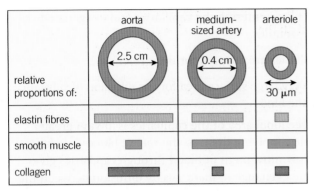

relative proportions of:	aorta	medium-sized artery	arteriole
	2.5 cm	0.4 cm	30 μm
elastin fibres			
smooth muscle			
collagen			

▲ Figure 2 *The differing proportions of components of the artery wall depending on how far the artery is from the heart and thus its role in the body*

Synoptic link

You learnt about the structure of elastin and collagen in Topic 3.6 The structure of proteins.

✚ Collagen, elastin, and aortic aneurysms

An aneurysm is a bulge or weakness in a blood vessel. The most common places for aneurysms are in the aorta and in the arteries of the brain. Most people do not know they have an aneurysm until it bursts. This is very serious and can be fatal. High blood pressure is one factor that increases the risk of an aneurysm. However, scientists have also discovered changes in the proportion of collagen to elastin in the aorta wall. The ratio of collagen : elastin in a normal aorta is 1.85 : 1. In a small aneurysm it increases to around 3.75 : 1 and in large aortic aneurysms it is 7.91 : 1. Research is continuing to see if this apparent link is real – and, if so, whether it can be used to predict who is at risk so they can have regular aortic screening.

1 What hypothesis for the formation of aneurysms can you develop from these data?
2 Suggest two possible ways in which patients might be treated to reduce the possibility of aneurysms developing or enlarging.

red blood cell

7-8 μm

Capillaries

The **capillaries** are microscopic blood vessels that link the arterioles with the venules. They form an extensive network through all the tissues of the body. The lumen of a capillary is so small that red blood cells (which have a diameter of only 7.5–8 μm) have to travel through in single file (Figure 3). Substances are exchanged through the capillary walls between the tissue cells and the blood. The gaps between the endothelial cells that make up the capillary walls in most areas of the body are relatively large. This is where many substances pass out of the capillaries into the fluid surrounding the cells. The exception is the capillaries in the central nervous system, which have very tight junctions between the cells.

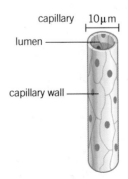

capillary 10μm

lumen

capillary wall

▲ Figure 3 *The structure of capillaries means that they are small enough to form the immense networks needed to exchange substances between the blood and the tissues*

In most organs of the body the blood entering the capillaries from the arterioles is oxygenated. By the time it leaves the capillaries for the venules it has less oxygen and more carbon dioxide (it is deoxygenated). Again, the lungs and the placenta are the exceptions, with deoxygenated blood entering the capillaries and oxygenated blood leaving in the venules.

Ways in which capillaries are adapted for their role:

● They provide a very large surface area for the diffusion of substances into and out of the blood.

● The total cross-sectional area of the capillaries is always greater than the arteriole supplying them so the rate of blood flow falls. The relatively slow movement of blood through capillaries gives more time for the exchange of materials by diffusion between the blood and the cells.

● The walls are a single endothelial cell thick, giving a very thin layer for diffusion.

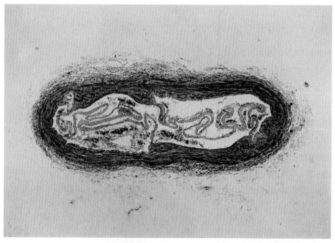

Veins and venules

The veins carry blood away from the cells of the body towards the heart and, with two exceptions, they carry deoxygenated blood. The pulmonary vein carries oxygenated blood from the lungs to the heart, and (during pregnancy) the umbilical vein carries oxygenated blood from the placenta to the fetus.

Deoxygenated blood flows from the capillaries into very small veins celled venules and then into larger veins. Finally it reaches the two main vessels carrying deoxygenated blood back to the heart – the inferior vena cava from the lower parts of the body and the superior vena cava from the head and upper body.

Veins do not have a pulse – the surges from the heart pumping are lost as the blood passes through the narrow capillaries. However, they do hold a large reservoir of blood – up to 60% of your blood volume is in your veins at any one time.

The blood pressure in the veins is very low compared with the pressure in the arteries. Medium-sized veins (the majority of the venous system) have valves to prevent the backflow of blood (see next page).

The walls contain lots of collagen and relatively little elastic fibre, and the vessels have a wide lumen and a smooth, thin lining (known as the endothelium) so the blood flows easily (Figure 4).

Venules link the capillaries with the veins. They have very thin walls with just a little smooth muscle. Several venules join to form a vein.

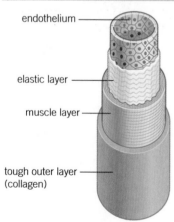

endothelium

elastic layer

muscle layer

tough outer layer (collagen)

▲ Figure 4 *Veins do not have to withstand high pressures like arteries do but they need a big capacity and this is reflected in their structure*

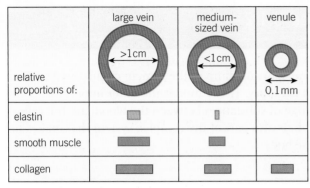

relative proportions of:	large vein	medium-sized vein	venule
elastin			
smooth muscle			
collagen			

▲ **Figure 5** *The differing proportions of components of the vein and venule walls depending on how far the vein is from the heart and thus its role in the body*

Deoxygenated blood in the veins must be returned to the heart to be pumped to the lungs and oxygenated again. However, the blood is under low pressure and needs to move against gravity. There are three main adaptations that enable the body to overcome this problem:

- The majority of the veins have one-way valves at intervals. These are flaps or infoldings of the inner lining of the vein. When blood flows in the direction of the heart, the valves open so the blood can pass through. If the blood starts to flow backwards, the valves close to prevent this from happening.
- Many of the bigger veins run between the big, active muscles in the body, for example in the arms and legs. When the muscles contract they squeeze the veins, forcing the blood towards the heart. The valves prevent backflow when the muscles relax.
- The breathing movements of the chest act as a pump. The pressure changes and the squeezing actions move blood in the veins of the chest and abdomen towards the heart.

In combination, these adaptations assist in the return of deoxygenated blood to the heart.

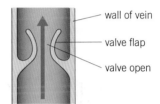

blood flowing towards the heart passes easily through the valves

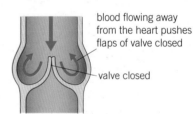

blood flowing away from the heart pushes valves closed and so blood is prevented from flowing any further in this direction

▲ **Figure 6** *The valves in a vein prevent the backflow of blood*

Summary questions

1 Veins have valves but arteries do not – explain why. *(3 marks)*

2 The structure of an arteriole is different from the structure of an artery. Describe the differences in structure and explain how they are related to the functions of the vessels. *(2 marks)*

3 a Explain how the different structures of large veins, medium veins and venules are related to their functions in the body. *(6 marks)*

 b Explain why the venules and veins in the lungs are so unusual. *(2 marks)*

8.3 Blood, tissue fluid, and lymph

Specification reference: 3.1.2

Blood is the main transport medium of the human circulatory system, but it is only part of the story. Tissue fluid is the other important player in the exchange of substances between the blood and the cells. A third liquid, lymph, is also part of the complex system that makes up the circulation of the body.

Blood

Blood consists of a yellow liquid – **plasma** – which carries a wide variety of other components including dissolved glucose and amino acids, mineral ions, hormones, and the large plasma proteins including albumin (important for maintaining the osmotic potential of the blood), fibrinogen (important in blood clotting) and globulins (involved in transport and the immune system). Plasma also transports red blood cells (which carry oxygen to the cells and also give the blood its red appearance) and the many different types of white blood cells. It also carries platelets. Platelets are fragments of large cells called megakaryocytes found in the red bone marrow, and they are involved in the clotting mechanism of the blood. Plasma makes up 55% of the blood by volume – and much of that volume is water. Only the plasma and the red blood cells are involved in the transport functions of the blood. The other components have different functions.

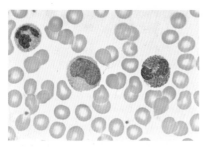
Functions of the blood

The composition of the blood is closely related to its functions in the body, many of which involve transport. They include transport of:

- oxygen to, and carbon dioxide from, the respiring cells
- digested food from the small intestine
- nitrogenous waste products from the cells to the excretory organs
- chemical messages (hormones)
- food molecules from storage compounds to the cells that need them
- platelets to damaged areas
- cells and antibodies involved in the immune response.

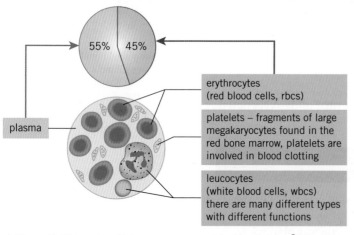

▲ Figure 2 The main cellular components of the blood. 1 mm³ of healthy blood contains around 5 million erythrocytes, 7000 leucocytes and 250 000 platelets, which are cell fragments

The blood also contributes to maintenance of a steady body temperature and acts as a buffer, minimising pH changes.

Learning outcomes

Demonstrate knowledge, understanding, and application of:

→ the differences in the composition of blood, tissue fluid and lymph

→ the formation of tissue fluid from plasma.

Synoptic link

You will learn more about clotting of the blood, phagocytosis, and the immune system in Topic 12.5, The specific immune system.

Tissue fluid

The substances dissolved in plasma can pass through the fenestrations in the capillary walls, with the exception of the large plasma proteins. The plasma proteins, particularly albumin, have an osmotic effect. They give the blood in the capillaries a relatively high solute potential (and so a relatively low water potential) compared with the surrounding fluid. As a result, water has a tendency to move into the blood in the capillaries from the surrounding fluid by osmosis. The tendency of water to move into the blood by osmosis is termed **oncotic pressure** and it is about −3.3 kPa.

However, as blood flows through the arterioles into the capillaries, it is still under pressure from the surge of blood that occurs every time the heart contracts. This is **hydrostatic pressure**. At the arterial end of the capillary, the hydrostatic pressure forcing fluid out of the capillaries is relatively high at about 4.6 kPa (Figure 3). It is higher than the oncotic pressure attracting water in by **osmosis**, so fluid is squeezed out of the capillaries. This fluid fills the spaces between the cells and is called **tissue fluid**. Tissue fluid has the same composition as the plasma, without the red blood cells and the plasma proteins. Diffusion takes place between the blood and the cells through the tissue fluid.

As the blood moves through the capillaries towards the venous system, the balance of forces changes. The hydrostatic pressure falls to around 2.3 kPa in the vessels as fluid has moved out and the pulse is completely lost. The oncotic pressure is still −3.3 kPa, so it is now stronger than the hydrostatic pressure, so water moves back into the capillaries by osmosis as it approaches the venous end of the capillaries. By the time the blood returns to the veins, 90% of the tissue fluid is back in the blood vessels.

Synoptic link

You learnt about hydrostatic pressure and osmosis in Topic 5.5, Osmosis.

Study tip

Filtration pressure = hydrostatic pressure − oncotic pressure

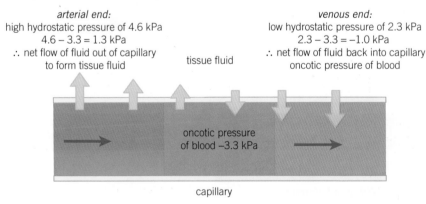

▲ **Figure 3** *Diagram showing differences in hydrostatic pressure at arterial and venous end and how this results in movement into or out of the capillary*

Lymph

Some of the tissue fluid does not return to the capillaries. 10% of the liquid that leaves the blood vessels drains into a system of blind-ended tubes called lymph capillaries, where it is known as **lymph**. Lymph is similar in composition to plasma and tissue fluid but has less oxygen and fewer nutrients. It also contains fatty acids, which have been absorbed into the lymph from the villi of the small intestine. The lymph capillaries join up to form larger vessels. The fluid is transported through them by the squeezing of the body muscles. One-way valves like those in veins prevent the backflow of lymph. Eventually the lymph returns to the blood, flowing into the right and left subclavian veins (under the clavicle, or collar bone).

Along the lymph vessels are the lymph nodes. **Lymphocytes** build up in the lymph node when necessary and produce antibodies, which are then passed into the blood. Lymph nodes also intercept bacteria and other debris from the lymph, which are ingested by phagocytes found in the nodes. The lymphatic system plays a major role in the defence mechanisms of the body.

Enlarged lymph nodes are a sign that the body is fighting off an invading pathogen. This is why doctors often examine the neck, armpits, stomach or groin of their patients – these are the sites of some of the major lymph nodes (which people often refer to as 'lymph glands').

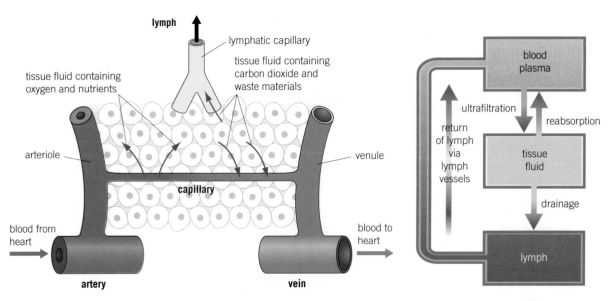

▲ Figure 4 *The human lymphatic system*

Summary questions

1 Describe the main functions of the blood. (*4 marks*)

2 a What are platelets? (*3 marks*)
   b What is the role of platelets in the body? (*3 marks*)
   c What percentage of the cells/cell fragments in the blood is made up of platelets? (*3 marks*)

3 Summarise the similarities and differences between plasma, tissue fluid, and lymph. (*5 marks*)

4 Explain how hydrostatic and oncotic pressure affect the movement of fluids into and out of capillaries. (*6 marks*)

8.4 Transport of oxygen and carbon dioxide in the blood

Specification reference: 3.1.2

The most specialised transport role of the blood is the transport of oxygen from the lungs to the cells of the body by the erythrocytes (red blood cells). The erythrocytes are also involved in the removal of carbon dioxide from the cells and its transport to the lungs for gaseous exchange.

Transporting oxygen

The erythrocytes are very specialised, with a number of adaptations to their main function of transporting oxygen.

Erythrocytes have a biconcave shape. This shape has a larger surface area than a simple disc structure or a sphere, increasing the surface area available for diffusion of gases. It also helps them to pass through narrow capillaries. In adults, erythrocytes are formed continuously in the red bone marrow. By the time mature erythrocytes enter the circulation they have lost their nuclei, which maximises the amount of haemoglobin that fits into the cells. It also limits their life, so they only last for about 120 days in the bloodstream.

Erythrocytes contain **haemoglobin**, the red pigment that carries oxygen and also gives them their colour. Haemoglobin is a very large globular conjugated protein made up of four peptide chains, each with an iron-containing haem prosthetic group. There are about 300 million haemoglobin molecules in each red blood cell and each haemoglobin molecule can bind to four oxygen molecules.

The oxygen binds quite loosely to the haemoglobin forming **oxyhaemoglobin**. The reaction is reversible.

$$Hb \quad + \quad 4O_2 \quad \rightleftharpoons \quad Hb(O_2)_4$$

haemoglobin + oxygen ⇌ oxyhaemoglobin

Carrying oxygen

When the erythrocytes enter the capillaries in the lungs, the oxygen levels in the cells are relatively low. This makes a steep concentration gradient between the inside of the erythrocytes and the air in the alveoli. Oxygen moves into the erythrocytes and binds with the haemoglobin. The arrangement of the haemoglobin molecule means that as soon as one oxygen molecule binds to a haem group, the molecule changes shape, making it easier for the next oxygen molecules to bind. This is known as positive cooperativity. Because the oxygen is bound to the haemoglobin, the free oxygen concentration in the erythrocyte stays low, so a steep diffusion gradient is maintained until all of the haemoglobin is saturated with oxygen.

When the blood reaches the body tissues, the situation is reversed. The concentration of oxygen in the cytoplasm of the body cells

Learning outcomes

Demonstrate knowledge, understanding, and application of:

→ the role of haemoglobin in transporting oxygen and carbon dioxide

→ the oxygen dissociation curve for fetal and adult haemoglobin.

Synoptic link

You learnt about haemoglobin as a globular conjugated protein in Topic 3.7, Types of proteins.

Study tip

Partial pressure is a useful way of talking about the concentration of a chemical when it is one of a mixture of gases. The whole mixture of gases has an overall pressure and each of the chemicals in the mixture can be thought of as contributing part of that pressure. If you were shut in an airtight room, as time went by the overall pressure of the air around you would not change but the partial pressure of oxygen (pO_2) would fall, and the partial pressure of carbon dioxide would rise (the partial pressure of nitrogen would not change). If you walk up a mountain the overall pressure of the air does decrease as the altitude increases. In this case the partial pressures of oxygen, carbon dioxide and nitrogen are all falling (but the proportions of each are not changing).

is lower than in the erythrocytes. As a result, oxygen moves out of the erythrocytes down a concentration gradient. Once the first oxygen molecule is released by the haemoglobin, the molecule again changes shape and it becomes easier to remove the remaining oxygen molecules.

An **oxygen dissociation curve** (Figure 1) is an important tool for understanding how the blood carries and releases oxygen. The percentage saturation haemoglobin in the blood is plotted against the partial pressure of oxygen (pO_2). Oxygen dissociation curves show the affinity of haemoglobin for oxygen. A very small change in the partial pressure of oxygen in the surroundings makes a significant difference to the saturation of the haemoglobin with oxygen, because once the first molecule becomes attached, the change in the shape of the haemoglobin molecule means other oxygen molecules are added rapidly. The curve levels out at the highest partial pressures of oxygen because all the haem groups are bound to oxygen and so the haemoglobin is saturated and cannot take up any more.

This means that at the high partial pressure of oxygen in the lungs the haemoglobin in the red blood cells is rapidly loaded with oxygen. Equally, a relatively small drop in oxygen levels in the respiring tissues means oxygen is released rapidly from the haemoglobin to diffuse into the cells. This effect is enhanced by the relatively low pH in the tissues compared with the lungs.

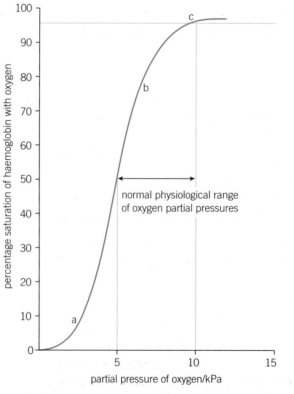

a   at low pO_2, few haem groups are bound to oxygen, so haemoglobin does not carry much oxygen

b   at higher pO_2, more haem groups are bound to oxygen, making it easier for more oxygen to be picked up

c   the haemoglobin becomes saturated at very high pO_2, as all the haem groups become bound

▲ **Figure 1** *The oxygen dissociation curve for human haemoglobin*

When you are not very active, only about 25% of the oxygen carried in your erythrocytes is released into the body cells. The rest acts as a reservoir for when the demands of the body increase suddenly.

The effect of carbon dioxide

As the partial pressure of carbon dioxide rises (in other words, at higher partial pressures of CO_2), haemoglobin gives up oxygen more easily (Figure 2). This change is known as the **Bohr effect**. The Bohr effect is important in the body because as a result:

- in active tissues with a high partial pressure of carbon dioxide, haemoglobin gives up its oxygen more readily

- in the lungs where the proportion of carbon dioxide in the air is relatively low, oxygen binds to the haemoglobin molecules easily.

Fetal haemoglobin

When a fetus is developing in the uterus it is completely dependent on its mother to supply it with oxygen. Oxygenated blood from the mother runs close to the deoxygenated fetal blood in the placenta. If the blood of the fetus had the same affinity for oxygen as the blood of the mother, then little or no oxygen would be transferred to the blood of the fetus. However, fetal haemoglobin has a higher affinity for oxygen than adult haemoglobin at each point along the dissociation curve (Figure 3). So it removes oxygen from the maternal blood as they move past each other.

Transporting carbon dioxide

Carbon dioxide is transported from the tissues to the lungs in three different ways:

- About 5% is carried dissolved in the plasma.

- 10–20% is combined with the amino groups in the polypeptide chains of haemoglobin to form a compound called **carbaminohaemoglobin**.

- 75–85% is converted into hydrogen carbonate ions (HCO_3^-) in the cytoplasm of the red blood cells.

Most of the carbon dioxide that diffuses into the blood from the cells is transported to the lungs in the form of hydrogen carbonate ions.

Carbon dioxide reacts slowly with water to form carbonic acid ($H_2CO_3^-$). The carbonic acid then dissociates to form hydrogen ions and hydrogen carbonate ions.

$$CO_2 + H_2O \rightleftharpoons H_2CO_3 \rightleftharpoons H^+ + HCO_3^-$$

In the blood plasma this reaction happens slowly. However, in the cytoplasm of the red blood cells there are high levels of the enzyme **carbonic anhydrase**. This enzyme catalyses the reversible reaction between carbon dioxide and water to form carbonic acid. The carbonic acid then dissociates to form hydrogen carbonate ions and hydrogen ions.

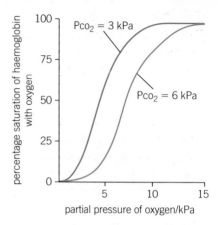

▲ **Figure 2** *The Bohr shift – as the proportion of carbon dioxide increases, the oxygen dissociation curve for haemoglobin moves to the right*

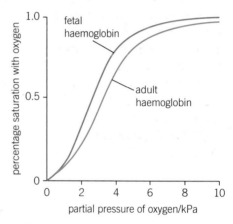

▲ **Figure 3** *The oxygen dissociation curves for adult and fetal haemoglobin show how the fetus can gain oxygen from the mother*

The negatively charged hydrogen carbonate ions move out of the erythrocytes into the plasma by diffusion down a concentration gradient and negatively charged chloride ions move into the erythrocytes, which maintains the electrical balance of the cell. This is known as the **chloride shift**.

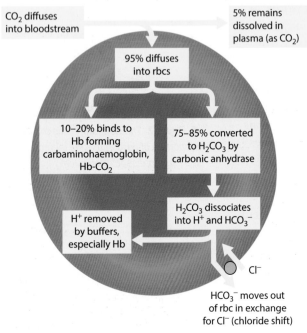

CO$_2$ diffuses into bloodstream

5% remains dissolved in plasma (as CO$_2$)

95% diffuses into rbcs

10–20% binds to Hb forming carbaminohaemoglobin, Hb-CO$_2$

75–85% converted to H$_2$CO$_3$ by carbonic anhydrase

H$_2$CO$_3$ dissociates into H$^+$ and HCO$_3^-$

H$^+$ removed by buffers, especially Hb

Cl$^-$

HCO$_3^-$ moves out of rbc in exchange for Cl$^-$ (chloride shift)

▲ **Figure 4** *The transport of carbon dioxide in the blood is a complex process*

By removing the carbon dioxide and converting it to hydrogen carbonate ions, the erythrocytes maintain a steep concentration gradient for carbon dioxide to diffuse from the respiring tissues into the erythrocytes.

When the blood reaches the lung tissue where there is a relatively low concentration of carbon dioxide, carbonic anhydrase catalyses the reverse reaction, breaking down carbonic acid into carbon dioxide and water. Hydrogen carbonate ions diffuse back into the erythrocytes and react with hydrogen ions to form more carbonic acid. When this is broken down by carbonic anhydrase it releases free carbon dioxide, which diffuses out of the blood into the lungs. Chloride ions diffuse out of the red blood cells back into the plasma down an electrochemical gradient.

Haemoglobin in the erythrocytes also plays a role in this process. It acts as a buffer and prevents changes in the pH by accepting free hydrogen ions in a reversible reaction to form **haemoglobinic acid**.

Summary questions

1   Explain how the structure of erythrocytes is adapted to their function in the body.   *(3 marks)*

2   a   Draw the oxygen dissociation curve for normal adult haemoglobin and fetal haemoglobin.   *(2 marks)*
    b   What is the difference in the oxygen saturation of adult and fetal haemoglobin at a partial pressure of oxygen of 6 kPa? Explain the importance of this difference in the survival of the fetus.   *(3 marks)*

3   Myoglobin is an oxygen-binding molecule found in the muscles.
    a   Sketch a graph to showing myoglobin oxygen affinity and haemoglobin oxygen affinity.   *(2 marks)*
    b   Explain the differences between the two curves.   *(2 marks)*

4   Use a flow chart to summarise how the carbon dioxide produced in the cells is carried to the lungs in the red blood cells.   *(6 marks)*

8.5 The heart

Specification reference: 3.1.2

The **heart** is the organ that moves the blood around the body. In some animal groups it is no more than a simple muscular tube. In mammals the heart is a complex, four-chambered muscular 'bag' found in the chest, enclosed by the ribs and sternum.

The human heart

The heart consists of two pumps, joined and working together. Deoxygenated blood from the body flows into the right side of the heart, which pumps it to the lungs. Oxygenated blood from the lungs returns to the left side of the heart, which pumps it to the body. The blood from the two sides of the heart does not mix.

The heart is made of cardiac muscle, which contracts and relaxes in a regular rhythm. It does not get fatigued and need to rest like skeletal muscle. The coronary arteries supply the cardiac muscle with the oxygenated blood it needs to keep contracting and relaxing all the time. The heart is surrounded by inelastic pericardial membranes, which help prevent the heart from over-distending with blood.

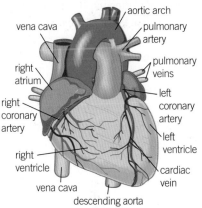

▲ **Figure 1** *Diagram showing the external structure of the heart*

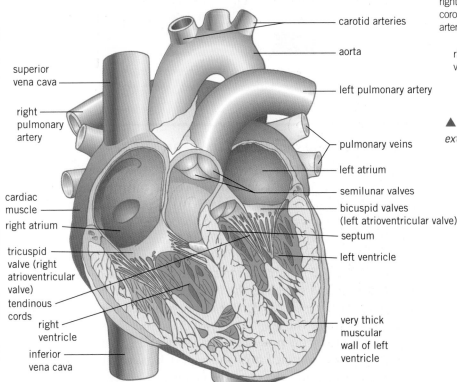

▲ **Figure 2** *The structure of the human heart is closely related to its function. In an average lifetime it will beat around 3×10^9 times, and each ventricle will pump around 200 million litres of blood*

Dissecting a heart

The heart of a sheep or a pig is similar in shape and size to a human heart and is often used in dissection. By careful examination of a heart you can identify many of the important structures in the mammalian heart – although the real thing is much more complicated than the standard diagram in Figure 3.

The external view of the heart enables you to see and trace the coronary arteries which supply the heart muscle with the blood it needs to beat. It is the narrowing or blockage of these blood vessels that cause the symptoms of coronary heart disease and even heart attacks.

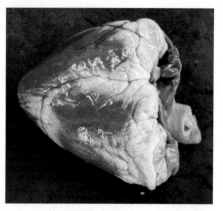

▲ **Figure 3** *Anterior view of a sheep's heart*

However hearts obtained from the butcher are often not intact. The major blood vessels will have been cut right back and often the atria have been removed – people don't want to eat all the tubes. So when you examine, dissect and draw a heart, you have to be aware of which, if any, parts are missing.

1 How does the dissected heart in Figure 4 differ from the diagram of the structures of the heart in Figure 2? Compare the two and explain the differences.

▲ **Figure 4** *A dissected sheep's heart, as you can see a dissected mammalian heart shows its more complex intricacies*

The structure and function of the heart

Deoxygenated blood enters the *right atrium* of the heart from the upper body and head in the superior vena cava, and from the lower body in the inferior vena cava, at relatively low pressure. The atria have thin muscular walls. As the blood flows in, slight pressure builds up until the atrio-ventricular valve (the tricuspid valve) opens to let blood pass into the right ventricle. When both the atrium and ventricle are filled with blood the atrium contracts, forcing all the blood into the **right ventricle** and stretching the ventricle walls. As the right ventricle starts to contract, the tricuspid valve closes, preventing any backflow of blood to the atrium. The tendinous cords make sure the valves are not turned inside out by the pressures exerted when the ventricle contracts. The right ventricle contracts fully and pumps deoxygenated blood through the semilunar valves into the pulmonary artery, which transports it to the capillary beds of the lungs. The semilunar valves prevent the backflow of blood into the heart.

At the same time oxygenated blood from the lungs enters the *left atrium* from the pulmonary vein. As pressure in the atrium builds the bicuspid valve opens between the left atrium and the

left ventricle so the ventricle also fills with oxygenated blood. When both the atrium and ventricle are full the atrium contracts, forcing all the oxygenated blood into the left ventricle. The left ventricle then contracts and pumps oxygenated blood through semilunar valves into the **aorta** and around the body. As the ventricle contracts the tricuspid valve closes, preventing any backflow of blood.

The muscular wall of the left side of the heart is much thicker than that of the right. The lungs are relatively close to the heart, and the lungs are also much smaller than the rest of the body so the right side of the heart has to pump the blood a relatively short distance and only has to overcome the resistance of the pulmonary circulation. The left side has to produce sufficient force to overcome the resistance of the aorta and the arterial systems of the whole body and move the blood under pressure to all the extremities of the body.

The septum is the inner dividing wall of the heart which prevents the mixing of deoxygenated and oxygenated blood.

The right and left side of the heart fill and empty together.

 ### A hole in the heart

The development of the septum is not completed until after birth. In the fetus the blood is oxygenated in the placenta, not in the lungs. As a result, all the blood in the heart is very similar and so mixes freely. In the days after birth the gap in the septum closes to ensure that the deoxygenated and oxygenated bloods are kept completely separate. Any gap remaining in the septum after the first few weeks of life is referred to as a 'hole in the heart' and it can often be heard with a stethoscope as a heart murmur. Many people have a small hole in their septum without knowing about it. However, if the hole is large it can lead to severe health problems unless it is diagnosed and repaired by surgery.

1 Explain why a small hole in the septum of the heart might not have any adverse effects but a large hole will cause health problems.

The cardiac cycle and the heartbeat

The **cardiac cycle** describes the events in a single heartbeat, which lasts about 0.8 seconds in a human adult.

- In **diastole** the heart relaxes. The atria and then the ventricles fill with blood. The volume and pressure of the blood in the heart build as the heart fills, but the pressure in the arteries is at a minimum.

- In **systole** the atria contract (atrial systole), closely followed by the ventricles (ventricular systole). The pressure inside the heart increases dramatically and blood is forced out of the right side of the heart to the lungs and from the left side to the main body circulation. The volume and pressure of the blood in the heart are low at the end of systole, and the blood pressure in the arteries is at a maximum.

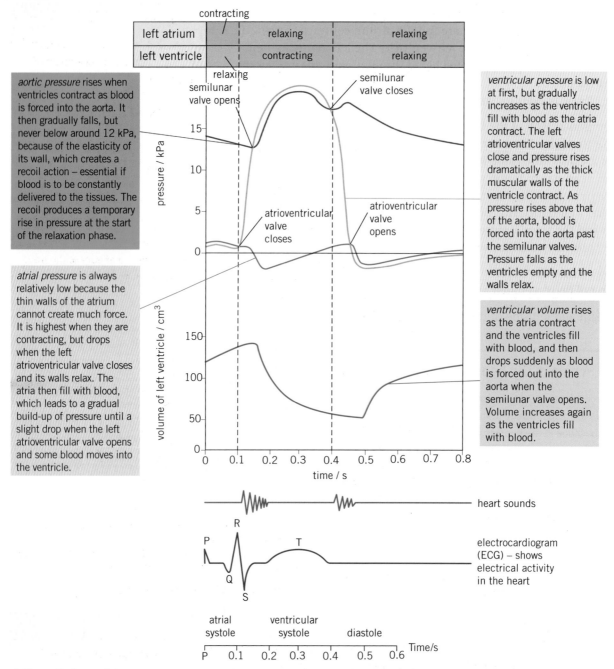

▲ Figure 5 *Some of the pressure changes in the heart during the cardiac cycle, as well as the heart sounds and an ECG trace*

Heart sounds

The sounds of the heartbeat, which can be heard through a stethoscope, are made by blood pressure closing the heart valves. The two sounds of a heartbeat are described as 'lub-dub'. The first sound comes as the blood is forced against the atrio-ventricular valves as the ventricles contract, and the second sound comes as a backflow of blood closes the semilunar valves in the aorta and pulmonary artery as the ventricles relax.

The basic rhythm of the heart

Cardiac muscle is **myogenic** – it has its own intrinsic rhythm at around 60 beats per minute (bpm). This prevents the body wasting resources maintaining the basic heart rate. The average resting heart rate of an adult is higher, at around 70 bpm. This is because other factors including exercise, excitement, and stress also affect our heart rate.

The basic rhythm of the heart is maintained by a wave of electrical excitation, rather like a nerve impulse (Figure 6).

- A wave of electrical excitation begins in the pacemaker area called the **sino-atrial node (SAN)**, causing the atria to contract and so initiating the heartbeat. A layer of non-conducting tissue prevents the excitation passing directly to the ventricles.

- The electrical activity from the SAN is picked up by the **atrio-ventricular node (AVN)**. The AVN imposes a slight delay before stimulating the **bundle of His**, a bundle of conducting tissue made up of fibres (**Purkyne fibres**), which penetrate through the septum between the ventricles.

- The bundle of His splits into two branches and conducts the wave of excitation to the apex (bottom) of the heart.

- At the apex the Purkyne fibres spread out through the walls of the ventricles on both sides. The spread of excitation triggers the contraction of the ventricles, starting at the apex. Contraction starting at the apex allows more efficient emptying of the ventricles.

The way in which the wave of excitation spreads through the heart from the SAN, with AVN delay, makes sure that the atria have stopped contracting before the ventricles start.

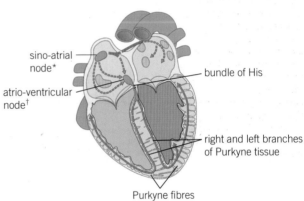

sino-atrial node*

atrio-ventricular node†

bundle of His

right and left branches of Purkyne tissue

Purkyne fibres

*The sino-atrial node (SAN) or natural pacemaker sets up a wave of electrical excitation.

†The atrio-ventricular node (AVN) is excited by the SAN. From here the excitation passes into the Purkyne tissue.

▲ **Figure 6** *The SAN initiates the heartbeat*

Electrocardiograms

You can measure the spread of electrical excitation through the heart as a way of recording what happens as it contracts. This recording of the electrical activity of the heart is called an **electrocardiogram (ECG)**. An ECG doesn't directly measure the electrical activity of your heart. It measures tiny electrical differences in your skin, which result from the electrical activity of the heart.

To pick up these tiny changes, electrodes are stuck painlessly to clean skin to get the good contacts needed for reliable results. The signal from each of the electrodes is fed into the machine, which produces an ECG. A normal ECG is shown in Figure 5 and Figure 7(a). ECGs are used to help diagnose heart problems. For example, if someone is having a heart attack, recognisable changes take place in the electrical activity of their heart, which can be used to diagnose the problem and treat it correctly and fast.

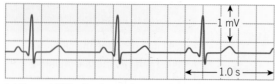

(a) Normal ECG – beats evenly spaced, rate 60–100/min

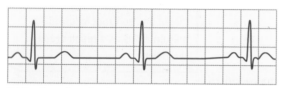

(b) Bradycardia – slow heart rate – beats evenly spaced, rate <60/min

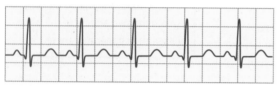

(c) Tachycardia – fast heart rate – beats evenly spaced, rate >100/min

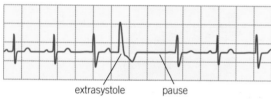

extrasystole   pause

(d) Ectopic heart beat – altered rhythm, extra beat followed by longer than normal gap before the next beat

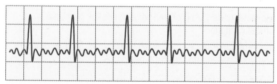

(e) Atrial fibrillation – abnormal irregular rhythm from atria, ventricles lose regular rhythm

▲ **Figure 7** *Normal and abnormal ECGs*

Heart rhythm abnormalities that commonly show up on ECGs include:

- **Tachycardia** – when the heartbeat is very rapid, over 100 bpm. This is often normal, for instance when you exercise, if you have a fever, if you are frightened or angry. If it is abnormal it may be caused by problems in the electrical control of the heart and may need to be treated by medication or by surgery.

- **Bradycardia** – when the heart rate slows down to below 60 bpm. Many people have bradycardia because they are fit – training makes the heart beat more slowly and efficiently. Severe bradycardia can be serious and may need an artificial pacemaker to keep the heart beating steadily.

- **Ectopic heartbeat** – extra heartbeats that are out of the normal rhythm. Most people have at least one a day. They are usually normal but they can be linked to serious conditions when they are very frequent.

- **Atrial fibrillation** – this is an example of an **arrhythmia**, which means an abnormal rhythm of the heart. Rapid electrical impulses are generated in the atria. They contract very fast (fibrillate) up to 400 times a minute. However, they don't contract properly and only some of the impulses are passed on to the ventricles, which contract much less often. As a result the heart does not pump blood very effectively.

Blood pressure

The blood travels through the arterial system at pressures that vary as the ventricles contract. The blood pressure is also affected by the diameter of the blood vessels themselves. Narrowing the arteries is one way in which the body affects and controls local blood flow, but permanent changes can cause severe health problems.

Most people will have their blood pressure taken at some point in their lives. Blood pressure is expressed as two figures, the first higher than the second. But what is being measured? Traditionally blood pressure is measured using a manual sphygmomanometer. A cuff, which is connected to a mercury manometer (a way of measuring pressure using the height of a column of mercury), is placed around the upper arm. The cuff is then inflated until the blood supply to the lower arm is completely cut off.

A stethoscope is positioned over the blood vessels at the elbow. Air is slowly let out of the cuff. The pressure at which the blood sounds first reappear as a slight tapping sound is recorded. The first blood to get through the cuff is that under the highest pressure – in other words, when the left ventricle of the heart is contracting strongly. The height of the mercury at this point gives the *systolic blood pressure* in mmHg (the height of the mercury column). The blood sounds return to normal at the point when even the lowest pressure during diastole is sufficient to get through the cuff. This gives the *diastolic blood pressure*. A reading of 120/80 mmHg is regarded as being normal. More recently a simpler, digital sphygmomanometer is often used – but the same principles apply. The stethoscope is simply built into the cuff applied around the arm.

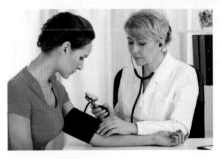

▲ **Figure 8** *Measuring blood pressure with a aneroid sphygmomanometer*

1   Blood pressure is used as an indicator of the health of both the heart and the blood vessels. A weakened heart may produce a low blood pressure, whereas damaged blood vessels that are closing up or becoming less elastic will give a raised blood pressure. Explain how these symptoms might be produced.

Summary questions

1   Explain why healthy coronary arteries are important for maintaining a regular heart rhythm.   (*2 marks*)

2   a   What causes the heart sounds?   (*2 marks*)
    b   Explain the relationship between the heart sounds and the events of the cardiac cycle.   (*6 marks*)

3   Explain the following responses:
    a   Bradycardia is common in diving mammals such as whales and seals.   (*5 marks*)
    b   Many people experience tachycardia when they travel to high altitudes.   (*5 marks*)

4   Look carefully at Figure 7:
    a   Work out the heart rate of the individuals with i) a normal heart rhythm ii) bradycardia and iii) tachycardia.   (*4 marks*)
    b   What is the percentage decrease or increase in the heart rate over the normal rate in these patients?   (*3 marks*)

Practice questions

1 In mammals, the lungs are adapted to enable efficient gaseous exchange.

The table below lists some of the adaptations of the lungs.

Complete the table explaining how each adaptation improves the efficiency of gaseous exchange.

Adaptation	How this adaptation improves efficiency of gaseous exchange
squamous epithelium	
large number of alveoli	
good blood supply	
good ventilation	

(4 marks)

OCR F211/01 2013

2 Animals and fish live in completely different environments. However their cells respire in the same way and depend on a constant supply of oxygen.

a Explain how both gills and lungs are designed to maximise the rate of diffusion in gaseous exchange. *(4 marks)*

b Using the simple representations of the two systems below, discuss the different way that gills and lungs are designed to maximise the rate of gaseous exchange. *(3 marks)*

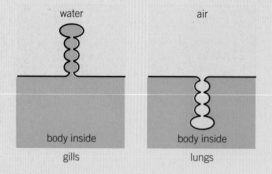

3 The diagram below summarises the formation of tissue fluid.

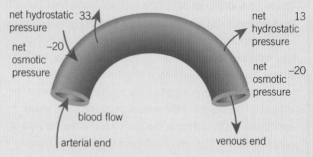

a (i) Calculate the net filtration pressures (NFP) at either end of the capillary in the diagram above:

arterial end

venous end *(2 marks)*

Interstitial fluid is constantly being formed at the arterial ends of capillary networks. Approximately 90% of this fluid is reabsorbed at the venous end.

(ii) Describe what happens to the other 10% and explain what could happen if this process did not occur.

(4 marks)

b Transport systems are essential in multicellular organisms to overcome the limitations of diffusion.

Using a mammalian transport system as an example, explain why diffusion is still an essential process. *(3 marks)*

c The diagram shows the circulatory system of a mollusc.

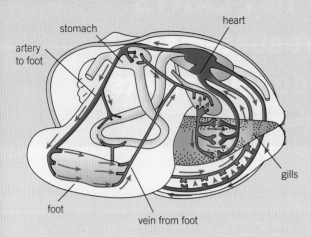

stomach
heart
artery to foot
gills
foot
vein from foot

State, with reasons, which type of circulatory system is present in a mollusc. *(4 marks)*

4 Varicose veins occur when the valves in the veins fail to work correctly and blood starts to pool in the veins. This occurs most commonly in the legs and feet.

a State the function of the valves in the veins. *(1 mark)*

b Suggest why you do not see 'varicose arteries'. *(2 marks)*

c (i) Explain why haemoglobin is known as a conjugate protein. *(2 marks)*

(ii) Explain why an oxygen dissociation curve is sigmoidal (s shaped). *(4 marks)*

d Name the molecule formed when carbon dioxide binds to haemoglobin. *(1 mark)*

e Myoglobin is a molecule with one haem group that binds to oxygen and only releases it at low oxygen levels. It effectively acts as store of oxygen in muscles.

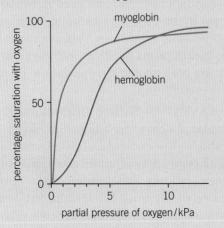

(i) Explain why the position of the dissociation curve for myoglobin is different than that of haemoglobin. *(2 marks)*

(ii) Suggest why the curve for myoglobin is not sigmoidal. *(2 marks)*

(iii) Draw a curve on the graph for fetal haemoglobin. *(2 marks)*

5 The majority of carbon dioxide released by respiring cells diffuses into red blood cells. Carbon dioxide can bind to haemoglobin forming a compound called carbaminohaemoglobin. About 10% of carbon dioxide is transported to the lungs by this method.

The remaining carbon dioxide undergoes an enzyme controlled reaction once inside a red blood cell.

a State what type of molecule forms an enzyme. *(2 marks)*

b Describe what an enzyme does. *(2 marks)*

c Outline the fate of carbon dioxide after it has entered a red blood but does not bind to haemoglobin. *(3 marks)*

TRANSPORT IN PLANTS

9.1 Transport systems in dicotyledonous plants

Specification reference: 3.1.3

▲ **Figure 1** *This saguaro cactus* (Carnegiea gigantea) *in the Sonoran Desert is as tall as many trees*

Synoptic link

You will learn about the classification and scientific naming of species in Topic 10.1, Classification.

Plants have transport systems to move substances between leaves, stems, and roots. These transport systems work at tremendous pressures. The pressure in the phloem, one of the main transport tissues of a plant, is around 2000 kPa. For comparison, the systolic blood in your main arteries is at a pressure of around 16 kPa, while the steam turbines in a power station work at around 4000 kPa. The higher pressures in plants, however, are confined to much smaller spaces than in arteries and turbines.

The need for plant transport systems

There are three main reasons why multicellular plants need transport systems:

● Metabolic demands – The cells of the green parts of the plant make their own glucose and oxygen by photosynthesis – but many internal and underground parts of the plant do not photosynthesise. They need oxygen and glucose transported to them and the waste products of cell metabolism removed. Hormones made in one part of a plant need transporting to the areas where they have an effect. Mineral ions absorbed by the roots need to be transported to all cells to make the proteins required for enzymes and the structure of the cell.

● Size – Some plants are very small but because plants continue to grow throughout their lives, many perennial plants (plants that live a long time and reproduce year after year) are large and some of them are enormous. The tallest trees in the world include the coastal redwood (*Sequoia sempervirens*) and giant redwood (*Sequoiadendron giganteum*) in the USA (up to around 115 m tall) and the mountain ash (*Eucalyptus regnans*) in Australia (up to around 114 m tall). This means plants need very effective transport systems to move substances both up and down from the tip of the roots to the topmost leaves and stems.

● Surface area : volume ratio (SA : V) – Surface area : volume ratios are not simple in plants. Leaves are adapted to have a relatively large SA : V ratio for the exchange of gases with the air. However, the size and complexity of multicellular plants means that when the stems, trunks, and roots are taken into account they still have a relatively small SA : V ratio. This means they cannot rely on diffusion alone to supply their cells with everything they need.

Transport systems in dicotyledonous plants

Dicotyledonous plants (dicots) make seeds that contain two cotyledons, organs that act as food stores for the developing embryo plant and form the first leaves when the seed germinates. There are herbaceous dicots, with soft tissues and a relatively short life cycle (leaves and stems that die down at the end of the growing season to the soil level), and woody (arborescent) dicots, which have hard, lignified tissues and a long life cycle (in some cases hundreds of years). You will be looking at herbaceous dicots in this section.

Dicotyledonous plants have a series of transport vessels running through the stem, roots, and leaves. This is known as the **vascular system**. In herbaceous dicots this is made up of two main types of transport vessels, the xylem and the phloem, described later in this topic. These transport tissues are arranged together in **vascular bundles** in the leaves, stems, and roots of herbaceous dicots. The pattern of the vascular tissue is easily recognised and is shown in transections (TS) in Table 1.

Synoptic link

It will help to look back at your work on surface area : volume ratios in Topic 7.1, Specialised exchange surfaces and Topic 8.1, Transport systems in multicellular animals.

▼ Table 1

TS stem of young herbaceous plant 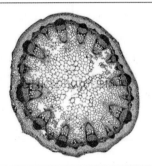	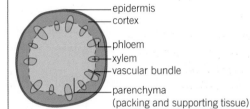 In the stem, the vascular bundles are around the edge to give strength and support.
TS root of young herbaceous plant 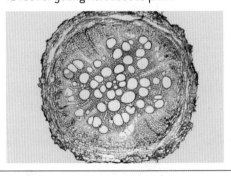	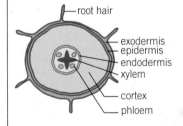 In the roots, the vascular bundles are in the middle to help the plant withstand the tugging strains that result as the stems and leaves are blown in the wind.
TS dicot leaf 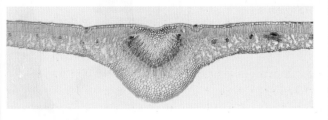	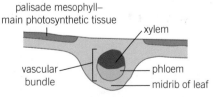 In the leaves, the midrib of a dicot leaf is the main vein carrying the vascular tissue through the organ. It also helps to support the structure of the leaf. Many small, branching veins spread through the leaf functioning both in transport and support.

Observing xylem vessels in living plant stems

Xylem vessels can be seen clearly stained in transverse and longitudinal sections of plant stems and roots on prepared slides. However, it is also possible to observe these vessels in living tissue.

If plant material, for example celery stalks with plenty of leaves, flowers such as a gerberas, or the roots of germinating seeds, are put in water containing a strongly coloured dye for at least 24 hours, you can remove the plant from the dye, rinse it and look for the xylem vessels which should have been stained by the dye.

- In one specimen, make clean transverse cuts across the stem with a sharp blade on a white tile. Take great care with the blade.
- Observe and draw the position of the xylem vessels which should show up as coloured spots.
- In another specimen make a careful longitudinal cut through a region where you expect there to be xylem vessels.
- Observe and draw the xylem vessels which may show up as coloured lines.

1   How does the information from your dissection compare with the micrographs of transverse and longitudinal sections of a dicot stem in Table 1 on the previous page?
2   What are the limitations of this method of observing the transport tissues in plants?

The structure and functions of the xylem

The **xylem** is a largely non-living tissue that has two main functions in a plant – the transport of water and mineral ions, and support. The flow of materials in the xylem is up from the roots to the shoots and leaves. Xylem is made up of several types of cells, most of which are dead when they are functioning in the plant. The xylem vessels are the main structures. They are long, hollow structures made by several columns of cells fusing together end to end.

There are two other tissues associated with xylem in herbaceous dicots. Thick-walled xylem parenchyma packs around the xylem vessels, storing food, and containing tannin deposits. Tannin is a bitter, astringent-tasting chemical that protects plant tissues from attack by herbivores. Xylem

Xylem vessel showing lignin spirals

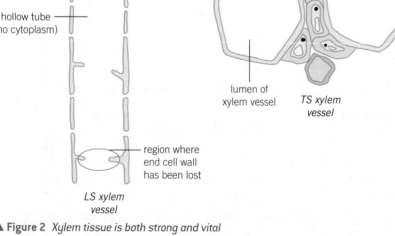

spirals of lignin running around the lumen of the xylem, lignin helps reinforce the xylem vessels so that they do not collapse under the transpiration pull

▲ **Figure 2** *Xylem tissue is both strong and vital for the transport of water and minerals around plants, lignified xylem vessels not shown* fibres are long cells with lignified secondary walls that provide extra mechanical strength but do not transport water. Lignin can be laid down in the walls of the xylem vessels in several different ways. It can form rings, spirals or relatively solid tubes with lots of small unlignified areas called bordered pits. This is where water leaves the xylem and moves into other cells of the plant.

The structure and functions of the phloem

Phloem is a living tissue that transports food in the form of organic solutes around the plant from the leaves where they are made by photosynthesis. The phloem supplies the cells with the sugars and amino acids needed for cellular respiration and for the synthesis of all other useful molecules. The flow of materials in the phloem can go both up and down the plant.

The main transporting vessels of the phloem are the **sieve tube elements**. Like xylem, the phloem sieve tubes are made up of many cells joined end to end to form a long, hollow structure. Unlike xylem tissue, the phloem tubes are not lignified. In the areas between the cells, the walls become perforated to form **sieve plates**, which look like sieves and let the phloem contents flow through. As the large pores appear in these cell walls, the tonoplast (vacuole membrane), the nucleus and some of the other organelles break down. The phloem becomes a tube filled with phloem sap and the mature phloem cells have no nucleus.

Closely linked to the sieve tube elements are **companion cells**, which form with them. These cells are linked to the sieve tube elements by many plasmodesmata – microscopic channels through the cellulose cell walls linking the cytoplasm of adjacent cells. They maintain their nucleus and all their organelles. The companion cells are very active cells and it is thought that they function as a 'life support system' for the sieve tube cells, which have lost most of their normal cell functions.

Phloem tissue also contains supporting tissues including fibres and sclereids, cells with extremely thick cell walls.

Synoptic link
You learnt about the cells of the xylem and phloem in Topic 6.4, The organisation and specialisation of cells.

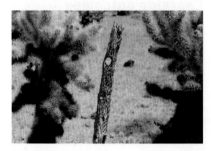
▲ Figure 3 Plants such as these desert cholla grow big but do not form wood like a tree, so they are very dependent on the complex lignified xylem structures to act as a supporting skeleton

Study tip
Be clear about the structural differences between xylem and phloem. Neither xylem vessels nor mature sieve tubes have nuclei, but xylem vessels are dead and phloem vessels are living tissue.

Summary questions

1 Explain why multicellular plants need transport systems. (4 marks)
2 State three differences between transport systems in multicellular plants and multicellular animals. (3 marks)
3 State the positioning of the transport tissue in herbaceous dicot stems, roots and leaves, and explain how this positioning is related to their functions. (6 marks)
4 Compare and contrast the structure and function of the main cell types in xylem and phloem tissues. (6 marks)

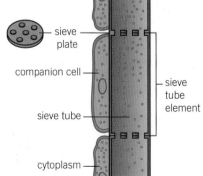
▲ Figure 4 Phloem tissue transports carbohydrates and other solutes around a plant

9.2 Water transport in multicellular plants

Specification reference: 3.1.3

Learning outcomes

Demonstrate knowledge, understanding, and application of:

→ the transport of water into the plant, through the plant and to the air surrounding the leaves

→ the mechanisms of water movement in plants.

Synoptic link

You learnt about osmosis, water potential and turgor pressure in Topic 5.5, Osmosis.

Synoptic link

You learnt about meristems in plant roots and shoots (the tissues of a plant where growth can take place) in Topic 6.5, Stem cells.

Unlike some other multicellular organisms, plants do not have a muscular heart beating to move fluids through the vessels of the xylem and phloem. In this topic you are going to discover exactly how they manage to move the substances they need through bodies that may be many metres tall and weigh many tonnes.

Water transport in plants

Water is key in both the structure and in the metabolism of plants:

- Turgor pressure (or hydrostatic pressure) as a result of osmosis in plant cells provides a hydrostatic skeleton to support the stems and leaves. So, for example, the turgor pressure in leaf cells is around 1.5 MPa – that is 11 251 mm Hg (unit of pressure still commonly used in medicine) (human systolic blood pressure is around 120 mm Hg).

- Turgor also drives cell expansion – it is the force that enables plant roots to force their way through tarmac and concrete.

- The loss of water by evaporation helps to keep plants cool.

- Mineral ions and the products of photosynthesis are transported in aqueous solutions.

- Water is a raw material for photosynthesis.

To understand the role of water in plants we need to look at how water moves into a plant from the soil and how it moves around the plant.

Movement of water into the root

Root hair cells are the exchange surface in plants where water is taken into the body of the plant from the soil. A root hair is a long, thin extension from a root hair cell, a specialised epidermal cell found near the growing root tip.

Root hairs are well adapted as exchange surfaces:

- Their microscopic size means they can penetrate easily between soil particles.

- Each microscopic hair has a large SA : V ratio and there are thousands on each growing root tip.

- Each hair has a thin surface layer (just the cell wall and cell-surface membrane) through which diffusion and osmosis can take place quickly.

- The concentration of solutes in the cytoplasm of root hair cells maintains a water potential gradient between the soil water and the cell.

Soil water has a very low concentration of dissolved minerals so it has a very high water potential. The cytoplasm and vacuolar sap of the root hair cell (and the other root cells) contain many different solvents including sugars, mineral ions, and amino acids so the water potential in the cell is lower. As a result water moves into the root hair cells by osmosis.

Movement of water across the root

Once the water has moved into the root hair cell it continues to move across the root to the xylem in one of two different pathways:

The symplast pathway

Water moves through the **symplast** – the continuous cytoplasm of the living plant cells that is connected through the plasmodesmata – by osmosis. The root hair cell has a higher water potential than the next cell along. This is the result of water diffusing in from the soil, which has made the cytoplasm more dilute. So water moves from the root hair cell into the next door cell by osmosis. This process continues from cell to cell across the root until the xylem is reached.

As water leaves the root hair cell by osmosis, the water potential of the cytoplasm falls again, and this maintains a steep water potential gradient to ensure that as much water as possible continues to move into the cell from the soil.

The apoplast pathway

This is the movement of water through the **apoplast** – the cell walls and the intercellular spaces. Water fills the spaces between the loose, open network of fibres in the cellulose cell wall. As water molecules move into the xylem, more water molecules are pulled through the apoplast behind them due to the cohesive forces between the water molecules. The pull from water moving into the xylem and up the plant along with the cohesive forces between the water molecules creates a tension that means there is a continuous flow of water through the open structure of the cellulose wall, which offers little or no resistance.

Movement of water into the xylem

Water moves across the root in the apoplast and symplast pathways until it reaches the endodermis – the layer of cells surrounding the vascular tissue (xylem and phloem) of the roots (Figure 4). The endodermis is particularly noticeable in the roots because of the effect of the Casparian strip. The Casparian strip is a band of waxy material called suberin that runs around each of the endodermal cells forming a waterproof layer (Figure 3). At this point, water in the apoplast pathway can go no further and it is forced into the cytoplasm of the cell, joining the water in the symplast pathway. This diversion to the cytoplasm is significant as to get to get there, water must pass through the selectively permeable cell surface membranes, this excludes any potentially-toxic solutes in the soil water from reaching living tissues, as the membranes would have no carrier proteins to admit them. Once forced into the cytoplasm the water joins the symplast pathway.

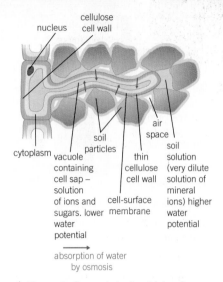

▲ **Figure 1** *Osmosis is the driving force behind the absorption of water by a root hair cell*

Synoptic link

You learned about water potential in Topic 5.5, Osmosis.

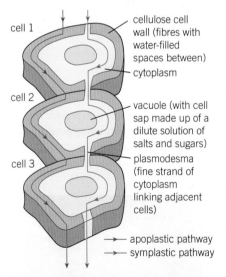

▲ **Figure 2** *The apoplast and symplast pathways along which water moves across the root*

Synoptic link

You learnt about the properties of water in Topic 3.2, Water.

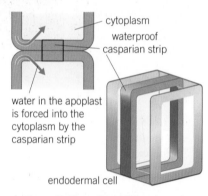

▲ Figure 3 *The effect of the casparian strip on the movement of water across the endodermis*

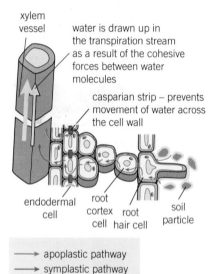

apoplastic pathway
symplastic pathway

▲ Figure 4 *The root cells provide a very efficient system for moving water from the soil into the xylem*

▲ Figure 5 *Guttation – evidence for root pressure independent of transpiration*

The solute concentration in the cytoplasm of the endodermal cells is relatively dilute compared to the cells in the xylem. In addition, it appears that the endodermal cells move mineral ions into the xylem by active transport. As a result the water potential of the xylem cells is much lower than the water potential of the endodermal cells. This increases the rate of water moving into the xylem by osmosis down a water potential gradient from the endodermis through the symplast pathway.

Once inside the vascular bundle, water returns to the apoplast pathway to enter the xylem itself and move up the plant. The active pumping of minerals into the xylem to produce movement of water by osmosis results in **root pressure** and it is independent of any effects of transpiration. Root pressure gives water a push up the xylem, but under most circumstances it is not the major factor in the movement of water up from the roots to the leaves.

Evidence for the role of active transport in root pressure

It has taken some time and many different investigations to determine the role of active transport in moving water from the root endodermis into the xylem. There are several strands of evidence supporting the current model:

- Some poisons, such as cyanide, affect the mitochondria and prevent the production of ATP. If cyanide is applied to root cells so there is no energy supply, the root pressure disappears.
- Root pressure increases with a rise in temperature and falls with a fall in temperature, suggesting chemical reactions are involved.
- If levels of oxygen or respiratory substrates fall, root pressure falls.
- Xylem sap may exude from the cut end of stems at certain times. In the natural world, xylem sap is forced out of special pores at the ends of leaves in some conditions – for example overnight, when transpiration is low. This is known as *guttation*.

Summary questions

1 Explain how a root hair cell is adapted for its role in the uptake of water from the soil.   *(4 marks)*

2 Explain the differences between the apoplast and symplast pathways for water movement across the root of a plant into the xylem.   *(6 marks)*

3 a Suggest why the effect of temperature on root pressure is not sufficient to prove that active transport is involved in the development of root pressure.   *(2 marks)*

   b Explain why the effects of cyanide, oxygen levels, and respiratory substrates on root pressure are taken as evidence that active transport is involved in the development of root pressure.   *(5 marks)*

Photosynthesis, the process by which green plants make their own food, takes place mainly in the leaves. Carbon dioxide (CO_2) and water (H_2O) are both needed so, for successful photosynthesis to take place in a leaf, water must be transported there from the roots and carbon dioxide must be taken into the cells of the leaf from the air.

Carbon dioxide diffuses into the leaf cells down a concentration gradient from the air spaces within the leaf. In a process of gaseous exchange, oxygen (O_2) also moves out of the leaf cells into the air spaces by diffusion down a concentration gradient (oxygen is a waste product of photosynthesis). At the same time water evaporates from the surfaces of the leaf cells into the air spaces.

The process of transpiration

Leaves have a very large surface area for capturing sunlight and carrying out photosynthesis. Their surfaces are covered with a waxy cuticle that makes them waterproof. This is an important adaptation that prevents the leaf cells losing water rapidly and constantly by evaporation from their surfaces. However, it is also important that gases can move into and out of the air spaces of the leaf so that photosynthesis is possible.

Carbon dioxide moves from the air into the leaf and oxygen moves out of the leaf by diffusion down concentration gradients through microscopic pores in the leaf (usually on the underside of the leaf) called **stomata** (singular stoma). The stomata can be opened and closed by **guard cells**, which surround the stomatal opening (further detail is given later in this topic).

When the stomata are open to allow an exchange of carbon dioxide and oxygen between the air inside the leaf and the external air, water vapour also moves out by diffusion and is lost. This loss of water vapour from the leaves and stems of plants is called **transpiration**. Transpiration is an inevitable consequence of gaseous exchange.

Stomata open and close to control the amount of water lost by a plant, but during the day a plant needs to take in carbon dioxide for photosynthesis and at night when no oxygen is being produced by photosynthesis it needs to take in oxygen for cellular respiration, so at least some stomata need to be open all the time.

It has been estimated that an acre of corn loses around 11 500–15 000 litres of water through transpiration every day, whilst a single large tree can lose more than 700 litres a day.

The transpiration stream

As you learn in Topic 9.2, water enters the roots of the plant by osmosis and is transported up in the xylem until it reaches the leaves. Here, it moves by osmosis across membranes and by diffusion in the apoplast pathway from the xylem through the cells of

Learning outcomes

Demonstrate knowledge, understanding, and application of:

→ the process of transpiration and the environmental factors that affect transpiration rate

→ practical investigations to estimate transpiration rates

→ the transport of water into the plant, through the plant and to the air surrounding the leaves.

the leaf where it evaporates from the freely permeable cellulose cell walls of the mesophyll cells in the leaves into the air spaces. The water vapour then moves into the external air through the stomata along a diffusion gradient. This is the **transpiration stream**.

The transpiration stream moves water up from the roots of a plant to the highest leaves – a height which can be up to 100 m or more in the tallest trees. As you learn in Topic 9.1, xylem vessels are non-living, hollow tubes so the process must be passive. How does it work?

- Water molecules evaporate from the surface of mesophyll cells into the air spaces in the leaf and move out of the stomata into the surrounding air by diffusion down a concentration gradient.

- The loss of water by evaporation from a mesophyll cell lowers the water potential of the cell, so water moves into the cell from an adjacent cell by osmosis, along both apoplast and symplast pathways.

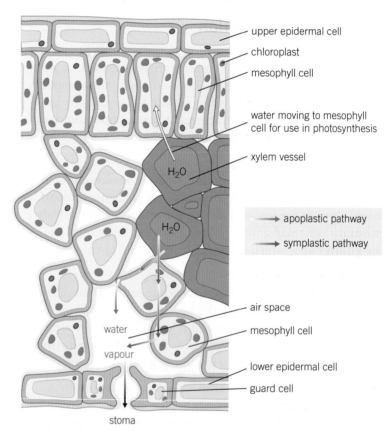

▲ **Figure 1** *Movement of water across a leaf*

- This is repeated across the leaf to the xylem. Water moves out of the xylem by osmosis into the cells of the leaf.

- Water molecules form hydrogen bonds with the carbohydrates in the walls of the narrow xylem vessels – this is known as adhesion. Water molecules also form hydrogen bonds with each other and so tend to stick together – this is known as cohesion. The combined effects of adhesion and cohesion result in water exhibiting capillary action. This is the process by which water can rise up a narrow tube against the force of gravity. Water is drawn up the xylem in a continuous stream to replace the water lost by evaporation. This is the *transpiration pull*.

● The transpiration pull results in a tension in the xylem, which in turn helps to move water across the roots from the soil.

This model of water moving from the soil in a continuous stream up the xylem and across the leaf is known as the **cohesion-tension theory**.

Synoptic link

You learn about the adhesive and cohesive properties of water and capillary action in Topic 3.2, Water.

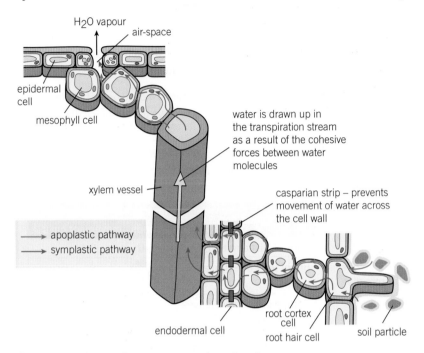

▲ **Figure 2** *A model of water transport through a plant*

Evidence for the cohesion-tension theory

Several pieces of evidence support the cohesion-tension theory for the movement of water up the xylem of a plant. These include the following:

● Changes in the diameter of trees. When transpiration is at its height during the day, the tension in the xylem vessels is at its highest too. As a result the tree shrinks in diameter. At night, when transpiration is at its lowest, the tension in the xylem vessels is at its lowest and the diameter of the tree increases. This can be tested by measuring the circumference of a suitably sized tree at different times of the day.

● When a xylem vessel is broken – for example when you cut flower stems to put them in water – in most circumstances air is drawn in to the xylem rather than water leaking out.

● If a xylem vessel is broken and air is pulled in as described in the previous bullet, the plant can no longer move water up the stem as the continuous stream of water molecules held together by cohesive forces has been broken.

In summary, transpiration delivers water, and the mineral ions dissolved in that water, to the cells where they are needed. The evaporation of water from the leaf cell surfaces also helps to cool the

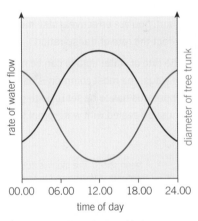

▲ **Figure 3** *The relationship between water flow in the xylem and the diameter of a tree trunk*

Study tip

Read questions about water movement in plants carefully. A question about how water in the xylem in the root reaches the cells of the leaves of a plant does not require you to give any information about the uptake of water from the soil by the root hairs or about the movement of water across the root.

leaf down and prevent heat damage (refer back to Topic 3.2, Water). However, transpiration, specifically the water loss, is also a problem for a plant because the amount of water available is often limited. In high intensity sunlight when the plant is photosynthesising rapidly, there will be a high rate of gaseous exchange, the stomata will all be open and the plant may lose so much water through transpiration that the supply cannot meet the demand.

 ## Measuring transpiration

It is very difficult to make direct measurements of transpiration because of the practical difficulties with condensing and collecting all of the water that evaporates from the surfaces of the leaves and stems of a plant without also collecting water that evaporates from the soil surface. It is also very difficult to separate water vapour from transpiration and water vapour produced as a waste product of respiration.

However, it is relatively easy to measure the uptake of water by a plant. As around 99% of the water taken up by a plant is then lost by transpiration, water uptake gives us a good working model of transpiration losses. By measuring factors that affect the uptake of water by a plant, you are effectively also measuring the factors that affect the rate of transpiration.

The rate of water uptake can be measured in a variety of ways. The most common is using a **potometer**. The apparatus has to be set up with great care, and all joints must be sealed with waterproof jelly to make sure that

any water loss measured is as a result of transpiration from the stem and leaves (Figure 4).

Rate of water uptake = distance moved by air bubble/ time taken for air bubble to move that distance. The units are cm s^{-1}.

A plant was placed in a potometer in bright light. The time taken for the bubble to move 5 cm was recorded under different conditions:

Conditions	Time in seconds	Rate of bubble movement cm s^{-1}
Normal lab conditions	35.0	0.14
Bright light directed at leaves	18.0	0.28
Bottom surfaces of leaves covered with vaseline	175.0	0.028

Data like these can be used to help demonstrate the effect of different factors on transpiration

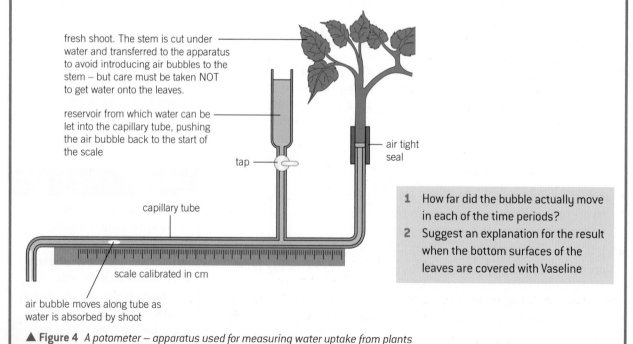

1 How far did the bubble actually move in each of the time periods?
2 Suggest an explanation for the result when the bottom surfaces of the leaves are covered with Vaseline

▲ **Figure 4** *A potometer – apparatus used for measuring water uptake from plants*

Stomata – controlling the rate of transpiration

As you learn at the start of this topic, the main way in which the rate of transpiration is controlled by the plant is by the opening and closing of the stomatal pores. This is a turgor-driven process. When turgor is low the asymmetric configuration of the guard cell walls closes the pore. When the environmental conditions are favourable guard cells pump in solutes by active transport, increasing their turgor. Cellulose hoops prevent the cells from swelling in width, so they extend lengthways. Because the inner wall of the guard cell is less flexible than the outer wall, the cells become bean-shaped and open the pore. When water becomes scarce, hormonal signals from the roots can trigger turgor loss from the guard cells, which close the stomatal pore and so conserve water.

Factors affecting transpiration

Any factor affecting the rate of water loss from the leaves of a plant will affect the rate of transpiration. Factors that affect water loss from the leaf must either act on the opening/closing of the stomata, the rate of evaporation from the surfaces of the leaf cells or the diffusion gradient between the air spaces in the leaves and the air surrounding the leaf. The effects of some of these factors are described below:

- Light is required for photosynthesis and in the light the stomata open for the gas exchange needed. In the dark, most of the stomata will close. Increasing light intensity gives increasing numbers of open stomata, increasing the rate of water vapour diffusing out and therefore increasing the evaporation from the surfaces of the leaf. So, increasing light intensity increases the rate of transpiration (Figure 6).

- Relative humidity is a measure of the amount of water vapour in the air (humidity) compared to the total concentration of water the air can hold. A very high relative humidity will lower the rate of transpiration because of the reduced water vapour potential gradient between the inside of the leaf and the outside air. Very dry air has the opposite effect and increases the rate of transpiration.

- Temperature affects the rate of transpiration in two ways:
  - An increase in temperature increases the kinetic energy of the water molecules and therefore increases the rate of evaporation from the spongy mesophyll cells into the air spaces of the leaf.
  - An increase in temperature increases the concentration of water vapour that the external air can hold before it becomes saturated (so decreases its relative humidity and its water potential).

Both factors increase the diffusion gradient between the air inside and outside the leaf, thus increasing the rate of transpiration.

- Air movement – Each leaf has a layer of still air around it trapped by the shape of the leaf and features such as hairs on the surface of

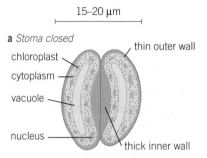
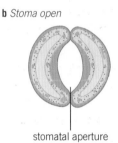

15–20 μm

a *Stoma closed*
chloroplast — thin outer wall
cytoplasm
vacuole
nucleus — thick inner wall

b *Stoma open*

stomatal aperture

▲ **Figure 5** *The opening and closing of the stomata is vital to the control of transpiration in a plant*

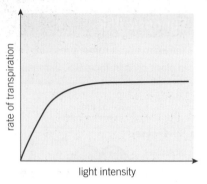

▲ **Figure 6** *Graph to show the effect of light intensity on transpiration*

the leaf decrease air movement close to the leaf. The water vapour that diffuses out of the leaf accumulates here and so the water vapour potential around the stomata increases, in turn reducing the diffusion gradient. Anything that increases the diffusion gradient will increase the rate of transpiration. So air movement or wind will increase the rate of transpiration, and conversely a long period of still air will reduce transpiration (Figure 7).

● Soil-water availability – The amount of water available in the soil can affect transpiration rate. If it is very dry the plant will be under water stress and the rate of transpiration will be reduced.

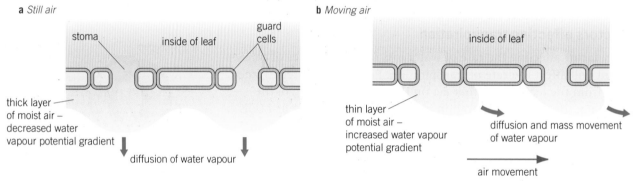

▲ **Figure 7** *The effect of air movements on the rate of transpiration*

Summary questions

1 Explain the difference between transpiration and the transpiration stream. *(2 marks)*

2 Compare root pressure and the transpiration pull. *(3 marks)*

3 Describe how you might use a potometer to investigate the effect of air movements on transpiration rates. *(6 marks)*

4 In an investigation into water uptake by a shoot using a bubble potometer, a group recorded how far the air bubble moved in a set time period of 40 seconds, in different conditions.
   Their results are as follows:
   a In the dark: 0.4 cm *(5 marks)*
   b In normal daylight: 5.6 cm *(5 marks)*
   c With bright lights shining on the plant shoot: 8.0 cm *(5 marks)*
   d With bright light shining and a fan blowing on the shoot: 12 cm *(6 marks)*

   Calculate the rate of water movement under each set of conditions and explain each result.

5 a Describe the cohesion-tension theory of transpiration. *(6 marks)*
   b Use the theory to explain:
     i the changes in diameter measured in the trunk of a tree between midday and midnight *(4 marks)*
     ii the fact that cut flowers placed in water may droop and die very quickly. *(4 marks)*

9.4 Translocation

Specification reference: 3.1.3

The leaves of a plant produce large amounts of glucose, which is needed for respiration by all the cells of the plant. The glucose is converted to sucrose for transport. When it reaches the cells where it is needed it is converted back to glucose for respiration, or to starch for storage, or used to produce the amino acids and other compounds needed within the cell.

From source to sink

Plants transport organic compounds in the phloem from **sources** to **sinks** (the tissues that need them) in a process called **translocation**. In many plants translocation is an active process that requires energy to take place and substances can be transported up or down the plant. The products of photosynthesis that are transported are known as **assimilates**. Although glucose is made in the process of photosynthesis, the main assimilate transported around the plant is **sucrose**. The sucrose content of most cell sap is only around 0.5%, but it can be 20–30% of the phloem sap content.

The main **sources** of assimilates in a plant are:

- green leaves and green stems
- storage organs such as tubers and tap roots that are unloading their stores at the beginning of a growth period
- food stores in seeds when they germinate.

Some of these need to transport resources downwards, and some need to move materials up the plant.

The main **sinks** in a plant include:

- roots that are growing and/or actively absorbing mineral ions
- meristems that are actively dividing
- any parts of the plant that are laying down food stores, such as developing seeds, fruits or storage organs.

▼ **Table 1** *Carbohydrates in cyclamen. The data in this table suggest that sucrose is transported in the phloem*

Plant part	Mean carbohydrate content ($\mu g\,g^{-1}$ fresh mass ± standard error of mean)	
	sucrose	starch
leaf blade	1312 ± 212	62 ± 25
vascular bundle in the leaf stalk, consisting of xylem and phloem	5757 ± 1190	<18
tissue surrounding the vascular bundle in the leaf stalk	417 ± 96	<18
buds, roots, and tubers (underground storage organs)	2260 ± 926	152 ± 242

The process of translocation

Translocation is a vital and effective process – a large tree can transport around 250 kg of sucrose down its trunk in a year, and substances move at speeds of around 0.15–7 metres per hour. The details of how substances are moved in the phloem of plants are still the subject of active investigation but the main steps are described here:

Phloem loading

In many plants the soluble products of photosynthesis are moved into the phloem from the sources by an active process. Sucrose is the main carbohydrate transported – it is not used in metabolism as readily as glucose and is therefore less likely to be metabolised during the transport process.

There are two main ways in which plants load assimilates into the **phloem** (phloem loading) for transport. One is largely passive, the other is active. Active phloem loading by the apoplast route is the most widely studied.

The symplast route

In some species of plants the sucrose from the source moves through the cytoplasm of the mesophyll cells and on into the sieve tubes by diffusion through the plasmodesmata (known as the **symplast** route). Although phloem loading and translocation are often referred to as active processes, this route is largely *passive*. The sucrose ends up in the sieve elements and water follows by osmosis. This creates a pressure of water that moves the sucrose through the phloem by mass flow.

The apoplast route

In many plant species sucrose from the source travels through the cell walls and inter-cell spaces to the companion cells and sieve elements (known as the **apoplast** route) by diffusion down a concentration gradient, maintained by the removal of sucrose into the phloem vessels. In the companion cells sucrose is moved into the cytoplasm across the cell membrane in an *active process*. Hydrogen ions (H^+) are actively pumped out of the companion cell into the surrounding tissue using ATP. The hydrogen ions return to the companion cell down a concentration gradient via a co-transport protein. Sucrose is the molecule that is co-transported. This increases the sucrose concentration in the companion cells and in the sieve elements through the many plasmodesmata between the two linked cells.

Companion cells have many infoldings in their cell membranes to give an increased surface area for the active transport of sucrose into the cell cytoplasm. They also have many mitochondria to supply the ATP needed for the transport pumps.

As a result of the build up of sucrose in the companion cell and sieve tube element, water also moves in by osmosis. This leads to a build up of turgor pressure due to the rigid cell walls. The water carrying the assimilates moves into the tubes of the sieve elements, reducing the pressure in the companion cells, and moves up or down the plant by mass flow to areas of lower pressure (the sinks).

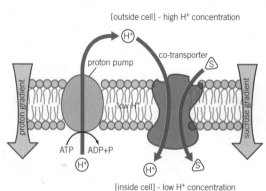

▲ **Figure 1** *The active movement of sucrose (S) into a companion cell or sieve tube across the cell membrane*

Solute accumulation in source phloem leads to an increase in turgor pressure that forces sap to regions of lower pressure in the sinks. The pressure generated in the phloem is around 2 MPa (15 000 mm Hg) – considerably higher than the 0.016 MPa (120 mm Hg) of pressure in a human artery. These pressure differences in plants can transport solutes and water rapidly over many metres. Solutes are translocated either up or down the plant, depending on the positions of the source.

Phloem unloading

The sucrose is unloaded from the phloem at any point into the cells that need it. The main mechanism of phloem unloading seems to be by diffusion of the sucrose from the phloem into the surrounding cells. The sucrose rapidly moves on into other cells by diffusion or is converted into another substance (for example glucose for respiration, starch for storage) so that a concentration gradient of sucrose is maintained between the contents of the phloem and the surrounding cells.

The loss of the solutes from the phloem leads to a rise in the water potential of the phloem. Water moves out into the surrounding cells by osmosis. Some of the water that carried the solute to the sink is drawn into the transpiration stream in the xylem.

Looking at the evidence

There is still a lot of research to be done to determine all the details of translocation. But there is a body of evidence that supports the main principles:

- Advances in microscopy allow us to see the adaptations of the companion cells for active transport.

- If the mitochondria of the companion cells are poisoned, translocation stops.

- The flow of sugars in the phloem is about 10 000 times faster than it would be by diffusion alone, suggesting an active process is driving the mass flow.

- Aphids can be used to demonstrate the translocation of organic solutes in the phloem. Using evidence from aphid studies, it has been shown that there is a positive pressure in the phloem that forces the sap out through the stylet. The pressure and therefore the flow rate in the phloem is lower closer to the sink than it is near the source. The concentration of sucrose in the phloem sap is also higher near to the source than near the sink.

However, some questions remain. Not all solutes in the phloem move at the same rate. On the other hand, sucrose always seems to move at the same rate regardless of the concentration at the sink. And no one is yet completely sure about the role of the sieve plates in the process.

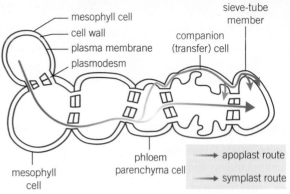

▲ **Figure 2** *The apoplast and symplast routes by which sucrose is loaded into the phloem*

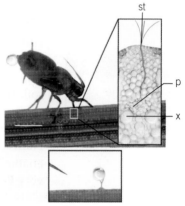

▲ **Figure 3** *Aphids (order Hemiptera) are a simple but very effective research tool for plant biologists. Top: Aphids penetrate the plant tissue with their mouth parts or stylet (st) to reach the phloem (p). Bottom: If the aphid is anaesthetised and removed from the stylet, phloem continues to flow out of the stylet due to pressure from the phloem contents. The rate of flow and composition of sap can be analysed*

Summary questions

1 Explain the difference between a source and a sink. (*4 marks*)

2 Describe how aphids can be used to demonstrate translocation, including the effect light intensity on the process. (*5 marks*)

3 a Explain the role of active transport in translocation in the phloem. (*5 marks*)
  b Describe the evidence that supports this model. (*5 marks*)

9.5 Plant adaptations to water availability

Specification reference: 3.1.3

Synoptic link

You will learn more about the adaptations of marram grass in Topic 10.7, Adaptations.

Synoptic link

You will learn about the classification and naming of species in Topic 10.1, Classification.

Synoptic link

You will learn about habitat biodiversity in Topic 11.1, Biodiversity.

Land plants exist in a state of constant compromise between getting the carbon dioxide they need for photosynthesis and losing the water they need for turgor pressure and transport. They must have a large SA : V ratio for gaseous exchange and the capture of light for photosynthesis, but this greatly increases their risk of water loss by transpiration.

Xerophytes

Most plants have adaptations to conserve water. These include a waxy cuticle to reduce transpiration from the leaf surfaces, stomata found mainly on the underside of the leaf that can be closed to prevent the loss of water vapour, and roots that grow down to the water in the soil. However, in habitats where water is often in very short supply, this is not enough. In hot conditions, particularly hot, dry, and breezy conditions – water will evaporate from the leaf surfaces very rapidly. Plants in dry habitats have evolved a wide range of adaptations that enable them to live and reproduce in places where water availability is very low indeed. They are known as **xerophytes**.

Conifers (class Pinopsida) are xerophytes. So is marram grass (*Ammophila* spp.), a plant found widely on sand dunes and coastal areas, in dry and salty conditions.

Many plants that survive in very cold and icy conditions are also xerophytes – the water in the ground is not freely available to them because it is frozen.

Perhaps the best known xerophytes are the cacti (members of the plant family Cactaceae).

▲ Figure 1 *Successful desert plants like these have many adaptations to a xerophytic way of life because the gap between rains may be months or even years so water is rarely available*

Ways of conserving water

Xerophytes use a range of strategies for conserving water:

- A thick waxy cuticle – in most plants up to 10% of the water loss by transpiration is actually through the cuticle. Some plants have a particularly thick waxy cuticle to help minimise water loss. This adaptation is common in evergreen plants and helps them survive both hot dry summers and cold winters when water can be hard to absorb from the frozen ground. Holly (*Ilex* spp.) is an example commonly seen in the UK.

- Sunken stomata – many xerophytes have their stomata located in pits, which reduce air movement, producing a microclimate of still, humid (moist) air that reduces the water vapour potential gradient and so reduces transpiration. These are seen clearly in xerophyes such as marram grass, cacti, and conifers.

- Reduced numbers of stomata – Many xerophytes have reduced numbers of stomata, which reduce their water loss by transpiration but also reduce their gas exchange capabilities.

- Reduced leaves – by reducing the leaf area, water loss can be greatly reduced. The leaves of conifers are reduced to thin needles. These narrow leaves, which are almost circular in cross-section, have a greatly reduced SA : V ratio, minimising the amount of water lost in transpiration.

- Hairy leaves – some xerophytes have very hairy leaves that, like the spines of some cacti, create a microclimate of still, humid air, reducing the water vapour potential gradient and minimising the loss of water by transpiration from the surface of the leaf. Some plants – such as marram grass – even have microhairs in the sunken stomatal pits (Figure 4).

- Curled leaves – another adaptation that greatly reduces water loss by transpiration, especially in combination with other adaptations, is the growth of curled or rolled leaves. This confines all of the stomata within a microenvironment of still, humid air to reduce diffusion of water vapour from the stomata. Marram grass is a good example of a plant with this strategy.

- Succulents – succulent plants store water in specialised parenchyma tissue in their stems and roots. They get their name because, unlike other plants, they often have a swollen or fleshy appearance. Water is stored when it is in plentiful supply and then used in times of drought. *Salicornia* spp. (edible samphire), which grows on UK salt marshes, and desert cacti are examples of succulents, as are aloes, which include *Aloe vera*, a plant often used in cosmetics.

- Leaf loss – some plants prevent water loss through their leaves by simply losing their leaves when water is not available. Palo verde (*Parkinsonia* spp.) is a desert tree that loses all of its leaves in the long dry seasons. The trunk and branches turn green and photosynthesise with minimal water loss to keep it alive. Its name is derived from the Spanish words meaning 'green pole'.

Synoptic link

You learned about the structure and function of stomata in Topic 6.4, The organisation and specialisation of cells and Topic 9.3, Transpiration.

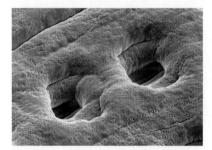

▲ **Figure 2** *The stomata of this sitka spruce (*Picea stichensis*) are in sunken pits and the thick waxy cuticle shows up as a pale green layer – two clear adaptations to minimise water loss*

▲ **Figure 3** *Reducing the leaves to spines in this Barrel cactus prevents transpiration from the leaves but also helps create a microclimate to prevent water loss from the stem*

▲ **Figure 4** *This Borage flower growing in the desert needs these very hairy leaves to prevent water loss and survive*

Study tip

Clearly relate the structural adaptations of the leaves of xerophytic plants to their functions in reducing the water vapour potential gradient and so reducing evaporation from the stomata, thus reducing transpiration.

▲ **Figure 5** *The folds in the stem of this saguaro cactus (Carnegiea gigantea) allow it to expand in the rare times when water is freely available. A fully hydrated saguaro can weigh over a tonne.*

▲ **Figure 6** *A leafless palo verde (Parkinsonia sp.) not only photosynthesises but also flowers without any leaves*

Synoptic link

You learnt about disaccharides in Topic 3.3, Carbohydrates.

- Root adaptations – many xerophytes have root adaptations that help them to get as much water as possible from the soil. Long tap roots growing deep into the ground can penetrate several metres, so they can access water that is a long way below the surface. A mass of widespread, shallow roots with a large surface area able to absorb any available water before a rain shower evaporates is another adaptation. Many cacti show both of these adaptations, including the giant saguaro (*Carnegiea gigantea*), which can get enough water to grow to around 12–18 metres tall and live for around 200 years. The root system of marram grass consists of long vertical roots that penetrate metres into the sand. They also have a mat of horizontal rhizomes (modified stems) from which many more roots develop to form an extensive network that helps to change their environment and enable the sand to hold more water.

- Avoiding the problems – some plants are adapted to cope with the problems of low water availability by avoiding the situation entirely. Plants may lose their leaves and become dormant, or die completely, leaving seeds behind to germinate and grow rapidly when rain falls again. Others survive as storage organs such as bulbs (onions, daffodils), corms (crocuses) or tubers (potatoes, dahlias). A few plants can withstand complete dehydration and recover – they appear dead but when it rains the cells recover, the plant becomes turgid and green again and begins to photosynthesise. The ability to survive in this way is linked to the disaccharide trehalose, which appears to enable to the cells to survive unharmed.

 Investigating stomatal numbers

You can compare the numbers of stomata on the leaves and stems of plants by taking an impression of the epidermis of the leaf or the stem and looking at it under a microscope.

You can use clear nail varnish, Germolene New Skin or DIY water based varnish.

You need to observe the numbers of stomata, and whether they are open or closed over the same area each time so take care to use the same magnification with your microscope. You may choose to use a graticule.

This technique can be used to compare stomatal numbers in different areas of a plant and in different types of plants. It can also be used to investigate the opening and closing of the stomata under different conditions.

1   How would you predict the number and position of the stomata might vary between xerophytes and normal plants?
2   Suggest why it can be difficult to investigate the stomata of cacti, even though they are some of the most effective xerophytes.

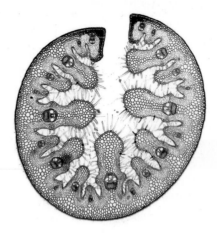

▲ **Figure 7** *Marram grass shows many adaptations to reduced water availability including both vertical and horizontal roots as well as the stomatal pits, hairs, and curled leaves visible in the light micrograph (× 10 magnification)*

Hydrophytes

Not all plants have to conserve water. In fact the **hydrophytes** – plants that actually live in water (submerged, on the surface or at the edges of bodies of water) – need special adaptations to cope with growing in water or in permanently saturated soil.

Examples of hydrophytes include water lilies (plants of the family Nymphaeaceae) and water cress (*Nasturtium officinale*), which grow at the surface, duckweeds (genus *Lemna*), which are submerged or free-floating plants, and marginals such bulrushes (*Typha latifolia*) and yellow iris (*Iris pseudacorus*), which grow at the edge of the water.

It is important in surface water plants that the leaves float so they are near the surface of the water to get the light needed for photosynthesis.

Water-logging is a major problem for all hydrophytes. The air spaces of the plant need to be full of air, not water, for the plant to survive.

▲ **Figure 8** *Hydrophytes like this water lily (family Nymphaeaceae) have no need to conserve water, but they have other problems to overcome*

Adaptations of hydrophytes:

- Very thin or no waxy cuticle – hydrophytes do not need to conserve water as there is always plenty available so water loss by transpiration is not an issue.

- Many always-open stomata on the upper surfaces – maximising the number of stomata maximises gaseous exchange. Unlike other plants there is no risk to the plant of loss of turgor as there is always an abundance of water available, so the stomata are usually open all the time for gaseous exchange and the guard cells are inactive. In plants with floating leaves such as water lilies the stomata need to be on the upper surface of the leaf so they are in contact with the air.

- Reduced structure to the plant – the water supports the leaves and flowers so there is no need for strong supporting structures.

- Wide, flat leaves – some hydrophytes, including the water lilies, have wide, flat leaves that spread across the surface of the water to capture as much light as possible.

- Small roots – water can diffuse directly into stem and leaf tissue so there is less need for uptake by roots.

- Large surface areas of stems and roots under water – this maximises the area for photosynthesis and for oxygen to diffuse into submerged plants.

- Air sacs – some hydrophytes have air sacs to enable the leaves and/or flowers to float to the surface of the water.

- Aerenchyma – specialised parenchyma (packing) tissue forms in the leaves, stems and roots of hydrophytes. It has many large air spaces, which seem to be formed at least in part by apoptosis (programmed cell death) in normal parenchyma. It has several different functions within the plants, including:

  — making the leaves and stems more buoyant

  — forming a low-resistance internal pathway for the movement of substances such as oxygen to tissues below the water. This helps the plant to cope with anoxic (extreme low oxygen conditions) conditions in the mud, by transporting oxygen to the tissues.

Aerenchyma is found in crop species that grow in water, such as rice (*Oryza sativa* and *Oryza glaberrima*). Studies suggest that aerenchyma may provide a low resistance pathway by which methane produced by the rice plants can be vented into the atmosphere. This is part of a major problem. Atmospheric methane, which contributes of the greenhouse effect and the resulting climate change, has doubled over the past two centuries and flooded rice paddies represent a major source.

In situations where there is plenty of water – for example in mangrove swamps, the roots can become waterlogged. It is air rather than water that is in short supply. Special aerial roots called pneumatophores grow upwards into the air. They have many lenticels, which allow the entry of air into the woody tissue.

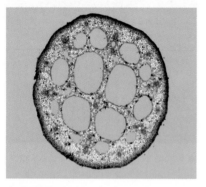

▲ **Figure 9** *Light micrograph showing aerenchyma tissue in the stem of a water lily leaf (Nymphaea alba) – compare this to the TS stem of a land plant you looked at in Topic 9.1 Transport systems in dicotyledonous plants.*

Summary questions

1 State three structural adaptations seen in the leaves of xerophytes and explain how these are related to their functions. *(3 marks)*

2 State three structural adaptations of hydrophytes and explain how they are related to their functions. *(3 marks)*

3 Compare the challenges of a plant of living in dry conditions to a plant living in water or waterlogged conditions. *(4 marks)*

4 Based on your knowledge of plant adaptations to water availability, suggest two characteristics that might be targeted by scientists in their search for suitable genes to use in future crops to help them withstand drought and two characteristics that might help them withstand flooding. Explain the advantages they confer on the plants and why you have chosen those characteristics. *(6 marks)*

Practice questions

1 The term vascular is derived from the Latin word vas meaning vessel.

   a State what is meant by the term vascular bundle. *(2 marks)*

   b Draw what you would expect to see in a dicotyledonous plant stem. *(2 marks)*

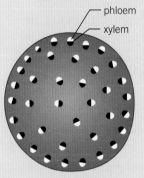
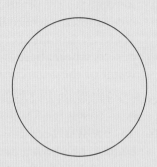

phloem
xylem

monocot stem

   c Describe and explain what happens when a ring of bark is removed from a tree. *(4 marks)*

2 The movement of water and solutes through a plant is known as the transpiration stream. It is the way essential minerals are transported from roots to leaves.

   a (i) Define transpiration. *(2 marks)*

      (ii) Explain why transpiration is an inevitable consequence of photosynthesis. *(4 marks)*

   b (i) Name the piece of apparatus used to measure transpiration. *(1 mark)*

      (ii) Explain why the measurements obtained from this apparatus are only an estimate of the rate of transpiration. *(2 marks)*

   c Table 1 below shows the readings obtained using the apparatus named in bi for a plant exposed to different environments.

The distance moved by a column of water in a capillary tube was measured every three minutes for 30 minutes as the plant was exposed to each environment.

Water loss ml/m²											
Time/minutes		0	3	6	9	12	15	18	21	24	27	30
treatment	fan											
	mist	0.00	4.17	4.17	4.17	4.17	2.08	0.00	2.08	2.08	0.00	2.08

      (i) Given the following information, copy and complete the first table and fill in the missing figures. *(2 marks)*

Water loss ml											
Time/minutes fan	0	3	6	9	12	15	18	21	24	27	30
	0.00	0.01	0.01	0.01	0.01	0.01	0.02	0.01	0.01	0.01	0.00

Mass of leaves = 1.1 g

Leaf Surface Area = 0.0044 m²

      (ii) Suggest how the mass of leaves could have been used to calculate the surface area of the leaves. *(3 marks)*

      (iii) State the independent variable, the dependent variable and two variables that would have been controlled in the experiment. *(4 marks)*

      (iv) Plot a graph to display the results. *(4 marks)*

      (v) Identify any anomalous results giving the reason for your choice. *(2 marks)*

      (vi) Explain the trends shown on the graph. *(4 marks)*

   d Describe how water is moving at A, B and C in the diagram below.

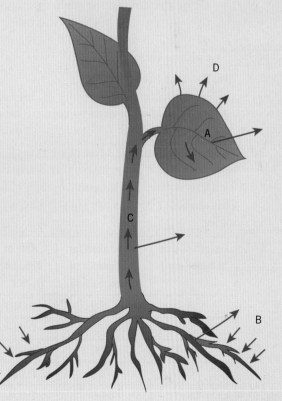

(6 marks)

3 In an experiment to measure the rate of diffusion, a student placed cubes of agar jelly containing an indicator into dilute hydrochloric acid. The indicator changes from pink to colourless in acidic conditions.

The student used cubes of different sizes and recorded the time taken for the pink colour of each cube to disappear completely.

The students results are recorded in the table

Length of side of cube (mm)	Surface area of cube (mm^2)	Volume of cube (mm^3)	Surface area to volume ratio	Time taken for pink colour to disappear (s)	Rate of diffusion (mm s^{-1})
2	24	8	3.0:0	50	0.020
5	150	125	1.2:1	120	0.021
10	600	1000		300	0.017
20	2400	8000	0.3:1	700	0.014
30	5400	27000	0.2:1	1200	0.013

a Define diffusion. *(1 mark)*

b Calculate the surface area to volume ratio of the cube with 10 mm sides.

Show your working. *(2 marks)*

c **(i)** Using the data in the table describe the relationship between the rate of diffusion and the surface area to volume ratio. *(2 marks)*

     (ii) Explain the significance of the relationship between rate of diffusion and the surface area to volume ratio for large plants. *(2 marks)*

d Another student used the same raw data obtained in the experiment but calculated a different rate of diffusion for each cube. This student's results are shown in the table.

Length of side of cube (mm)	Time taken for pink colour to disappear (s)	Rate of diffusion (mm s^{-1})
2	50	0.040
5	120	0.042
10	300	0.033
20	700	0.029
30	1200	0.025

In this student's table, the calculation of the rate of diffusion is incorrect.

     (i) suggest the method used to calculate the rate of diffusion in the table *(1 mark)*

     (ii) state why the method in ei is not correct *(1 mark)*

OCR F211/01 2013

4 Xerophytes are plants adapted to living in environments where little water is available. These environments can be as varied as deserts where there is a complete lack of water or the antarctic where all the water is present as ice or snow.

a Explain what is meant by the term xerophyte. *(1 mark)*

b Describe how a cactus plant is adapted to the environment in which it lives. *(3 marks)*

5 **a** Explain the difference between active transport and facilitated transport. *(3 marks)*

b Describe the difference in the direction of movement in the phloem and xylem. *(2 marks)*

c The following statements are evidence that support the explanation of translocation in phloem.

1 *there is an exudation of solution from the phloem when the stem is cut or punctured by the mouthparts of an aphid*

2 *concentration gradients of organic solutes are proved to be present between the sink and the source.*

3 *when viruses or growth chemicals are applied to a well-illuminated leaf, they are translocated downwards to the roots. This does not happen if the leaves are shaded.*

Explain how each statement supports this explanation. *(6 marks)*

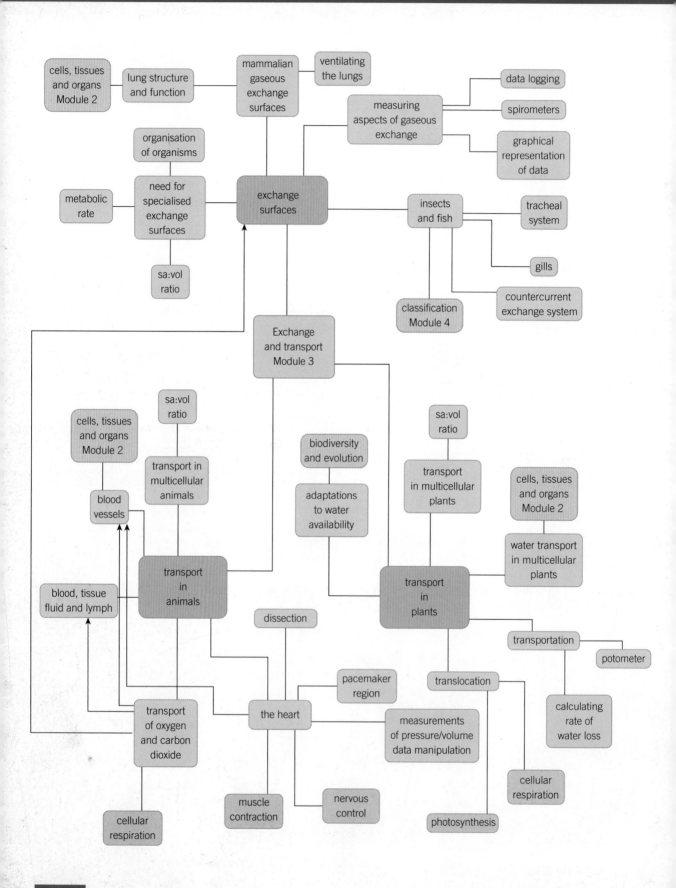

Application

Aphids feed by tapping into the phloem of plants. They push their stylets through the tissues of the plant stem or leaf until they reach the phloem. The stylet has a food canal to transport the sugary sap from the phloem into the insect. It also has a saliva canal to carry saliva down the stylet into the plant.

How does the plant respond to this invasion? There is now evidence that plants have a number of protein–based systems which act in seconds to block off damaged phloem vessels and prevent the loss of sap.

● the plastids may burst and release their contents which then coagulate and block the sieve elements

● certain proteins associated with the endoplasmic reticulum of the sieve elements will coagulate if the cells are damaged, blocking the vessel

● some plants, for example, broad beans, contain special proteins called *forisomes* in the phloem. As soon as a sieve element is damaged, calcium ions flow into the cells and the forisomes expand and block the vessel.

Researcher have hypothesised that the saliva pumped down the stylet of an aphid into the phloem of a plant helps prevent this coagulation from taking place.

▲ **Figure 1** *Why doesn't a lawn 'bleed' to death every time it is cut?*

1   When a lawn is cut, the ends of the leaves of the grass are chopped off, exposing the cut surfaces of the phloem. Suggest why the grass plants do not die as a result of losing all of the assimilates from photosynthesis through the cut transport vessels.

2   Why is it so important for aphids to prevent the protein coagulation which occurs in damaged phloem vessels if they are to successfully transport of food into their guts?

3   Find out more – produce a powerpoint presentation EITHER on phloem tissue and the mechanisms which protect the plant when the phloem is damaged OR looking at aphids and the ways in which they overcome the defence mechanisms of the plant phloem. Make sure you use lots of resources to help compile your presentation.

Extension

In mammals such as human beings, the blood is transported around the body in the blood vessels. If the blood vessels are damaged, there is a clotting mechanism which involves both proteins and calcium ions in a cascade which rapidly forms a clot to block the damaged blood vessel and prevent blood loss.

1   Investigate the clotting mechanism of the blood. Produce large, illustrated flow diagrams to compare the way in which the plant transport system prevents the loss of sugar-rich sap when the phloem is damaged with the clotting mechanism seen in the human circulatory system when a blood vessel is damaged. Write a commentary to highlight the similarities and differences between the two systems.

MODULE 4

Biodiversity, evolution, and disease

Chapters in this module

10 Classification and evolution

11 Biodiversity

12 Communicable diseases

Introduction

In this module you will learn about the vast biodiversity of organisms – how they are classified and the ways in which biodiversity can be measured. Classification is an attempt to impose a hierarchy on the complex and dynamic variety of life on Earth. The way in which organisms are classified has changed many times as our understanding of biological molecules and genetics increases.

The module also serves as an introduction to ecology, emphasising practical techniques used to study biodiversity and an appreciation of the need to maintain biodiversity.

Finally you will gain an understanding of the variety of organisms that are pathogenic and the way in which plants and animals have evolved defences to deal with disease. The impact of the evolution of pathogens on the treatment of disease is also considered.

Classification and evolution introduces you to the current system of classification used by scientists. It also explains historically how organisms were classified, and why the system has changed as our knowledge of the biology of organisms develops. It also covers how organisms are adapted to their environment and how, as a result of naturally occurring variation organisms have evolved, and continue to evolve.

Biodiversity is an important indicator in the study of habitats. You will learn how to sample habitats to measure and monitor biodiversity. You will also study the importance of maintaining biodiversity for ecological, economic and aesthetic reasons. To ensure biodiversity is maintained you will learn about how conservation action must be taken at local, national and global levels.

Communicable diseases explores how organisms are surrounded by pathogens and have evolved defences against them. You will discover how plants defend themselves and the role of the mammalian immune system. You will also learn how medical intervention can be used to support these natural defences such as the role of vaccinations and antibiotics.

Knowledge and understanding checklist

From your Key Stage 4 study you should be able to answer the following questions. Work through each point, using your Key Stage 4 notes and other resources. There is also support available on Kerboodle.

☐ Describe how to carry out a field investigation into the distribution and abundance of organisms in an ecosystem and explain how to determine their numbers in a given area.

☐ Describe both positive and negative human interactions within ecosystems and explain their impact on biodiversity.

☐ Explain some of the benefits and challenges of maintaining local and global biodiversity.

☐ Explain how evolution occurs through natural selection of variants that give rise to phenotypes best suited to their environment and may result in the formation of new species.

☐ Describe the evidence for evolution, including fossils and antibiotic resistance in bacteria.

☐ Describe the impact of developments in biology on classification systems.

☐ Explain how communicable diseases are spread in animals and plants.

☐ Describe the non-specific defence systems and the role of the immune system in the human bodies defence against disease.

☐ Explain the use of vaccines and medicines in the prevention and treatment of disease.

Maths skills checklist

In this module, you will need to use the following maths skills.

☐ **Standard deviation.** You will need to calculate the standard deviation to measure the spread in a set of data.

☐ **Student's t test.** You will need to use this test to compare the means of data values of two populations.

☐ **Correlation coefficient.** You will need to use this test to consider the relationship between two sets of data. This will determine if and how the data is correlated.

☐ **Simpson's Index of Diversity.** You will use this formula to measure biodiversity in a habitat. The higher the value of Simpson's Index of Diversity, the more diverse the habitat.

☐ **Proportion of polymorphic gene loci.** You will use this formula to measure genetic biodiversity.

MyMaths.co.uk
Bringing Maths Alive

10 CLASSIFICATION AND EVOLUTION

10.1 Classification

Specification reference: 4.2.2

No one knows how many different types of organism currently exist on the Earth. Through studying evolutionary relationships and performing mathematical calculations, in 2011 a team of scientists from the UK, USA, and Canada arrived at a widely accepted estimate of 8.7 million. However, the vast majority of these organisms have not been identified, and cataloguing them all could take more than 1000 years. The team also warned that many species would become extinct before they were studied.

Classification systems

Classification is the name given to the process by which living organisms are sorted into groups. The organisms within each group share similar features.

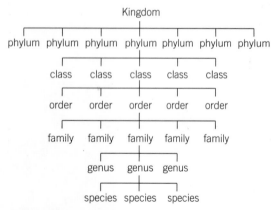

▲ **Figure 1** *This diagram shows how the seven taxonomic groups are arranged into a hierarchy. An organism can only belong to one group at each level of the hierarchy*

A number of different classification systems exist. Until recently the most widely used system contained seven groups ordered in a hierarchy – these are referred to as **taxonomic groups**. The seven groups are: **kingdom**, phylum (plural phyla), class, order, family, genus (plural genera) and **species** (Figure 1). Kingdoms are the biggest and broadest taxonomic group, with species being the smallest and most specific classification. Similar or related groups at one hierarchical level are combined into more inclusive groups at the next higher level.

Hierarchical classification systems are often referred to as Linnaean classification, after the 18th century Swedish botanist, Carl Linnaeus who was the first to propose such a system.

Based on recent studies of genetic material many scientists now add a further level of classification into the hierarchy. It is known as a *domain* (you will find out more about the three domain system of classification in Topic 10.2, The five kingdoms). This level of classification is placed at the top of the hierarchy. As new scientific discoveries are made (for example, through genome sequencing), the current system of classification may change again.

The development of classification systems

The system of classification of living organisms provides a good example of how our scientific knowledge and understanding has developed over time, as new information is gathered or discovered. Advances in scientific techniques have provided more detailed information on the genetic and biological make up of organisms, which has led to several revisions in the way organisms are classified. Our current system of classification may well change in the future, as further discoveries are made.

Some of the key steps in the development of the current system of classification are shown here:

1 Explain why the classification system we use today is different from the one originally proposed by Aristotle around the year 350 BC.

Ernst Haeckel

Konrad Gesner, a Swiss botanist, produces a four volume encyclopaedia of the then-known animal world, consisting of over 4500 pages

Ernst Haeckel, a German biologist and naturalist, proposes the third kingdom of protoctista, in addition to Linnaeus' animal and plant kingdoms

Following genetic analysis, Carl Woese introduces the six kingdom classification model, sub-dividing the prokaryotes into archaebacteria and eubacteria

1551 — 1866 — 1977

350 BC — 1758 — 1969

Aristotle compiles his book 'A history of animals', the first comprehensive study of animals

Carl Linnaeus publishes the tenth edition of his book Systema Naturae, now considered the starting point of binomial nomenclature

A five kingdom classification system is proposed by Robert Whittaker, consisting of prokaryotae, protoctista, fungi, plantae, and animalia

Why do scientists classify organisms?

- To identify species – by using a clearly defined system of classification, the species an organism belongs to can be easily identified.
- To predict characteristics – if several members in a group have a specific characteristic, it is likely that another species in the group will have the same characteristic.
- To find evolutionary links – species in the same group probably share characteristics because they have evolved from a common ancestor.

By using a single classification system, scientists worldwide can share their research. Links between different organisms can be seen, even if they live on different continents. Remember, though, that classification systems have been created to order observed organisms. This form of hierarchical organisation is not defined by 'nature'.

How are organisms classified?

The classification system begins by separating organisms into the three domains – Archaea, Bacteria, and Eukarya (discussed further in Topic 10.2, The five kingdoms). These are the broadest groups. As you move down the hierarchy there are more groups at each level, but fewer organisms in each group. The organisms in each group become more similar and share more of the same characteristics.

The system ends with organisms being classified as individual species. These are the smallest units of classification – each group contains only one type of organism. A species is defined as a group of organisms that are able to reproduce to produce fertile offspring. For example, donkeys can reproduce with other donkeys, the offspring of which can subsequently breed. Likewise, horses can breed with other horses to produce fertile offspring. However, when a horse is bred with a donkey, the offspring produced (a mule or a hinny) is infertile. Therefore, donkeys and horses are classified as belonging to different species. Mules or hinnies are not a species.

Mules and hinnies are infertile because their cells contain an odd number of chromosomes (63). This means that meiosis and gamete production cannot take place correctly as all chromosomes must pair up. This chromosome number is created because horses have 64 chromosomes (32 pairs) whereas donkeys have 62 chromosomes (31 pairs).

Synoptic link

You learnt about meiosis in Topic 6.3, Meiosis.

Study tip

Do not confuse viable offspring with fertile offspring. Viable means the organism produced survives, but does not mean that it is capable of producing offspring.

▲ **Figure 2** *Mules (left) are produced by crossing a horse (*Equus caballus*, middle) and a donkey (*Equus asinus*, right). Because a mule is infertile, it is not classified as a species.*

To show how the system works, the classification of three organisms is given in Table 1.

▼ **Table 1** *Classification of three organisms, you do not need to learn these examples*

Level of hierarchy	Brewer's yeast	English oak tree	European badger
Domain	Eukaryote	Eukaryote	Eukaryote
Kingdom	Fungi	Plantae	Animalia
Phylum	Ascomycota	Angiosperms	Chordata
Class	Saccharomycetes	Eudicots	Mammalia
Order	Saccharomycetales	Fagales	Carnivora
Family	Saccharomycetaceae	Fagaceae	Mustelidae
Genus	*Saccharomyces*	*Quercus*	*Meles*
Species	*cerevisiae*	*robur*	*meles*

Classification of humans

You belong to a species named *Homo sapiens*. This is the scientific name for humans. Humans are classified as shown in Table 2.

Naming organisms

Before classification systems were widely used, many organisms were given names according to certain physical characteristics, behaviour or habitat. Examples are 'blackbirds' for their colour, 'song thrushes' for their song and 'fieldfares' for their habitat. These are called their 'common names'.

This was not a very useful system for scientists working internationally, as organisms may have more than one common name, and different names in different languages. Another problem is that common names do not provide information about relationships between organisms. For example, the blackbird, song thrush, and fieldfare all belong to the genus *Turdus*, meaning that they have all evolved from a common ancestor, but you wouldn't know this from their common names, nor necessarily from their observable characteristics.

▲ **Figure 3** *The blackbird* (<u>Turdus</u> <u>merula</u>), *song thrush* (<u>Turdus</u> <u>philomelos</u>) *and fieldfare* (<u>Turdus</u> <u>pilaris</u>) *all belong to the genus* <u>Turdus</u>

To ensure scientists the world over are discussing the same organism we now use a system developed in the 18th century, also by Carl Linnaeus, a Swedish botanist. This system is known as **binomial nomenclature**.

All species are given a scientific name consisting of two parts:

- The first word indicates the organism's genus. It is called the generic name; you can think of this as being equivalent to your surname or family name, as it is shared by close relatives.

- The second word indicates the organism's species. It is called the specific name.

- Unlike people, no two species have the same generic and specific name. Two different species could have the same specific name, however their genus would be different. An example of this is *Anolis cuvieri* (a lizard) and *Oplurus cuvieri* (a bird). The only link between them is that they are both named after the famous French naturalist and zoologist Georges Cuvier (1769–1832). Many of these scientific names derive from Latin.

When naming an organism using its scientific name the word should be presented in italics. As it is difficult to handwrite in italics, the

▼ **Table 2** *Classification of humans*

Level of hierarchy	Human
Domain	Eukarya
Kingdom	Animalia
Phylum	Chordata
Class	Mammalia
Order	Primates
Family	Hominidae
Genus (plural – genera)	*Homo*
Species	*sapiens*

Study tip

The abbreviation for species 'sp' is used after genus, when not identifying the species fully. For example, you may only know the willow tree in your garden to be *Salix* sp.

The plural 'spp.' is used to refer to multiple species within a genus.

▲ **Figure 4** *This is a jungle cat (*Felis chaus*). It is closely related to the domestic cat (*Felis catus*) and both belong to the genus* Felis

standard procedure in handwritten documents is to underline the name. The name should be written in lowercase, with the exception of the first letter of the genus name, which should be uppercase.

Some examples of scientific names are included in Table 3. Split the name into two parts and you can easily work out which genus and species the organism belongs to.

▼ **Table 3** *Examples of scientific names*

Common name	Scientific name	Genus	Species
dog	*Canis familiaris*	*Canis*	*familiaris*
lion	*Panthera leo*	*Panthera*	*leo*
daisy	*Bellis perennis*	*Bellis*	*perennis*
Christmas tree (Norway spruce)	*Picea abies*	*Picea*	*abies*
E.coli	*Escherichia coli*	*Escherichia*	*coli*

Summary questions

▲ **Figure 5** *The domestic cat is also related to lions, tigers and leopards, but not as closely as the jungle cat. All these cats belong to the family Felidae*

1   State two reasons why classification is important.   *(2 marks)*

2   Ligers are the offspring of male lions (*Panthera leo*) and female tigers (*Panthera tigris*). Suggest two reasons why ligers are not classified as a species, but their parents are.   *(2 marks)*

3   *Erithacus rubecula* is the scientific name for the robin. Complete the table below to show its full classification   *(3 marks)*

Kingdom	Animalia
a	Chordata
Class	Aves
Order	Passeriformes
Family	Muscicapidae
Genus	b
Species	c

4   The loganberry (*Rubus loganobaccus*) is the fertile offspring of the blackberry (*Rubus ursinus*) and the raspberry (*Rubus idaeus*). Explain why the loganberry is difficult to classify into a taxonomic group.   *(3 marks)*

Study tip

As an organism's scientific name may be long or difficult to pronounce, the genus name is often shortened to the first letter after the first mention in full. For example, baker's yeast, *Saccharomyces cerevisiae* is often shortened to *S. cerevisiae* and the bacterium *Escherichia coli* to *E. coli*.

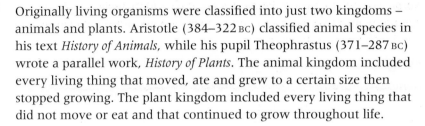

Originally living organisms were classified into just two kingdoms – animals and plants. Aristotle (384–322 BC) classified animal species in his text *History of Animals*, while his pupil Theophrastus (371–287 BC) wrote a parallel work, *History of Plants*. The animal kingdom included every living thing that moved, ate and grew to a certain size then stopped growing. The plant kingdom included every living thing that did not move or eat and that continued to grow throughout life.

As more was discovered about organisms and more species were discovered, it became increasingly difficult to divide living organisms into just two kingdoms. For example, the introduction of the microscope in the 16th to 17th century enabled scientists to study the cells of an organism and showed that bacteria have a very different cell structure to that of other organisms. From the 1960s, scientists classified organisms into five kingdoms. This classification system was introduced by Robert Whittaker, an American plant ecologist, based on the principles developed by Carl Linnaeus.

What are the five kingdoms?

Living organisms can be classified into five kingdoms:

- **Prokaryotae** (bacteria) — the **prokaryotes**
- **Protoctista** (the unicellular eukaryotes) ⎫
- **Fungi** (e.g., yeasts, moulds, and mushrooms) ⎬ the **eukaryotes**
- Plantae (the plants) ⎪
- Animalia (the animals) ⎭

Organisms were originally classified into these kingdoms based on similarities in their observable features, as described below.

Prokaryotae
General features:

Examples include the bacteria *Escherichia coli*, *Staphylococcus aureus*, and *Bacillus anthracis*.

- unicellular
- no nucleus or other membrane-bound organelles – a ring of 'naked' DNA – small ribosomes
- no visible feeding mechanism – nutrients are absorbed through the cell wall or produced internally by photosynthesis.

Protoctista
General features:

Examples include species belonging to the genera *Paramecium* and *Amoeba*.

- (mainly) unicellular
- a nucleus and other membrane-bound organelles
- some have chloroplasts

Learning outcomes

Demonstrate knowledge, understanding, and application of:

→ the features used to classify organisms into the five kingdoms.

→ evidence that has been used more recently in order to clarify relationships including evidence that has led to the classification of organisms into the three domains of life.

Synoptic link

You learnt about the difference in the structure of prokaryotic cells and eukaryotic cells in Topic 2.4, Cell structure and Topic 2.6, Prokaryotic and eukaryotic cells.

▲ **Figure 1** *This is* <u>Streptococcus thermophilus</u>, *a type of bacteria used in yoghurt production. All bacteria belong to the Prokaryotae kingdom. Scanning electron micrograph, × 8000 magnification.*

▲ **Figure 2** *This is an amoeba (Amoeba proteus), it is member of the Protoctista kingdom. Light micrograph × 100 magnification*

▲ **Figure 3** *Two different types of fungi. Top: Scanning electron micrograph of microscopic yeast, Saccharomyces cerevisiae × 2000 magnification. Bottom: Large mushrooms*

- some are sessile, but others move by cilia, flagella, or by amoeboid mechanisms
- nutrients are acquired by photosynthesis (**autotrophic** feeders), ingestion of other organisms (**heterotrophic** feeders), or both – some are parasitic.

Fungi

General features:

Examples include mushrooms, moulds, and yeast.

- unicellular or multicellular
- a nucleus and other membrane-bound organelles and a cell wall mainly composed of chitin
- no chloroplasts or chlorophyll
- no mechanisms for locomotion
- most have a body or mycelium made of threads or hyphae
- nutrients are acquired by absorption – mainly from decaying material – they are **saprophytic** feeders – some are parasitic
- most store their food as glycogen.

Plantae

With over 250 000 species, the plant kingdom is the second largest of the kingdoms.

Examples include flowering plants such as roses, trees such as oak, and grasses.

General features:

- multicellular
- a nucleus and other membrane-bound organelles including chloroplasts, and a cell wall mainly composed of cellulose
- all contain chlorophyll
- most do not move, although gametes of some plants move using cilia or flagella
- nutrients are acquired by photosynthesis – they are **autotrophic** feeders – organisms that make their own food
- store food as starch.

Animalia

The animal kingdom is the largest kingdom with over 1 million known species.

General features:

Examples include mammals such as cats, reptiles such as lizards, birds, insects, molluscs, worms, sponges, and anemones.

- multicellular
- a nucleus and other membrane-bound organelles (no cell walls)
- no chloroplasts
- move with the aid of cilia, flagella, or contractile proteins, sometimes in the form of muscular organs

- nutrients are acquired by ingestion – they are heterotrophic feeders
- food stored as glycogen.

Recent changes to classification systems

As scientists learn more about organisms, classification systems change. Originally classification systems were based on observable features. Through the study of genetics and other biological molecules, scientists are now able to study the evolutionary relationships between organisms. These links can then be used to classify organisms.

When organisms evolve, their internal and external features change, as does their DNA. This is because their DNA determines the proteins that are made, which in turn determines the organism's characteristics. In order for their characteristics to have changed, their DNA must also have changed. By comparing the similarities in the DNA and proteins of different species, scientists can discover the evolutionary relationships between them. You will learn more about DNA sequencing and its use in studying evolutionary relationships in Topic 10.4, Evidence for evolution.

An example of a protein that has changed in structure is haemoglobin. Haemoglobin has four polypeptide chains, each made up of a fixed number of amino acids. The haemoglobin of humans differs from chimpanzees in only one amino acid, from gorillas in three amino acids and from gibbons in eight amino acids. As the structure of haemoglobin is remarkably similar, it indicates a common ancestry between the various primate groups.

> **Synoptic link**
>
> You learnt about haemoglobin in Topic 3.7, Types of protein ans Topic 8.4, Transport of oxygen and carbon dioxide in the blood.

Are there now six kingdoms?

The current classification system used by scientists is known as the 'Three Domain System', and was proposed by Carl Woese, an American microbiologist in 1977, reusing the word 'Kingdom'. In 1990 it was renamed 'Domain'. Domains are a further level of classification at the top of the hierarchy.

Woese's system groups organisms using differences in the sequences of nucleotides in the cells' ribosomal RNA (rRNA), as well as the cells' membrane lipid structure and their sensitivity to antibiotics. Observation of these differences was made possible through advances in scientific techniques.

> **Synoptic link**
>
> You learnt about ribosomal RNA and nucleotides in Topic 3.8 Nucleic acids.

Under the Three Domain System, organisms are classified into three domains and six kingdoms. The three domains are Archaea, Bacteria, and Eukarya. The organisms in the different domains contain a unique form of rRNA and different ribosomes:

- Eukarya – have 80s ribosomes
  - RNA polymerase (responsible for most mRNA transcription) contains 12 proteins.

- Archaea – have 70s ribosomes
  ○ RNA polymerase of different organisms contains between eight and 10 proteins and is very similar to eukaryotic ribosome.
- Bacteria – have 70s ribosomes
  ○ RNA polymerase contains five proteins.

The organisation of this system is shown in Figure 4.

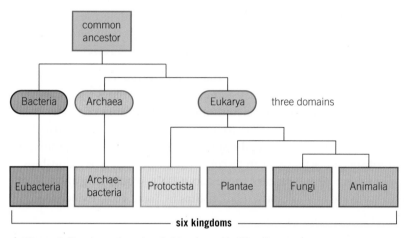

▲ **Figure 4** *The three domain, six kingdom classification system*

In Woese's system the Prokaryotae kingdom becomes divided into two kingdoms – Archaebacteria and Eubacteria. The six kingdoms are therefore: Archaebacteria, Eubacteria, Protoctista, Fungi, Plantae and Animalia.

Although both Archaebacteria and Eubacteria are single-celled prokaryotes, Eubacteria are classified in their own kingdom because their chemical makeup is different from Archaebacteria. For example, they contain peptidoglycan (a polymer of sugars and amino acids) in their cell wall whereas Archaebacteria do not.

Archaebacteria

Archaebacteria, also known as ancient bacteria, can live in extreme environments. These include hot thermal vents, anaerobic conditions, and highly acidic environments. For example, methanogens live in anaerobic environments such as sewage treatment plants and make methane.

Eubacteria

Eubacteria, also known as true bacteria, are found in all environments and are the ones you will be most familiar with. Most bacteria are of the Eubacteria kingdom.

Some scientists still use the traditional five kingdom system, but since Archaebacteria have been found to be different chemically from Eubacteria, most scientists now use the three domain, six kingdom system.

▲ **Figure 5** *The hot springs of Yellow stone National Park, USA, were among the first places Archaebacteria were discovered. These types of bacteria are called thermophiles. They can survive in extreme heat*

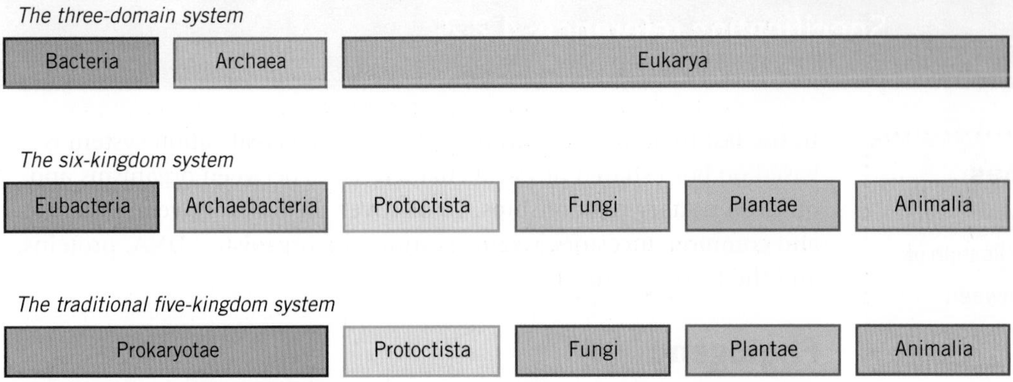

The three-domain system

Bacteria	Archaea	Eukarya

The six-kingdom system

Eubacteria	Archaebacteria	Protoctista	Fungi	Plantae	Animalia

The traditional five-kingdom system

Prokaryotae	Protoctista	Fungi	Plantae	Animalia

▲ **Figure 6** *This diagram shows how the three commonly used classification systems are related*

Summary questions

1   State two differences between fungi and plants.   (*2 marks*)

2   Using the information in this topic, classify the following organisms into the correct kingdom using your knowledge of the Five kingdom system of classification:
   a   *Escherichia coli* – an organism that lives in the human intestine. It is unicellular and has no nucleus.
   b   *Saccharomyces cerevisiae* – an organism used in the manufacture of beer. It is unicellular and has a nucleus and a cell wall made of chitin.
   c   *Euglena* – an organism that lives in fresh water. It is unicellular and contains chloroplasts.   (*3 marks*)

3   Explain why prokaryotes are now classified as two separate domains.   (*3 marks*)

4   Describe how and why classification systems have changed over time.   (*6 marks*)

10.3 Phylogeny

Specification reference: 4.2.2

Learning outcomes

Demonstrate knowledge, understanding, and application of:

→ the relationship between classification and phylogeny.

In the last topic you studied how the current classification system is based on both shared physical characteristics between organisms and on evolutionary relationships. To discover the links between organisms and common ancestors, scientists study the organisms' DNA, proteins, and the fossil record.

Phylogeny

Phylogeny is the name given to the *evolutionary* relationships between organisms. The study of the evolutionary history of groups of organisms is known as phylogenetics. It reveals which group a particular organism is related to, and how closely related these organisms are. You will learn more about the evidence that scientists use to study evolutionary relationships in the next topic.

Classification can occur without any knowledge of phylogeny, as occurred in the past. However, it is the objective of many scientists to develop a classification system that also correctly takes into account the phylogeny of an organism.

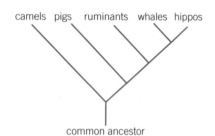

The closer the branches, the closer the evolutionary relationship. Hippos and whales are more closely related than hippos and ruminants.

▲ **Figure 1** *A phylogenetic tree showing the evolutionary relationships between certain mammals*

Phylogenetic trees

A phylogenetic tree (or evolutionary tree) is a diagram used to represent the evolutionary relationships between organisms. They are branched diagrams, which show that different species have evolved from a common ancestor.

The diagram is similar in structure to that of a branching tree – the earliest species is found at the base of the tree and the most recent species are found at the tips of the branches.

Phylogenetic trees are produced by looking at similarities and differences in species' physical characteristics and genetic makeup. Much of the evidence has been gained from fossils.

How do you interpret phylogenetic trees?

The tips of the phylogenetic tree represent groups of descendent organisms (often species). The nodes on the tree (the points where the new lines branch off) represent the common ancestors of those descendants. Two descendants that split from the same node are called sister groups. The closer the branches of the tree are, the closer the evolutionary relationship.

Study Figure 2. Begin by looking at the base of the tree. The organism at this point is the common ancestor of all the organisms on the tree. The letters A–F represent six different species that have evolved from this ancestor.

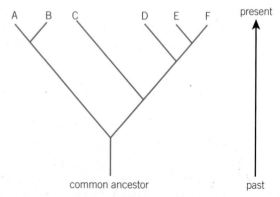

▲ **Figure 2** *A phylogenetic tree structure*

Then look at the top of the tree. You will see that species A and B are sister groups as these share a common ancestor. Species E and F are also sister groups that share their own common ancestor, which itself shared a common ancestor with species D further back in time. Further back in time again, C shared a common ancestor with D, E and F.

1 Another species exists (species X) which shares a common ancestor with species D. Add a line onto Figure 1 to show the correct evolutionary relationships of species X.
2 Species Y is now extinct but shared a common ancestor with species A and Species B. Add another line onto Figure 1 line to show where this species should be placed on the phylogenetic tree.

Advantages of phylogenetic classification

Phylogeny can be done without reference to Linnaean classification. Classification uses knowledge of phylogeny in order to confirm the classification groups are correct or causes them to be changed. For example, a dolphin has many of the same characteristics as a fish, so in theory a dolphin could be classified as a fish. However, knowledge of the phylogeny of dolphins confirms its classification as a mammal.

Other advantages:

● Phylogeny produces a continuous tree whereas classification requires discrete taxonomical groups. Scientists are not forced to put organisms into a specific group that they do not quite fit.

● The hierarchal nature of Linnaean classification can be misleading as it implies different groups within the same rank are equivalent. For example, the cats (Felidae) and the orchids (Orchidaceae) are both families. However, the two groups are not comparable – one has a longer history than the other (cats have existed for around 30 million years, but orchids have been in existence for over 100 million years). The two families also have different levels of diversity (with approximately 35 cat species and 20 000 orchid species) and different degrees of biological differentiation (many orchids of different genera are able to hybridise, but cats cannot).

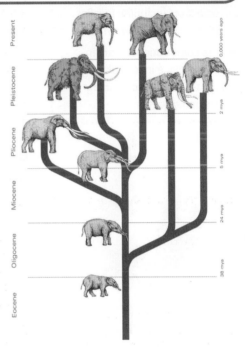

▲ **Figure 3** All species of elephant evolved from a palaeomastodon, which lived around 40 million years ago. Over time their appearance changed as the elephant adapted to the current environment. Today, only two species of elephant survive – the Indian and African elephant. The timeline stops for the other species of elephant as they are now extinct

Summary questions

1 State the main difference between early classification systems and systems based on phylogeny. (*1 mark*)

2 Describe the advantages of phylogenetic classification over the Linnaean system. (*3 marks*)

3 ⚙ Use the information in the phylogenetic tree to answer the following questions:
a State which group of organisms is most closely related to lizards. (*1 mark*)
b Explain why dinosaurs are listed lower in the diagram than the other organisms. (*2 marks*)

c Explain how the diagram shows that birds are more closely related to crocodiles than turtles. (*2 marks*)

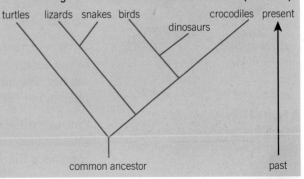

10.4 Evidence for evolution

Specification reference: 4.2.2

Learning outcomes

Demonstrate knowledge, understanding, and application of:

→ the evidence for the theory of evolution by natural selection, including fossils, DNA and molecular evidence

→ the contribution of Darwin and Wallace in formulating the theory of evolution by natural selection.

You are probably aware that Charles Darwin is credited with formulating the theory of **evolution**. However, it was studying the work of a number of other scientists as well as his own observations that led him to develop his theory.

Evolution is the theory that describes the way in which organisms evolve, or change, over many many years as a result of natural selection. Darwin realised that organisms best suited to their environment are more likely to survive and reproduce, passing on their characteristics to their offspring. Gradually, a species changes over time to have a more advantageous phenotype for the environment in which it lives. We now know that the advantageous characteristics are passed on from one generation to the next by genes in DNA molecules.

Developing the theory of evolution

When Charles Darwin was born in 1809, most people in Europe believed, in a literal sense, in the Christian Bible. They believed God directly created all life on Earth, including human beings. The Bible doesn't state how far in the past this occurred – in Darwin's day the common belief was that this creation had occurred only a few thousand years before.

In 1831 aboard the HMS Beagle, Darwin read '*Principles of Geology*'. This book was written by his friend Charles Lyell, a Scottish geologist. He suggested that fossils were actually evidence of animals that had lived millions of years ago. We now have scientific evidence that supports this.

In it Lyell also popularised the principle of uniformitarianism (the concept itself was originally proposed by another Scottish geologist, James Hutton). This is the idea that in the past, the Earth was shaped by forces that you can still see in action today, such as sedimentation in rivers, wind erosion, and deposition of ash and lava from volcanic eruptions. In emphasising these natural processes, he challenged the claims of earlier geologists who had tried to explain geological formations as a result of biblical events such as floods. This concept prompted Darwin to think of evolution as a slow process, one in which small changes gradually accumulate over very long periods of time.

Darwin carried out some of his most famous observations on finches in the Galapagos Islands. He noticed that different islands had different finches. The birds were similar in many ways and thus must be closely related, but their beaks and claws were different shapes and sizes.

Through these observations Darwin realised that the design of the finches' beaks was linked to the foods available on each island. He concluded that a bird born with a beak more suited to the food

▲ **Figure 1** *Charles Darwin, author of* 'On the Origin of Species'

Synoptic link

You learnt about DNA as the molecule of heredity in Topic 3.8, Nucleic acids and 3.9, DNA replication and the genetic code.

▲ **Figure 2** *Darwin 'noticed that the shape of the finches' beaks were adapted to the food available on the island*

available would survive longer than a bird whose beak was less suited. Therefore, it would have more offspring, passing on its characteristic beak. Over time the finch population on that island would all share this characteristic.

Throughout his trip Darwin sent specimens of organisms back to the UK for other scientists to preserve and classify. This enabled scientists not only to see specimens first hand but also enabled them to spot characteristics and links between organisms that Darwin had not. For example, Darwin did not notice that the tortoises (which the Galapagos islands are named after) present on different islands were different subspecies. Before this was pointed out to him he had simply stacked their shells randomly in the hold.

Upon his return to England, Darwin spent many years developing ideas. He also carried out experimental breeding of pigeons to gain direct evidence that his ideas might work.

At the same time as Darwin was developing his ideas, another scientist, Alfred Wallace, was working on his own theory of evolution in Borneo. In 1858 he sent his ideas to Darwin for peer review before its publication. As Wallace's ideas were so similar to Darwin's, they proposed the theory of evolution through a joint presentation of two scientific papers to the Linnean Society of London on 1st July 1858.

A year later in 1859, Darwin published '*On the Origin of Species*'. It was in this book that he named the theory that he and Wallace had presented independently as the theory of evolution by **natural selection** (see Topic 10.7, Adaptations).

The book was extremely controversial at the time. The theory of evolution conflicted with the religious view that God had created all of the animals and plants on Earth in their current form, and only about six thousand years ago. A further implication of Darwin's theory is that humans are simply a type of animal evolved from apes, which conflicted with the widely held Christian belief that God created 'man' in his own image.

Darwin's theory split the scientific community before his idea became generally agreed. Darwin's theory of evolution is now widely accepted, however, even today, debate with religious groups continues.

Evidence for evolution

Scientists use a number of sources to study the process of evolution. These include:

- palaeontology – the study of fossils and the fossil record
- comparative anatomy – the study of similarities and differences between organisms' anatomy
- comparative biochemistry – similarities and differences between the chemical makeup of organisms.

▲ **Figure 3** *One of the most common fossils found in the UK is an ammonite, an organism that lived in the sea. These organisms became extinct about 65 million years ago. Radioisotope dating is used to determine the age of the rock strata and the fossils found within the layer*

Palaeontology

Fossils are formed when animal and plant remains are preserved in rocks. Over long periods of time, sediment is deposited on the earth to form layers (strata) of rock. Different layers correspond to different geological eras, the most recent layer being found on the top. Within the different rock strata the fossils found are quite different, forming a sequence from oldest to youngest, which shows that organisms have gradually changed over time. This is known as the fossil record.

Evidence provided by the fossil record:

- Fossils of the simplest organisms such as bacteria and simple algae are found in the oldest rocks, whilst fossils of more complex organisms such as vertebrates are found in more recent rocks. This supports the evolutionary theory that simple life forms gradually evolved over an extremely long time period into more complex ones.

- The sequence in which the organisms are found matches their ecological links to each other. For example, plant fossils appear before animal fossils. This is consistent with the fact that animals require plants to survive.

- By studying similarities in the anatomy of fossil organisms, scientists can show how closely related organisms have evolved from the same ancestor. For example zebras and horses, members of the genus *Equus*, are closely related to the rhinoceros of the family Rhinocerotidae. An extensive fossil record of these organisms exists, which spans over 60 million years and links them to the common ancestor *Hyracotherium*. This lineage has been based on structural similarities between their skull (including teeth) and skeleton, in particular the feet (Figure 4).

- Fossils allow relationships between extinct and living (extant) organisms to be investigated.

The fossil record is, however, not complete. For example, many organisms are soft-bodied and decompose quickly before they have a chance to fossilise. The conditions needed for fossils to form are not often present. Many other fossils have been destroyed by the Earth's movements, such as volcanoes, or still lie undiscovered.

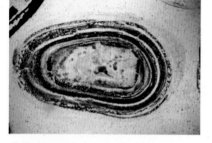

▲ **Figure 4** *You may be surprised to learn that bacteria can become fossils. The cyanobacteria (blue-green algae) have left behind a fossil record. The oldest cyanobacteria-like fossils known are nearly 3.5 billion years old, among the oldest fossils currently known*

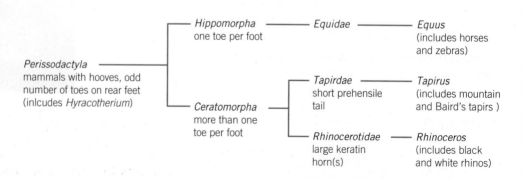

▲ **Figure 5** *This evolutionary tree shows that zebras and horses are closely related to the rhinoceros*

Comparative anatomy

As the fossil record is incomplete, scientists look for other sources of evidence to determine evolutionary relationships. Comparative anatomy is the study of similarities and differences in the anatomy of different living species.

Homologous structures

A **homologous structure** is a structure that appears superficially different (and may perform different functions) in different organisms, but has the same underlying structure. An example is the pentadactyl limb of vertebrates.

Vertebrate limbs are used for a wide variety of functions such as running, jumping, and flying. You would expect the bone structure of these limbs in a flying vertebrate to be very different from that in a walking vertebrate or a swimming vertebrate. However, the basic structures of all vertebrate limbs are actually very similar (Figure 6) – the same bones are adapted to carry out the whole range of different functions. An explanation is that all vertebrates have evolved from a common ancestor, therefore vertebrate limbs have all evolved from the same structure.

The presence of homologous structures provides evidence for **divergent evolution**. This describes how, from a common ancestor, different species have evolved, each with a different set of adaptive features. This type of evolution will occur when closely related species diversify to adapt to new habitats as a result of migration or loss of habitat.

Comparative biochemistry

Comparative biochemistry is the study of similarities and differences in the proteins and other molecules that control life processes. Although these molecules can change over time, some important molecules are

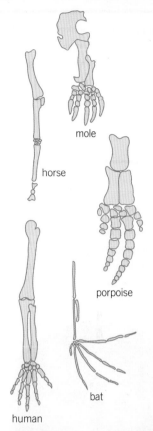

▲ **Figure 6** *These are the pentadactyl limbs from a number of organisms (not to scale). Scientists believe that their function has altered as a result of evolution from a common ancestor. The changes in bone structures of organisms living in the past can be studied through the fossil record*

 ## Evolutionary embryology

Embryology is the study of embryos. It is another source of evidence to show evolutionary relationships. An embryo is an unborn (or unhatched) animal in its earliest phases of development. Embryos of many different animals look very similar and it is often difficult to tell them apart. This shows that the animals develop in a similar way, implying that the processes of embryonic development have a common origin and the animals share common ancestry but have gradually evolved different traits.

Many traits of one type of animal appear in the embryo of another type of animal. For example, fish, and human embryos both have gill slits. In fish these develop into gills, but in humans they disappear before birth.

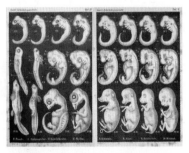

▲ **Figure 7** *Darwin used Haeckel's drawings of embryos as evidence for evolution*

Darwin considered the evidence from embryology to be 'by far the strongest single class of facts in favour of' his theory. He studied a series of drawings produced by biologist Ernst Haeckel that depicted the growth of embryos from various classes of vertebrates. The pictures show that the embryos begin looking virtually identical (suggesting common ancestry), but as they develop, their appearances diverge to take the form of their particular group.

As new organs or structures evolved, these features develop at the end of an organism's embryonic development. As a result, an organism's evolutionary history can be traced in the development of its embryos.

> 1 Explain whether embryos could form part of the fossil record.

Synoptic link

You learnt about ribosomal RNA and its role in protein synthesis in Topic 3.10, Protein synthesis.

highly conserved (remain almost unchanged) among species. Slight changes that occur in these molecules can help identify evolutionary links. Two of the most common molecules studied are cytochrome c, a protein involved in respiration, and ribosomal RNA.

The hypothesis of neutral evolution states that most of the variability in the structure of a molecule does not affect its function. This is because most of the variability occurs outside of the molecule's functional regions. Changes that do not affect a molecule's function are called 'neutral'. Since they have no effect on function, their accumulation is not affected by natural selection. As a result, neutral substitutions occur at a fairly regular rate, although that rate is different for different molecules.

To discover how closely two species are related, the molecular sequence of a particular molecule is compared. (Scientists do this by looking at the order of DNA bases, or at the order of amino acids in a protein.) The number of differences that exist are plotted against the rate the molecule undergoes neutral base pair substitutions (which has been determined through studies). From this information scientists can estimate the point at which the two species last shared a common ancestor. Species that are closely related have the more similar DNA and proteins, whereas those that are distantly related have far fewer similarities. Ribosomal RNA has a very slow rate of substitution, so it is commonly used together with fossil information to determine relationships between ancient species.

▲ **Figure 8** *Humans and chimpanzees have very similar DNA sequences. They have been found to share at least 98% of their DNA. This provides evidence that chimpanzees are humans' closest living relatives*

Summary questions

1 Describe what is shown on a phylogenetic tree. *(2 marks)*

2 Describe two advantages and two disadvantages of using the fossil record as a source of evidence for evolution. *(4 marks)*

3 Describe how the work of three scientists was used in the development of the theory of evolution. *(6 marks)*

4 Explain how comparative biochemistry provides evidence of evolution. *(3 marks)*

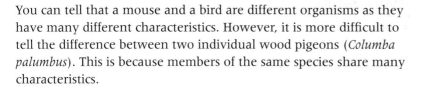

10.5 Types of variation

Specification reference: 4.2.2

You can tell that a mouse and a bird are different organisms as they have many different characteristics. However, it is more difficult to tell the difference between two individual wood pigeons (*Columba palumbus*). This is because members of the same species share many characteristics.

The differences in characteristics between organisms are called **variations**.

Types of variation

The widest type of variation is between members of different species – these differences are known as **interspecifc variation**. For example, a mouse has four legs, teeth, and fur whereas a bird has two legs, two wings, a beak and feathers.

Every organism in the world is different – even identical twins differ in some ways.

Differences between organisms within a species are called **intraspecific variation**. For example, people vary in height, build, hair colour, and intelligence.

▲ **Figure 2** *Dogs show immense variation within their species. For example, this Pug and Great Dane are both members of the species Canis familiaris. The differences between these individuals are examples of intraspecific variation. Due to humans selectively breeding dogs for their particular characteristics, the differences are more extreme than those that would occur naturally*

▲ **Figure 1** *These organisms all belong to the Plantae kingdom but they are members of different species. The differences between them are examples of interspecific variation*

Causes of variation

Two factors cause variation:

- An organism's genetic material – differences in the genetic material an organism inherits from its parents leads to **genetic variation**.

- The environment in which the organism lives – this causes environmental variation.

▲ **Figure 3** *These are wingcases of several fourteen-spot ladybirds (Propylea 14-punctata) collected from the same nettle plant. The variation in colouration patterns between the individuals is an example of intraspecific variation*

Synoptic link

You learnt about errors in DNA replication (mutations) in Topic 3.9, DNA replication and DNA replication and the genetic code in Topic 6.2, Mitosis.

Genetic causes of variation

Genetic variation is due to the genes (and alleles) an individual possesses. There are several causes for genetic variation being present within a population:

1   **Alleles** (variants) – genes have different alleles (alternative forms). With a gene for a particular characteristic, different alleles produce different effects. For example, the gene for human blood groups has three different alleles (A, B and O). Depending on the parental combination of these alleles (see point 4) four different blood groups can be produced (A, B, AB and O). Individuals in a species population may inherit different alleles of a gene.

2   **Mutations** – changes to the DNA sequence and therefore to genes can lead to changes in the proteins that are coded for. These protein changes can affect physical and metabolic characteristics. If a mutation occurs in somatic (body) cells, just the individual is affected. However, if a mutation occurs in the gametes it may be passed on to the organism's offspring. Both can result in variation.

3   **Meiosis** – gametes (sex cells – ovum and sperm) are produced by the process of meiosis in organisms that reproduce sexually. Each gamete receives half the genetic content of a parent cell. Before the nucleus divides and chromatids of a chromosome separate, the genetic material inherited from the two parents is 'mixed up' by **independent assortment** and **crossing over**. This leads to the gametes of an individual showing variation.

4   Sexual reproduction – the offspring produced from two individuals inherits genes (alleles) from each of the parents. Each individual produced therefore differs from the parents.

5   Chance – many different gametes are produced from the parental genome. During sexual reproduction it is a result of chance as to which two combine (often referred to as random fertilisation). The individuals produced therefore also differ from their siblings as each contains a unique combination of genetic material.

Synoptic link

You learnt about independent assortment and crossing over during meiosis in Topic 6.3, Meiosis.

Points 3, 4 and 5 are all aspects of sexual reproduction. As a result there is much greater variation in organisms that reproduce sexually than asexually. Asexual reproduction results in the production of clones (individuals that are genetically identical to their parents). Genetic variation can only be increased in these organisms as a result of mutation.

An example of a characteristic that is determined purely by genetic variation is your blood group. The genes passed onto you from your parents determine if your blood group will be type A, B, AB or O.

▲ **Figure 4** *These scars on the manatee's back are caused by damage from a boat propeller. They are an example of environmental variation*

Environmental causes of variation

All organisms are affected by the environment in which they live, although plants may be affected to a greater degree than animals due to their lack of mobility. For example, two rose bushes are planted in different positions in a garden. The one that has greater access to the sun will generally grow larger than one in a shadier position. As the plant cannot move to gain sunlight, it is more affected by the environment than an animal, which could move to another area to look for food or shelter.

An example of a characteristic that is determined purely by environmental variation is the presence (or absence) of any scars on your body. They will have occurred as a result of an accident or disease and have no genetic origin. Scars cannot be inherited from a parent.

▲ **Figure 5** *Hydrangeas (*Hydrangea* spp.) produce blue flowers in acidic soils and pink flowers in alkaline soils. This is an example of environmental variation*

Environmental and genetic causes

In most cases variation is caused by a combination of both environmental and genetic factors.

If you have very tall parents, you have most likely inherited the genes to also grow to a tall height. However, if you eat a very poor diet or suffer from disease you may only grow to below average height.

Another example of a characteristic that shows both environmental and genetic causes is your skin colour. This is determined by how much of the pigment, melanin, it contains. The more melanin present in your skin, the darker your skin is. Your skin colour at birth is determined purely by genetics – however, when you expose your skin to sunlight you produce more melanin to protect your skin from harmful UV rays. This results in your skin turning darker.

As many characteristics are caused by a combination of both genetic and environmental causes, it can be very difficult to investigate and draw conclusions about the causes of a variation in any particular case (however, see the Application on studying variation in identical twins). This is often referred to as the 'nature versus nurture' argument. For example, many studies have investigated the primary cause of variation in intelligence – genetics or environment? To date, no definitive conclusion has been reached.

Study tip

Examples of characteristics entirely due to the environment and without genetic influence are few.

For example, scarring as discussed – some people have skin that forms cheloid scars that are very obvious, and others have skin that heals easily with minimal scarring. So even scarring has genetic aspects.

However, the scars themselves cannot be inherited – so have only an environmental cause.

Studying variation in identical twins

Many studies have been carried out on identical twins to determine how much of a characteristic is a result of genetic variation, and how much is a result of the environment in which a person lives.

Identical twins are produced when an egg splits after fertilisation. At this point each twin contains identical genetic material, therefore they show no genetic variation.

If the twins are brought up in different environments, the results of environment on variation can clearly be seen. Even within the same environment, as the twins grow they will show some variation. The characteristics in which they show most variation must be influenced more greatly by the environment than by genes. Those in which they show least variation are controlled more by genes than environment.

One of the most famous case studies on identical twins is the 'Minnesota Study of Twins Reared Apart'. This study looked at the lives of several pairs of identical twins. One pair was known as the 'Jim twins'. Identical twins Jim Lewis and Jim Springer were four weeks old when they were separated, and adopted into different families. They were later reunited, aged 39. At this point, the similarities the twins shared amazed researchers at the University of Minnesota. Many physical characteristics were shared; both twins:

- were 6 feet tall
- had a body mass of 82 kg (13 stone)

- were fingernail biters
- suffered from migraine headaches.

This is perhaps not too surprising. It was also discovered that the twins shared a number of other astonishing similarities. Both twins:

- had owned a dog named Toy
- had been married twice (where both first wives were called Linda, and the second wives both called Betty)
- smoked the same brand of cigarettes
- had studied carpentry and mechanical drawing.

Of course, like other identical twins Jim Lewis and Jim Springer were not identical copies of each other. The two men styled their hair differently – one preferred the medium of speech to communicate, while the other preferred the written word.

1 Why are identical twins used in variation studies?
2 Why do differences between identical twins increase as they age?
3 Look at these data. What can you determine about the genetic and environmental causes of these characteristics?

Twin	Height	Eye colour	Ear piercing	Body mass
A	1.79 m	brown	yes	100 kg
B	1.81 m	brown	no	85 kg

Summary questions

1 State the difference between interspecific and intraspecific variation. *(1 mark)*

2 a Name two human characteristics with variation caused solely by the environment. *(1 mark)*
   b Name two human characteristics with variation caused solely by genetics. *(1 mark)*

3 Explain some of the causes of variation of human hair. *(3 marks)*

4 Explain why genetic variation is more common in organisms that reproduce sexually. *(3 marks)*

10.6 Representing variation graphically

Specification reference: 4.2.2

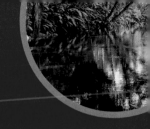

When studying variation, scientists take measurements of different characteristics within a species. To allow reliable conclusions to be formed, they need to collect measurements from large numbers of the population. In order to analyse and interpret this data it can be represented graphically. This allows any patterns to be seen clearly.

Based on the data collected, characteristics can be sorted into those that show **discontinuous variation** and those that show **continuous variation**.

Discontinuous variation

A characteristic that can only result in certain values is said to show discontinuous variation (or discrete variation). There can be no in-between values. Variation determined purely by genetic factors falls into this category. An animal's sex is an example of discontinuous variation as there are only two possible functional values – male or female.

An example of discontinuous variation in microorganisms is the shape of bacteria. They can be spherical (cocci), rods (bacilli), spiral (spirilla), comma (vibrios) or corkscrew shaped (spirochaetes).

Discontinuous variation is normally represented using a bar chart, but a pie chart may also be used. Human blood groups also show discontinuous variation. Like most other characteristics that show discontinuous variation, it is controlled by a single gene, the ABO gene.

Continuous variation

A characteristic that can take any value within a range is said to show continuous variation. There is a graduation in values from one extreme to the other of a characteristic – this is known as a continuum. The height and mass of plants and animals are examples of such characteristics.

Characteristics that show continuous variation are not controlled by a single gene but a number of genes (polygenes). They are also often influenced by environmental factors.

Data on characteristics that show continuous variation are collected in a frequency table (Table 1). These data are then plotted onto a histogram (Figure 1). Normally a curve is then drawn onto the graph to show the trend.

Normal distribution curves

When continuous variation data are plotted onto a graph, they usually result in the production of a bell-shaped curve known as a **normal**

▼ **Table 1** *Frequency of heights (measured to the nearest 2 cm)*

Height/cm	Frequency
140	0
144	1
148	23
152	90
156	261
160	393
164	458
168	413
172	177
176	63
180	17
184	4
188	1
190	0
192	0

▲ **Figure 2** *The surface area of the leaves on this tree show continuous variation – depending on their position on the tree they have all grown to slightly different sizes*

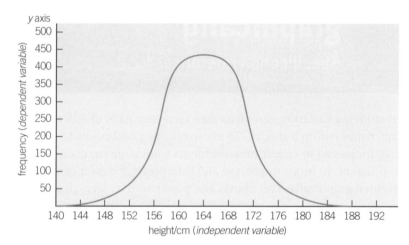

▲ **Figure 1** *This graph shows the frequency against height for a sample of humans. The height of the human population ranges from the shortest person in the world to the tallest person. A person's height can take any value in between. This is an example of continuous variation*

distribution curve (Figure 3). The data is said to be normally distributed.

Characteristics of a normal distribution:

- The mean, mode, and median are the same.
- The distribution has a characteristic 'bell shape', which is symmetrical about the mean.
- 50% of values are less than the mean and 50% are greater than the mean.
- Most values lie close to the mean value – the number of individuals at the extremes are low.

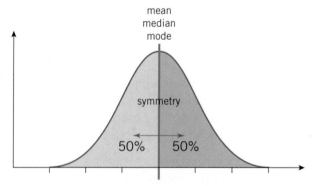

▲ **Figure 3** *A normal distribution curve*

Standard deviation

The standard deviation is a measure of how spread out the data is. The greater the standard deviation is, the greater the spread of the data. In terms of variation, a characteristic which has a high standard deviation has a large amount of variation.

When you calculate the standard deviation of data that display a normal distribution you will generally find that:

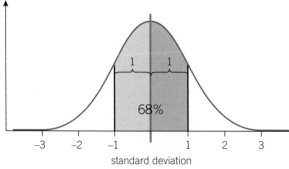

68% of values are within
1 standard deviation of the mean

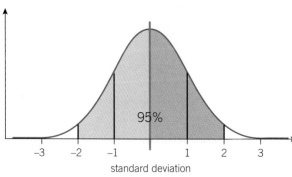

95% of values are within
2 standard deviations of the mean

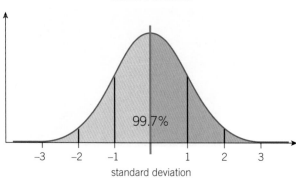

99.7% of values are within
3 standard deviations of the mean

▲ **Figure 5** *The petiole (stalk)
length of ivy (*Hedera *spp.) leaves
show continuous variation*

▲ **Figure 4** *Graphs showing spread of data on a normal distribution*

 Worked example: Petiole length variation in ivy leaves

Standard deviation (denoted with the Greek
letter σ) is a measurement of the spread of data.

It is calculated using the following formula:

$$\sigma = \sqrt{\frac{\sum (x - \overline{x})^2}{n - 1}}$$

Σ = the sum (total) of

x = value measured

$\overline{x}$ = mean value

n = total number of values in the sample

Follow the worked example below to work out
the standard deviation of the length of petioles
(the stalk attaching the leaf to the stem) of a
sample of 10 ivy leaves. The data collected was:

Sample Number	1	2	3	4	5	6	7	8	9	10
Petiole length / mm	28	30	17	31	35	45	46	67	33	57

1 Calculate the mean value $\overline{x}$

$\overline{x} = \dfrac{\text{sum of the individual measurements}}{\text{number in the sample}}$

$= \dfrac{389}{10} = 38.9 \text{ mm}$

2 Subtract the mean value from each measured
value: $x - \overline{x}$

For example, the first measurement was
28 mm.

Measured value (x) – mean value $(\bar{x})$
$= 28 - 38.9 = -10.9\,\text{mm}$

For the other measurements you would calculate: $-8.9\,\text{mm}$, $-21.9\,\text{mm}$, $-7.9\,\text{mm}$, $-3.9\,\text{mm}$, $6.1\,\text{mm}$, $7.1\,\text{mm}$, $28.1\,\text{mm}$, $-5.9\,\text{mm}$, $18.1\,\text{mm}$

3 Square each of these values $(x - \bar{x})^2$
For example $-10.9^2 = 118.81$

For the other measurements you would calculate: $79.2\,\text{mm}$, $479.6\,\text{mm}$, $62.4\,\text{mm}$, $15.2\,\text{mm}$, $37.2\,\text{mm}$, $50.4\,\text{mm}$, $789.6\,\text{mm}$, $34.8\,\text{mm}$, $327.6\,\text{mm}$

4 Sum each of these values $\sum (x - \bar{x})^2$
$118.8 + 79.2 + 479.6 + 62.4 + 15.2 + 37.2 + 50.4 + 789.6 + 34.8 + 327.6 = 1994.8\,\text{mm}$

5 Divide this value by the sample size minus one $\dfrac{\sum(x-\bar{x})^2}{n-1}$

$$\frac{\sum(x-\bar{x})^2}{n-1} = \frac{1994.8}{10-1} = \frac{1994.8}{9} = 221.7$$

6 Find the square root of this value $\sqrt{\dfrac{\sum(x-\bar{x})^2}{n-1}}$

$$\sigma = \sqrt{\frac{\sum(x-\bar{x})^2}{n-1}} = \sqrt{221.7} = 14.9$$

Flagella length variation in *Salmonella*

Salmonella bacteria have a number of flagella to enable them to move. Five bacteria were chosen at random, and their longest flagellum measured. This is the data that was collected:

Bacterium	1	2	3	4	5
Longest flagellum (μm)	3.0	2.5	1.8	2.0	2.7

1 Calculate the mean value for the length of flagella in *Salmonella* bacteria.
2 Using the formula given in the worked example, calculate the standard deviation for the length of flagella in *Salmonella* bacteria. State your answer to two decimal places.
3 State the range of flagella lengths that 68% of the *Salmonella* population will have.
4 State and explain what type of variation is shown by the length of flagella in *Salmonella*.

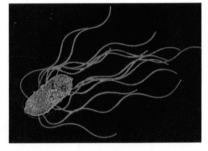

▲ **Figure 6** *This is a* Salmonella *bacteria (*Salmonella *sp.). The length of their flagella shows continuous variation*

Other statistical tests

Several statistical tests can be used by scientists to determine the significance of data collected. These tests can be used in a number of situations, for example when comparing variation within populations, or when comparing the effects of abiotic and biotic factors on organisms (Chapter 11, Biodiversity). These include:

- Student's *t* test – this is used to compare the means of data values of two populations
- Spearman's rank correlation coefficient – this is used to consider the relationship of between two sets of data.

Student's *t* test

Student's *t* test is used to compare the mean values of two sets of data. To use this test the data collected must be normally distributed and enough data should be collected to calculate a reliable mean. Different sample sizes may be used.

Study tip

A significant difference at *p* = 0.05 means that if the null hypothesis were correct (i.e., the samples or treatments do not differ) then we would expect to get a t value as great as this on exactly 5% of occasions. You can therefore be reasonably confident that the samples do differ from one another, but there is still nearly a 5% chance of this conclusion being wrong

If the calculated t value exceeds the tabulated value for *p* = 0.01, then there is a 99% chance of the means being significantly different (and a 99.9% chance if the calculated t value exceeds the tabulated value for *p* = 0.001). By convention, a difference between means at the 95% level is 'significant', a difference at 99% level is 'highly significant' and a difference at the 99.9% level is 'very highly significant'.

 Worked example: Comparing mean petiole length in ivy grown in the light and shade

Student's *t* test is is calculated using the following formula:

$$t = \frac{(\bar{x}_1 - \bar{x}_2)}{\sqrt{\left(\dfrac{\sigma_1^2}{n_1}\right) + \left(\dfrac{\sigma_2^2}{n_2}\right)}}$$

$\bar{x}_1$, $\bar{x}_2$ = mean of populations 1 and 2

σ_1, σ_2, = standard deviation of populations 1 and 2

n_1, n_2 = total number of values in samples 1 and 2

A sample of ten ivy leaves was collected from either side of a tall tree stump. Those collected from the south-facing side of the trunk were referred to as the 'light' leaves and those on the north-facing side of the stump, the 'shade' leaves. A group of students wanted to see if there was a significant difference in the size of the leaf's petiole, depending on whether ivy is grown in the light or the shade.

Before calculating Student's *t* test, the students had to produce a **null hypothesis**. This is a prediction that there is no significant difference between specified populations, and so any observed difference would be due to chance variation in the sample.

The data the students collected is summarised below:

Sample taken	Number in sample	Mean petiole length/mm	Standard deviation/mm
Light	10	38.9	14.9
Shade	10	52.8	15.1

The students then used the Student's *t* test to determine if there is statistical significance between the petiole length of ivy grown in the light, compared to those grown in the shade.

1   State the null hypothesis:

There will be no difference in the length of ivy petiole length of leaves growing in the light, compared with those in the shade.

2   Subtract the mean petiole length of sample two from sample one:

$\bar{x}_1 - \bar{x}_2 = 38.9 - 52.8 = (-)13.9$

Note: ignore minus signs.

3   For both populations, square the standard deviation and divide by the number in the sample:

Population 1: $\dfrac{\sigma_1^2}{n_1} = \dfrac{14.9^2}{10} = 22.201$

Population 2: $\dfrac{\sigma_2^2}{n_2} = \dfrac{15.1^2}{10} = 22.801$

4   Sum these values:

$\dfrac{\sigma_1^2}{n_1} + \dfrac{\sigma_2^2}{n_2} = 22.201 + 22.801 = 45.002$

5 Square root this value

$$\sqrt{\frac{\sigma_1^2}{n_1} + \frac{\sigma_2^2}{n_2}} = \sqrt{45.002} = 6.71$$

6 Calculate Student's *t* test

$$t = \frac{(\bar{x}_1 - \bar{x}_2)}{\sqrt{\frac{\sigma_1^2}{n_1} + \frac{\sigma_2^2}{n_2}}} \quad t = \frac{13.9}{6.71} = 2.07$$

To understand what this value means, you must look it up in the Student's *t* test significance tables (see appendix). First, calculate a quantity known as the 'degrees of freedom' (df) using the formula:

$df = (n_1 + n_2) - 2$ where n_1 = population 1, n_2 = population 2

In this example: $df = (n_1 + n_2) - 2$
$$= (10 + 10) - 2 = 18$$

Then look at the corresponding probability values. For the data to be considered significantly different from chance alone, the probability (p) must be 5% (0.05) or less.

At $df = 18$, the value of 2.07 falls between 5% and 10%.

The null hypothesis should be accepted, as we cannot be more that 95% confident that the results are not down to chance. We therefore cannot conclude that there is a significant difference between the petiole length of ivy grown in the light and in the shade.

Spearman's rank correlation coefficient

If two sets of data are related they are said to be correlated. Two sets of data can show:

- no correlation – no relationship between the data (Figure 9)
- positive correlation – as one set of data increases in value, the other set of data also increases in value (Figure 7)
- negative correlation – as one set of data increases in value, the other set of data decreases in value (Figure 8).

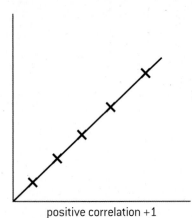

positive correlation +1

▲ **Figure 7** *Graph showing positive correlation*

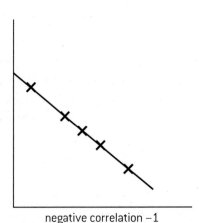

negative correlation −1

▲ **Figure 8** *Graph showing negative correlation*

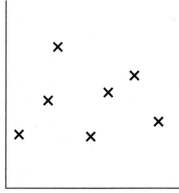

no correlation 0

▲ **Figure 9** *Graph showing no correlation*

Worked example – Using Spearman's rank correlation coefficient to compare ivy leaves' petiole length and leaf width

The correlation coefficient is calculated using the following formula:

$$r_s = 1 - \frac{6\Sigma d^2}{n(n^2-1)}$$

where:

r_s = correlation coefficient Σ = the sum (total) of

d = difference in ranks n = number of pairs of data

The group of students next wanted to find out if the petiole length of the ivy was related to the width of an ivy leaf. They took a sample of 10 ivy leaves from the north facing side of the stump. The data they collected is shown below.

Sample Number	1	2	3	4	5	6	7	8	9	10
Petiole length / mm	28	58	57	59	27	59	44	54	79	63
Leaf width / mm	38	66	64	66	30	65	48	54	78	62

The data for the two variables should be rank ordered, from lowest to highest. Using a table can help to make manipulating the data more straightforward.

Where identical values exist, the 'average rank' should be used. So, if two equal values appear at rank 5, both are assigned the rank 5.5 (between ranks 5 and 6).

petiole length / mm	leaf width / mm	Rank: petiole length	Rank: leaf width	Rank difference d	d^2
28	38	2	2	0	0
58	66	6	8.5	−2.5	6.25
57	64	5	6	−1	1
59	66	7.5	8.5	−1	1
27	30	1	1	0	0
59	65	7.5	7	0.5	0.25
44	48	3	3	0	0
54	54	4	4	0	0
79	78	10	10	0	0
63	62	9	5	4	16
					Σd^2 = 24.5

Substituting values from the table:

$$r_s = 1 - \frac{6\Sigma d^2}{n(n^2-1)} = 1 - \frac{(6 \times 24.5)}{10 \times (10^2 - 1)} = 1 - \left(\frac{147}{990}\right) = 0.852$$

where:

An r_s value of +1 shows a perfect positive correlation.

An r_s value of −1 shows a perfect negative correlation.

An r_s value of 0 shows no correlation.

Therefore in this example, petiole length and leaf width show an excellent correlation.

To work out the statistical strength of the correlation, the value should be looked up in the correlation coefficient critical value tables. Some tables refer to the number of data pairs (*n*); others ask you to calculate the degrees of freedom (df). The tables you will be using for your Spearman's rank correlation coefficient use *n*.

Then look at the probability values for this number of data pairs. As before, for the data to be considered significantly different from chance alone, the probability must be 5% (0.05) or less – a certainty of 95% or more.

	$p = 0.1$	$p = 0.05$	$p = 0.02$	$p = 0.01$
	10%	5%	2%	1%
n				
1	–	–	–	–
2	–	–	–	–
3	–	–	–	–
4	1.0000	–	–	–
5	0.9000	1.0000	1.0000	–
6	0.8286	0.8857	0.9429	1.0000
7	0.7143	0.7857	0.8929	0.9286
8	0.6429	0.7381	0.8333	0.8810
9	0.6000	0.7000	0.7833	0.8333
10	0.5636	0.6485	0.7455	0.7939
11	0.5364	0.6182	0.7091	0.7545
12	0.5035	0.5874	0.6783	0.7273

If p=0.01 then this correlation has only a 1% probability of having occurred by random chance. As the correlation is positive, we can conclude that the greater the petiole length, the greater the leaf width.

Summary questions

1 Sort the following list into those characteristics which show continuous variation and those which show discontinuous variation: (*2 marks*)
   a the presence in humans of lobed or lobeless ears
   b the size of an *E.coli* bacterium
   c the height of a group of seedlings, planted for a germination experiment
   d the number of spots present on a ladybird.

2 Describe the differences between the genetic and environmental control of characteristics that show discontinuous and continuous variation.   (*4 marks*)

3 Explain why a mean value should not normally be calculated for a characteristic showing discontinuous variation.   (*2 marks*)

4 Describe the pattern of variation that would be seen if the body mass of all wild rabbits was measured.   (*4 marks*)

5 The following data was collected from a student's fieldwork study:

Diameter of rose bush stem / mm	1	2	3	5	8	10	11	14
Number of thorns per unit length	8	11	9	12	12	27	23	30

   a Calculate Spearman's rank correlation coefficient for this set of data   (*6 marks*)
   b Evaluate the strength of the correlation calculated in part (a)   (*3 marks*)

10.7 Adaptations

Specification reference: 4.2.2

You should be familiar with the concept that organisms are adapted to the environment in which they live. Organisms can also be adapted to protect themselves from predators or attract a mate.

What are adaptations?

Adaptations are characteristics that increase an organism's chance of survival and reproduction in its environment. Adaptations can be divided into three groups:

- anatomical adaptations – physical features (internal and external)
- behavioural adaptations – the way an organism acts. These can be inherited or learnt from their parents.
- physiological adaptations – processes that take place inside an organism.

Many adaptations fall into more than one category. For example, the courtship behaviour of a peacock requires it to lift its huge, colourful tail to attract the peahen. This is an example of both a behavioural and anatomical adaptation.

Anatomical adaptations

Some examples of anatomical adaptations:

- Body covering – animals have a number of different body coverings such as hair, scales, spines, feathers, and shells. These can: help the organism to fly, such as feathers on birds – help it to stay warm, such as the thick hair on polar bears – provide protection, such as a snail's shell. Thick waxy layers on plants prevent water loss and spikes can deter herbivores and protect the tissues from sun damage.

- Camouflage – the outer colour of an animal allows it to blend into its environment, making it harder for predators to spot it. For example, the snowshoe hare is white in winter to match the snow, and turns brown in summer to blend in with the soil and rock environment in which it lives.

- Teeth – the shape and type of teeth present in an animal's jaw are related to its diet. Herbivores, such as sheep, have continuously growing molars for chewing tough grass and plants. Carnivores, such as tigers, have sharp large canines to kill prey and tear meat.

- Mimicry – copying another animal's appearance or sounds allows a harmless organism to fool predators into thinking it is poisonous or dangerous. For example, the harmless hoverfly mimics the markings of a wasp to deter predators.

▲ **Figure 1** *Otters* (Lutra *spp.*) *have webbed paws. This allows them to swim as well as walk. This is an example of an anatomical adaptation. It increases their chance of survival as they can live and hunt on land and in the water (× 10 magnification)*

▲ **Figure 2** *The harmless milk snake* (Lampropeltis triangulum, *top) mimics the markings of the deadly coral snake* (Micrurus alleni, *bottom)*

Synoptic link

You learnt about transpiration in Topic 9.3, Transpiration and about plant adaptations to prevent water loss in Topic 9.5, Plant adaptations to water availability.

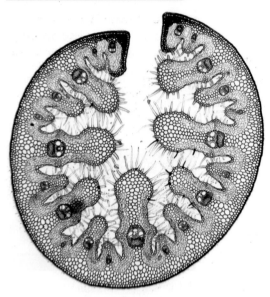

▲ **Figure 3** *A light micrograph of a curled leaf of marram grass plant (*Ammophila* *arenaria*). You can see the hairs on the inside surface of the leaf, slowing air movement and reducing water loss from the leaf (× 10 magnification)*

▲ **Figure 4** *This chimpanzee is using a twig/grass to get termites out of a termite mound*

Marram grass

Marram grass (*Ammophila* spp.) is commonly found on sand dunes around the UK. It is a xerophyte, a plant that has adapted to live in an environment with little water. Its adaptations reduce the rate of **transpiration** and include:

- curled (or rolled) leaves to minimise the surface area of moist tissue exposed to the air, and protect the leaves from the wind

  - hairs on the inside surface of the leaves to trap moist air close to the leaf, reducing the diffusion gradient
  - stomata sunk into pits, which make them less likely to open and lose water
  - a thick waxy cuticle on the leaves and stems, reducing water loss through evaporation.

Behavioural adaptations

Some examples of behavioural adaptations:

- Survival behaviours – for example, an opossum plays dead and a rabbit freezes when they think they have been seen.
- Courtship – many animals exhibit elaborate courtship behaviours to attract a mate. For example, scorpions perform a dance to attract a partner. This increases the organism's chance of reproducing.
- Seasonal behaviours – these adaptations enable organisms to cope with changes in their environment. They include:
  - migration – animals move from one region to another, and then back again when environmental conditions are more favourable. This may be for a better climate or a source of food
  - hibernation – a period of inactivity in which an animal's body temperature, heart rate and breathing rate slow down to conserve energy, reducing the animal's requirement for food. For example, brown bears hibernate during the winter.

Generally, behavioural adaptations fall into two main categories:

- Innate (or instinctive) behaviour – the ability to do this is inherited through genes. For example, the behaviour of spiders to build webs and woodlice to avoid light is innate. This allows the organism to survive in the habitat in which it lives.
- Learned behaviour – these adaptations are learnt from experience or from observing other animals. An example of learned behaviour is the use of tools. For example, sea otters use stones to hammer shells off rocks, and then to crack the hard shells open.

However, many behavioural adaptations are a combination of both innate and learned behaviours.

Physiological adaptations

Some examples of physiological adaptations:

- Poison production – many reptiles produce venom to kill their prey and many plants produce poisons in their leaves to protect themselves from being eaten.

- Antibiotic production – some bacteria produce antibiotics to kill other species of bacteria in the surrounding area.

- Water holding – the water-holding frog (*Cyclorana platycephala*) can store water in its body. This allows it to survive in the desert for more than a year without access to water. Many cacti and other desert plants can hold large amounts of water in their tissues.

Many other examples are less unusual, and include reflexes, blinking and temperature regulation.

▲ **Figure 5** *Aestivation is the name given to a period of inactivity in hot, dry places. The land snails aestivate to cope with periods of extreme dry heat. This is a combination of physiological and behavioural adaptations*

Anatomical adaptations provide evidence for convergent evolution

Analogous structures

Although the tail fins of a whale and a fish perform the same role, when you look at them in detail their structures are very different. These are known as **analogous structures** – they have adapted to perform the same function but have a different genetic origin.

Convergent evolution takes place when unrelated species begin to share similar traits. These similarities evolve because the organisms adapt to similar environments or other selection pressures (for an explanation of the term 'selection pressure', Topic 10.8, Changing population characteristics). The organisms live in a similar way to each other. Using our example of whales and fish, their similar characteristics have evolved over time to allow the organisms to move efficiently through water.

Marsupials in Australia and placental mammals in the Americas are an example of convergent evolution. Species in each continent resemble each other because they have adapted to fill similar niches.

In placental mammals, a placenta connects the embryo to its mother's circulatory system in the uterus. This nourishes the embryo, allowing it to reach a high level of maturity before birth. Marsupials also start life in the uterus, but then leave and enter the marsupium (pouch) while they are still embryos. They complete their development here by suckling milk.

These two subclasses of mammals separated from a common ancestor more than 100 million years ago. Each lineage then evolved independently. Despite this large temporal and geographical separation, marsupials in Australia and placental mammals in North America have produced varieties of species that bear a strong resemblance in overall shape, type of locomotion and feeding techniques. This is because they have adapted to similar climates and food supplies. However, these organisms have very different methods of reproduction. This is the feature that accurately reflects their distinct evolutionary relationships.

▲ **Figure 7** *(top) Flying phalanger (*Petaurus* sp.) – a marsupial mammal and (bottom) flying squirrel (family Pteromyini) – a placental mammal*

Examples include:

- marsupial and placental mice – both are small, agile climbers that live in dense ground cover and forage at night for small food items. The two mice are very similar in size and body shape

- flying phalangers and flying squirrels (Figure 7) – both are gliders that eat insects and plants. Their skin is stretched between their forelimbs and hind limbs to provide a large surface area for gliding from one tree to the next

- marsupial and placental moles (Figure 8) – both burrow through soft soil to find worms and grubs. They have a streamlined body shape and modified forelimbs for digging. They also have velvety fur, which allows smooth movement through the soil. However, they differ in fur colour – the marsupial mole ranges in colour from white to orange whereas the placental mole is grey.

Convergent evolution can also been seen in some plant species. For example, aloe and agave appear very similar as they have both adapted to survive in the desert. However, these species developed entirely separately from each other. Aloe are sometimes referred to as 'old world', having evolved in sub-Saharan Africa. Agave, by comparison, are 'new world', having evolved in Mexico and the southern United States.

▲ **FIGURE 8** *(top) Marsuipal mole (family Notoryctidae) found in Australia and (bottom) placental mole (family Talpidae) found in North America*

aloe

agave

▲ **Figure 9** *Aloe (Aloe spp.) and agave (Agave spp.) have a similar appearance. They provide an example of convergent evolution amongst plant species*

✚ Classification of giant pandas

Classification aims to place every organism into a particular taxonomical group. However, there are a number of organisms that do not fit easily into a group. An example of this is the giant panda (*Ailuropoda melanoleuca* – Figure 10, top).

Père Armand David, a catholic priest, was the first westerner to see a giant panda. He discovered the panda in 1869, and based on its appearance he concluded that it was related to a bear (family *Ursidae* – Figure 10,

bottom). He gave it a name that included the word *ursus* (the Latin word for bear).

A few years later, Alphonse Milne-Edwards, a French scientist, inspected the remains of a giant panda. He concluded that its anatomical structure was closer to the red panda (Figure 10, middle), a member of the raccoon family. He renamed the giant panda, and classified it into its own category. Many people disagreed with Milne-Edwards' conclusion because of its size. Red pandas have a mass of between 3 and 7 kg – the largest raccoons have a mass of around 30 kg. By comparison, the giant pandas can exceed 100 kg body mass.

The debate over the classification of the giant panda has continued for several decades.

Similarities to a red panda:

- Both eat bamboo and grip bamboo in the same manner.
- Both have similar snouts, teeth and paws.

Similarities to a bear:

- Both are a very similar shape and size.
- Both have shaggy fur.
- Both walk and climb in a similar manner.

Giant pandas and red pandas may have developed similar ways of eating bamboo separately as a result of convergent evolution. Equally, convergent evolution could explain their resemblance to bears.

1 Describe the adaptations of a giant panda.
2 Explain how you would classify a giant panda. Give reasons for your classification.
3 Suggest how recent biological techniques could be used to help classify the giant panda.

In the 1950s, the first molecular-level analysis of the giant panda occurred. Biologists used an immunological method to assess the closeness of bears to pandas. Through studying blood serum, they concluded that the 'serological affinities of the giant panda are with the bears rather than with the raccoons'. The giant panda is therefore a true bear and part of the Ursidae family, although it differentiated early in history from other bears.

Despite the shared name, habitat type, and diet, as well as a unique enlarged bone called the pseudo thumb (which helps them grip bamboo shoots), the giant panda and red panda are only distantly related. Molecular studies place the red panda in its own family – Ailuridae.

Summary questions

1 Classify the following adaptations into anatomical, behavioural or physiological adaptations:
melanin production; camouflage; migration; sharp canine teeth; production of toxins; courtship dance   (3 marks)

2 State the difference between analogous and homologous structures.   (1 mark)

3 Which of the following is an example of convergent evolution?   (1 mark)
Explain your answer.
a Insect wing and bird wing
b Bat wing and human arm

4 Select either a cactus or a hedgehog. For your chosen organism, state and explain how it is adapted to survive successfully in its habitat.   (4 marks)

5 State and explain how marsupial moles and placental moles provide evidence for convergent evolution.   (4 marks)

▲ Figure 10 *The giant panda (*Ailuropoda melanoleuca *– top) shares characteristics with both the red panda (*Ailurus fulgens *– middle) and bears (family Ursidae – for example this brown bear* Ursus arctos *– bottom). As a result, scientists have argued about how to classify giant pandas*

10.8 Changing population characteristics

Specification reference: 4.2.2

You have already learnt about the theory of evolution as a result of natural selection in Topic 10.4 Evidence for evolution. This process takes place over many, many generations; it generally takes several thousand years for a species to evolve. However, evolution is a dynamic process and is always occurring.

Natural selection

All organisms are exposed to **selection pressures**. These are factors that affect the organism's chances of survival or reproductive success (the ability to produce fertile offspring).

Organisms that are best adapted to their environment are more likely to survive and reproduce. As a result of natural selection these adaptations will become more common in the population. Organisms that are poorly adapted are less likely to survive and reproduce. Therefore their characteristics are not passed on to the next generation. As a result, less of the population will display these characteristics.

Natural selection follows a number of steps:

1   Organisms within a species show variation in their characteristics that are caused by differences in their genes (genetic variation). For example, they may have different alleles of a gene for a particular characteristic. New alleles can arise by mutation.

2   Organisms whose characteristics are best adapted to a selection pressure such as predation, competition (for mates and resources) or disease, have an increased chance of surviving and successfully reproducing. Less well-adapted organisms die or fail to reproduce. This process is known as 'survival of the fittest'.

3   Successful organisms pass the allele encoding the *advantageous characteristic* onto their offspring. Conversely, organisms that possess the non-advantageous allele are less likely to successfully pass it on.

4   This process is repeated for every generation. Over time, the proportion of individuals with the advantageous adaptation increases. Therefore the frequency of the allele that codes for this particular characteristic increases in the population's gene pool.

5   Over very long periods of time, many, many generations and often involving multiple genes, this process can lead to the evolution of a new species.

Modern examples of evolution

Antibiotic-resistant bacteria

Methicillin-resistant *Staphylococcus aureus* (MRSA) has developed resistance to many antibiotics. Bacteria reproduce very rapidly and so evolve in a relatively short time. When bacteria replicate, their DNA can be altered and this usually results in the bacteria dying. However, a mutation in some *S. aureus* arose that provided resistance to methicillin.

When the bacteria were exposed to this antibiotic, resistant individuals survived and reproduced, passing the allele for resistance on to their offspring. Non-resistant individuals died. Over time the number of resistant individuals in the population increased.

Peppered moths

Dramatic changes in the moth's environment in the 19th century caused changes in allele frequency in peppered moths (*Biston betularia*). Before the industrial revolution, most peppered moths in Britain were pale coloured. This provided camouflage against light-coloured tree bark, increasing their chance of survival. Those that were dark were easily spotted by birds and eaten. The different colourings are due to different alleles.

During the industrial revolution many trees became darker – partly due to being covered in soot, and partly due to the loss of lichen cover caused by increased atmospheric pollutants. The dark moths were now better adapted, as they were more highly camouflaged. More dark peppered moths survived and reproduced, increasing the frequency of dark moths (and the 'dark' allele) in the population. After a few years the number of dark peppered moths close to industrial towns and cities became much higher than pale peppered moths.

Since the Clean Air Act of 1956 steps have been taken to improve air quality in towns and cities, and to reduce the levels of pollution released from factories. The bark on the vast majority of trees in the UK is once again lighter coloured, and therefore the frequency of the pale allele in the moth gene pool has increased.

Sheep blowflies

Sheep blowflies (*Lucilia cuprina*) lay their eggs in faecal matter around a sheep's tail – the larvae then hatch and cause sores. This condition is known as 'flystrike', and if left untreated is normally fatal.

In the 1950s in Australia, the pesticide diazinon (an organophosphate pesticide) was used to kill the blow flies and prevent the condition. Within six years, blowflies had developed a high level of resistance to diazinon. Individual insects with resistance survived exposure to the insecticide, and passed on this characteristic through their alleles, allowing a resistant population to evolve.

To investigate how this evolution occurred so quickly, scientists extracted DNA from a sample of 70-year-old blowflies kept at

▲ **Figure 1** *Before the industrial revolution, pale peppered moths (*Biston betularia*) were better adapted to the environment. During the industrial revolution dark peppered moths were now better adapted – therefore the frequency of this characteristic increased in the population*

▲ **Figure 2** *Scientists have discovered that sheep blowflies (*Lucilia* spp.) have an inbuilt natural resistance to some organophosphate insecticides. This allowed their rapid evolution to become resistant to the pesticide diazinon*

the Australian National Insect Collection. Two Australian sheep blowflies were studied, *Lucilia cuprina* and the closely related *Lucilia sericata*. The researchers compared the blowflies' resistance genes before and after the introduction of the pesticide. Diazinon resistance was not found in the DNA of the 70-year-old flies, whereas it is present in the modern species. However, when they performed the same investigation with malathion (another organophosphate pesticide), they found resistance alleles in both the old and modern blowflies, showing there was pre-existing resistance to this chemical.

The scientists concluded that pre-adaptation contributed to the development of diazinon-resistance. Pre-adaptation is when an organism's existing trait is advantageous for a new situation. The alteration in the DNA that caused the pre-existing resistance allowed the flies to rapidly develop resistance to organophosphate chemicals in general, and ultimately a specific diazinon-resistance allele.

The existence of pre-adaptation in an organism may help researchers predict potential insecticide resistance in the future.

Flavobacterium

Most evolution occurs as a negative result of selection pressures. However, some organisms have evolved due to opportunities that have arisen in their environment. For example, scientists have found a new strain of *Flavobacterium* living in waste water from factories that produce nylon 6. Nylon 6 is used to make objects like toothbrushes and violin strings. This strain of bacteria has evolved to digest nylon and is therefore beneficial to humans as they help to clear up factory waste.

These bacteria use enzymes to digest the nylon known as nylonases. They are unlike any enzymes found in other strains of *Flavobacterium*, and they do not help the bacteria to digest any other known material. It is beneficial to the bacteria as it provides them with another source of nutrients.

Most scientists believe that the gene mutation that occurred to produce these enzymes was a result of a gene duplication, combined with a frameshift mutation (an insertion or deletion of DNA bases that causes the genetic code to be read incorrectly).

 ### Anolis lizards

When a few individuals of a species colonise a new area, their offspring initially experience a loss in genetic variation, often resulting in individuals that are physically and genetically different from their source population. This is known as the **founder effect**.

A 14-year experiment (led by Kolbe, a biologist at the University of Rhode Island) was carried out to study evolution. Pairs of *Anolis sagrei* were released across 14 small Caribbean islands that had no previous lizard populations. During the experiment, the lizard populations each became adapted to their respective environments through changes in their body shape driven by the flora in their environment. Several new species of lizards evolved.

1 State what is meant by the founder effect.
2 Explain why these particular islands were an ideal location for the experiment.

To determine how much of the evolution of the new species was due to the founder effect, and how much resulted from natural selection, Kolbe randomly selected pairs of Anolis lizards from the island of Iron Cay. He then released these organisms onto seven smaller islands that had no lizard population. Each island had the same types of insects, birds and short scrub vegetation but differed from Iron Cay, which is covered in forest.

Forest Anolis lizards have long hind limbs, which allow them to move quickly across thick branches, whereas short limbs give scrub-living lizards stability to walk along narrow perches. The scientists therefore predicted that the lizards in their experiment would develop shorter hind limbs than those of the lizards on Iron Cay.

After one year, the researchers noticed that the offspring of the experimental lizards had less genetic variability than the Iron Cay lizards – the founder effect. There were also significant differences in hind-limb length among the lizards on the islands. As the founder effect is a random process independent of the environment, there was no pattern to the length of the lizards' hind limbs. Over the next few years the lizards' hind limbs on all the experimental islands got shorter, making them better suited for their environment – natural selection.

▲ Figure 3 A male Anolis lizard (Anolis sagrei) displaying its eye-catching dewlap. When enlarged it makes the lizard appear much bigger than it really is. This mechanism is used to ward off predators and to attract females during the mating season

Kolbe concluded that both processes were evident during the experimental period.

3 Explain why scientists thought the lizards would develop short hind limbs.
4 Explain how the scientists showed whether the evolution was mainly a result of the founder effect or natural selection.
5 Some scientists were surprised that all the new populations of species survived as the presence of only a few individuals leads to inbreeding. Explain why this is disadvantageous.

Summary questions

1 State three selection pressures that may be experienced by a plant species.   (1 mark)

2 Describe the process of natural selection.   (3 marks)

3 DDT is a chemical insecticide that was used to kill mosquitoes to prevent the spread of malaria. Several years after its introduction large populations of mosquitoes became DDT resistant. Explain how this occurred.   (4 marks)

4 Using examples, state and explain the positive and negative effects on humans of recent examples of evolution in some species.   (6 marks)

Practice questions

1 As part of a sample collected from the Indian Ocean, scientists identified an organism as *Hydrophis spiralis*. Which of the following statements is/are correct about this organism?

Statement 1: The organism belongs to the genus *Hydrophis*

Statement 2: The organism belongs to the species *spiralis*

Statement 3: The organism belongs to the genus *spiralis*

  A 1, 2 and 3 are correct

  B 1 and 2 are correct

  C 2 and 3 are correct

  D Only 1 is correct   *(1 mark)*

2 Figure 1 shows an electron micrograph of an invertebrate known as 'water bear'.

▲ **Figure 1**

a Complete the following passage about the classification of water bears using the most appropriate terms.

The water bear, *Eschiniscus trisetosus* is a member of the genus...............and the family *Echiniscidae*. This family belongs to the....................Eschiniscoidea, which forms part of the class *Heterotardigrada*. Water bears, also known as tardigrades, are classified into a...... of their own called the *Tardigrada*. Tardigrades form part of the kingdom............... Within the domain........... . *(5 marks)*

b State the meaning of the term phylogeny and explain how phylogeny is related to classification. *(3 marks)*

c Water bears are extremely common in many habitats, including household gardens. However, they were not discovered until approximately 300 years ago.

Suggest reasons why they were not known before this time. *(2 marks)*

OCR June 2013 F212/01

3 Living organisms can be classified into five kingdoms, based on certain key characteristics.

a Table 1 shows some of the characteristics of the five kingdoms.

Copy and complete the table

kingdom	membrane-bound organelles	cell wall	type(s) of nutrition
prokaryote	absent	present – made of peptidoglycan	
	present	sometimes present – composition varies	heterotrophic and autotrophic
Fungi		present – made of chitin	heterotrophic
	present		autotrophic
animal		absent	heterotrophic

(6 marks)

b An unknown species is discovered. Its cells contain many nuclei scattered throughout the cytoplasm of thread-like structures.

Suggest the kingdom to which this species belongs *(1 mark)*

c Living organisms can also be classified into three groups called **domains.**

Outline the features of this system of classification compared with the five kingdom system. *(3 marks)*

OCR F212/01 2012

4 Adaptations are characteristics which increase an organism's chance of survival and reproduction in an environment.

a State the difference between a behavioural and a physiological adaptation. (*1 mark*)

b Marram grass is commonly found on sand dunes around the coast of Great Britain.

State and explain three anatomical adaptations which enable this plant to survive in an environment with little access to water. (*3 marks*)

c Anatomical adaptations provide evidence for convergent evolution. Explain what is meant by the term 'convergent evolution'. (*2 marks*)

OCR F212/01 2013

6 Bats are the only mammals that can truly fly. Many species of bat hunt flying insects at night. Bats are able to use sound waves (echolocation) in order to help them find their prey in the dark.

a Suggest how the ability to use echolocation may have evolved from an ancestor that did not have that ability (*4 marks*)

The pipistrelle is the most common species of bat in Europe. It was originally thought that all pipistrelles belonged to the same species, *Pipistrellus pipistrellus*. However, in the 1990s, it was decided that there were two species: the common pipistrelle, *Pipistrellus pipistrellus* and the soprano pipistrelle, *Pipistrellus pygmaeus*.

Data for both species are provided in Table 2

species	mean body mass (g)	mean wingspan (m)	range of echolocation (kHz)	colour
Common pipistrelle	5.5	0.22	42–47	medium to dark brown
Soprano pipistrelle	5.5	0.21	52–60	medium to dark brown

b (i) Name the genus to which the soprano pipistrelle belongs (*1 mark*)

(ii) Using the data in Table 2, suggest why pipistrelles were originally classified as one species. (*1 mark*)

(iv) Describe how it is possible to confirm, over a longer period of time, whether two organisms belong to different species or the same species. (*2 marks*)

c The soprano pipistrelle has an echolocation call that is 'high pitched' (between 52 and 60 kHz). The common pipistrelle has an echolocation call that is 'low pitched' (between 42 and 47 kHz).

Variation within and between species can be a result of genetic or environmental factors. Whatever the causes of variation, the type of variation displayed can occur in two different forms.

Using the pipistrelle as an example, describe the key features of both **forms** of variation. (*7 marks*)

OCR F212/01 2012

11 BIODIVERSITY
11.1 Biodiversity
Specification reference: 4.2.1

You may be familiar with the term **biodiversity** – the variety of living organisms present in an area. Biodiversity includes plants, animals, fungi, and other living things. In fact, it includes everything from gigantic redwood trees to single-celled algae.

The importance of biodiversity

Biodiversity is essential in maintaining a balanced ecosystem for all organisms. All species are interconnected – they depend on one another. For example, trees provide homes for animals. Animals eat plants, which in turn need fertile soil to grow. Fungi and other microorganisms help decompose dead plants and animals, returning nutrients to the soil. In regions of reduced biodiversity, these connections may not all be present, which eventually harms all species in the ecosystem.

We rely on balanced ecosystems as they provide us with the food, oxygen and other materials we need to survive. Unfortunately, many human activities, such as farming and clearing land for housing, can lead to a reduction in biodiversity.

Measuring biodiversity

Tropical, moist regions (that are warm all year round) have the most biodiversity. The UK's temperate climate (warm summers and cold winters) has less biodiversity. Very cold areas such as the Arctic, or very dry areas such as deserts, have the least biodiversity. Generally, the closer a region is to the Equator (the line of latitude of the Earth, halfway between the North Pole and South Pole), the greater the biodiversity. For example, over 40 000 plant species live in the Amazon rainforest, whereas less than 3000 live in Northern Canada.

Measuring biodiversity plays an important role in conservation. It informs scientists of the species that are present, thus providing a baseline for the level of biodiversity in an area. From this information, the effect of any changes to an environment can be measured. These may include the effect of human activity, disease or climate change, for example.

Before a major project is undertaken, such as building a new road or the creation of a new nature reserve, an Environmental Impact Assessment (EIA) is undertaken. This assessment attempts to predict the positive and negative effects of a project on the biodiversity in that area.

Biodiversity can be studied at different levels:
- habitat biodiversity
- species biodiversity
- genetic biodiversity.

▲ **Figure 1** *Coral reefs are amongst the most biodiverse ecosystems on the planet*

▲ **Figure 2** *There is very little biodiversity at the top of a high mountain*

Habitat biodiversity

Habitat biodiversity refers to the number of different habitats found within an area. Each habitat can support a number of different species. Therefore in general, the greater the habitat biodiversity, the greater the species biodiversity will be within that area.

The UK is home to large number of habitat types, including meadow, woodland, streams, and sand dunes. It has a large habitat biodiversity. By contrast Antarctica, covered almost entirely by an ice sheet, has a very low habitat biodiversity and very few species live in this region.

On a smaller scale, countryside that is habitat rich, perhaps with a river, woodland, hedgerows and wild grassland, will be more species rich than farmed countryside with large ploughed fields making up a single uniform habitat.

Species biodiversity

Species biodiversity has two different components:

- species richness – the number of different species living in a particular area, and
- species evenness – a comparison of the numbers of individuals of each species living in a community. (The community is all the populations of living organisms in a particular habitat.)

Therefore an area can differ in its species biodiversity even if it has the same number of species. For example, a cornfield and a grass meadow may both contain 20 species. However, in the cornfield, corn will make up 95% of the community with the remaining 5% made up of other organisms including weed plants, insects, mice, and birds. In the grass meadow the species will be more balanced in their populations.

Genetic biodiversity

Genetic biodiversity refers to the variety of genes that make up a species. Humans have about 25 000 genes, but some species of flowering plants have as many as 400 000 genes. Many of these genes are the same for all individuals within a species. However, for many genes, different versions (alleles) exist. This leads to genetic biodiversity within a species (you will learn more about genetic biodiversity in Topic 11.5, Calculating genetic biodiversity).

Genetic biodiversity within a species can lead to quite different characteristics being exhibited. For example, some genes are the same for all breeds of dog – these genes define the organism as a dog. Some of the genes have many alleles – they code for the wide variation in characteristics seen between different breeds of dog, for example coat colour and length.

Greater genetic biodiversity within a species allows for better adaptation to a changing environment, and is more likely to result in individuals who are resistant to disease.

▲ **Figure 3** *These two butterflies appear to belong to different species; however, both are examples of the Gaudy Commodore butterfly (Precis octavia). Genetic biodiversity within this species leads to different wing patterns and colours. The actual colours displayed depend on the season in which the butterflies are born*

Study tip

There are many key terms in this topic – make sure you are clear on their meaning. Why don't you try making your own biodiversity glossary?

Summary questions

1 State the difference between species richness and species evenness. *(2 marks)*

2 Compare the biodiversity of an arid desert and a temperate coastline. *(3 marks)*

3 Suggest why greater genetic biodiversity increases a species' chances of long-term survival. *(4 marks)*

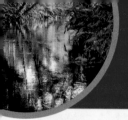

11.2 Types of sampling

Specification reference: 4.2.1

You can use a variety of techniques to measure and compare the biodiversity of different habitats. However, it is often impossible to count or measure all of the organisms present in an area, so sampling techniques are used.

What is sampling?

Sampling means taking measurements of a limited number of individual organisms present in a particular area.

Sampling can be used to estimate the *number* of organisms in an area without having to count them all. The number of individuals of a species present in an area is known as the *abundance* of the organism.

Sampling can also be used to measure a *particular characteristic* of an organism. For example, you cannot reliably determine the height of wheat by measuring one wheat plant in a farmer's field. However, if you measure the height of a number of plants and then calculate an average, your result is likely to be close to the average height of the entire crop.

After measuring a sample, you can use the results of the sample to make generalisations or estimates about the number of organisms, distribution of species or measured characteristic throughout the entire habitat.

Sampling can be done in two ways – random and non-random.

Random sampling

Random sampling means selecting individuals by chance. In a random sample, each individual in the population has an equal likelihood of selection, rather like picking names out of a hat.

To decide which organisms to study, random number tables or computers can be used. You have no involvement in deciding which organisms to investigate. For example, to take a random sample at a grass verge you could follow these steps:

1 Mark out a grid on the grass using two tape measures laid at right angles.

2 Use random numbers to determine the *x* coordinate and the *y* coordinate on your grid.

3 Take a sample at each of the coordinate pairs generated.

Non-random sampling

Non-random sampling is an alternative sampling method where the sample is not chosen at random. It can be divided into three main techniques:

● **Opportunistic** – this is the weakest form of sampling as it may not be representative of the population. Opportunistic sampling uses organisms that are conveniently available.

▲ **Figure 1** *This scientist is taking random soil samples in a field*

- **Stratified** – some populations can be divided into a number of strata (sub-groups) based on a particular characteristic. For instance, the population might be separated into males and females. A random sample is then taken from each of these strata proportional to its size.

- **Systematic** – in systematic sampling different areas within an overall habitat are identified, which are then sampled separately. For example, systematic sampling may be used to study how plant species change as you move inland from the sea. Systematic sampling is often carried out using a line or a belt transect. A **line transect** involves marking a line along the ground between two poles and taking samples at specified points, this can include describing all of the organisms which touch the line or distances of samples from the line. A **belt transect** provides more information; two parallel lines are marked, and samples are taken of the area between the two lines.

▲ **Figure 2** *These students are carrying out sampling by combining two different methods. The quadrat seen here is being used along a line transect. This is known as an interrupted belt transect. This is an example of systematic sampling*

Reliability

A sample is never entirely representative of the organisms present in a habitat. This may be due to the following:

- Sampling bias – the selection process may be biased. This may be by accident, or may occur deliberately. For example, you may choose to sample a particular area that has more flowers because it looks interesting. The effects of sampling bias can be reduced using random sampling, where human involvement in choosing the samples is removed.

- Chance – the organisms selected may, by chance, not be representative of the whole population. For example, a sample of five worms collected in a trap may be the five longest in the habitat. Chance can never be completely removed from the process, but its effect can be minimised by using a large sample size. The greater the number of individuals studied, the lower the probability that chance will influence the result. Therefore the larger the sample size, the more reliable the result.

Summary questions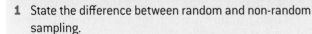

1   State the difference between random and non-random sampling. *(1 mark)*

2   Describe how you can increase the likelihood of a sample being a reliable representation of the population as a whole. *(4 marks)*

3   State and explain which type of sampling you would use to study:
    a   how organisms differ throughout the length of a stream *(2 marks)*
    b   the distribution of organisms on a school field. *(2 marks)*

11.3 Sampling techniques
Specification reference: 4.2.1

Learning outcomes

Demonstrate knowledge, understanding, and application of:

→ practical investigations collecting random and non-random samples in the field

→ how to measure species richness and species evenness in a habitat.

You can use many different techniques to sample the living organisms present in a habitat and the environment in which they live, as you saw in the previous topic. The technique you choose is dependent on the information you require. At each sampling point you would normally use more than one technique, so that a range of data can be collected.

Sampling animals

The following techniques can be used to collect living animals for study later. Remember, all living organisms must be handled carefully and for as short a time period as possible. As soon as any sample animals have been identified, counted and measured if required, they must be released back into the habitat at the point they were collected.

- A pooter is used to catch small insects. By sucking on a mouthpiece, insects are drawn into the holding chamber via the inlet tube. A filter before the mouthpiece prevents them from being sucked into the mouth.

- Sweep nets are used to catch insects in areas of long grass.

- Pitfall traps are used to catch small, crawling invertebrates such as beetles, spiders and slugs. A hole is dug in the ground, which insects fall into. It must be deep enough that they cannot crawl out and covered with a roof-structure propped above so that the trap does not fill with rainwater. The traps are normally left overnight, so that nocturnal species are also sampled.

- Tree beating is used to take samples of the invertebrates living in a tree or bush. A large white cloth is stretched out under the tree. The tree is shaken or beaten to dislodge the invertebrates. The animals will fall onto the sheet where they can be collected and studied.

- Kick sampling is used to study the organisms living in a river. The river bank and bed is 'kicked' for a period of time to disturb the substrate. A net is held just downstream for a set period of time in order to capture any organisms released into the flowing water.

▲ Figure 1 *This student is using a pooter to collect insects from a tree*

▲ Figure 2 *This is a pitfall trap. The glass perspex cover prevents rain entering the trap and potentially causing any trapped insects to drown*

Sampling plants

Plants are normally sampled using a **quadrat**, which can also be used to pinpoint an area in which the sample of plants should be collected. Quadrats can also be used to sample slow-moving animals such as limpets, barnacles, mussels, and sea anemones.

There are two main types of quadrat:

- Point quadrat – this consists of a frame containing a horizontal bar. At set intervals along the bar, long pins can be pushed through the bar to reach the ground. Each species of plant the pin touches is recorded (Figure 3).

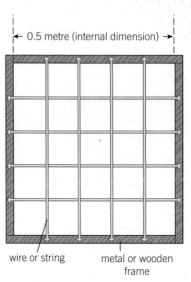

← 0.5 metre (internal dimension) →

wire or string   metal or wooden frame

▲ **Figure 4** *A frame quadrat*

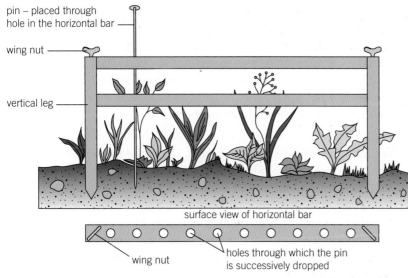

pin – placed through hole in the horizontal bar

wing nut

vertical leg

surface view of horizontal bar

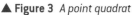

wing nut   holes through which the pin is successively dropped

▲ **Figure 3** *A point quadrat*

- Frame quadrat – this consists of a square frame divided into a grid of equal sections. The type and number of species within each section of the quadrat is recorded (Figure 4). Further details are given below.

To collect the most valid representative sample of an area, quadrats should be used following a random sampling technique (as discussed in Topic 11.2, Types of sampling). To study how the presence and distribution of organisms across an area of land varies, the quadrats can be placed systematically along a line or belt transect.

Measuring species richness

As you learnt in Topic 11.1, Biodiversity species richness is a measure of the number of different species living in a specific area. You should use a combination of the techniques described above to try to identify all the species present in a habitat. A list should be compiled of each species identified. The total number of species can then be calculated.

To enable scientists to accurately identify organisms, identification keys are often used. These may contain images to identify the organism, or a series of questions, which classify an organism into a particular species based on the presence of a number of identifiable characteristics.

Measuring species evenness

As you learnt in Topic 11.1, Biodiversity, species evenness refers to how close in numbers the populations of each species in an environment are. For example, 50 organisms are found living under

Synoptic link

You learned about how organisms are classified in Topic 10.1, Classification and 10.2, The five kingdoms.

a decaying log. Of these, 20 are woodlice, 15 are spiders, and 15 are centipedes – the community is quite evenly distributed between species. However, if the 50 insects comprised just 45 woodlice and 5 spiders, the community would be described as uneven.

Using frame quadrats

A frame quadrat is used to sample the population of plants living in a habitat. There are three main ways of doing this:

- Density – if individual large plants can be seen clearly, count the number of them in a 1m by 1m square quadrat. This will give you the density per square metre. This is an absolute measure, not an estimate as the following two methods.

- Frequency – this is used where individual members of a species are hard to count, like grass or moss. Using the small grids within a quadrat, count the number of squares a particular species is present in. For example, if clover is present in 65 out of 100 squares, the frequency of its occurrence is 65%. (Each square represents 1%.) Another commonly used quadrat contains 25 squares – in this case, each square represents 4% of the study area. Therefore if, during a sampling of grassland, eight quadrat squares contained buttercups, the frequency of occurrence would be 32%.

- Percentage cover – this is used for speed as lots of data can be collected quickly. It is useful when a particular species is abundant or difficult to count. It is an estimate by eye of the area within a quadrat that a particular plant species covers.

For each approach, samples should be taken at a number of different points. The larger the number of samples taken, the more reliable your results. You should then calculate the mean of the individual quadrat results to get an average value for a particular organism per m^2 (To calculate the mean value, sum the individual quadrat results, then divide by the number of samples taken). To work out the total population of an organism in an area that has been sampled, multiply the mean value per m^2 by the total area.

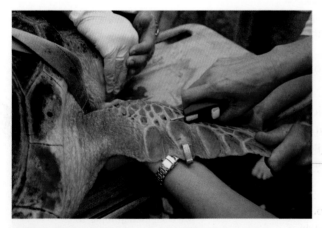

▲ **Figure 5** *An electronic tag is being fitted to this green sea turtle (Chelonia mydas). This is as part of a study to estimate population size*

Estimating animal population size

As animals are constantly moving through a habitat and others may be hidden, it can be difficult to accurately determine their population size. A technique known as capture-mark-release-recapture is often used to estimate a population size. This involves capturing as many individuals of a species in an area as possible. The organisms are marked and then released back into the community. Time is allowed for the organisms to redistribute themselves throughout the habitat before another sample of animals is collected. By comparing the number of marked individuals with the number of unmarked individuals in the second sample, scientists can estimate population size. The greater the number of marked individuals recaptured, the smaller the population.

The species evenness in an area can then be calculated by comparing the total number of each organism present. Populations of plants or animals that are similar in size or density represent an even community and hence a high species evenness. Species evenness can also be expressed as a ratio between the numbers of each organism present.

Measuring abiotic factors

Abiotic factors are the non-living conditions in a habitat. They have a direct effect on the living organisms that reside there. Examples are the amount of light and water available. To enable them to draw conclusions about the organisms present and the conditions they need for survival, scientists normally measure these conditions at every sampling point.

Table 1 summarises the ways in which common abiotic factors can be measured.

▼ **Table 1**

Abiotic factor	Sensor used	Example unit of measurement
wind speed	anemometer	$m\,s^{-1}$
light intensity	light meter	lx
relative humidity	humidity sensor	$mg\,dm^{-3}$
pH	pH probe	pH
temperature	temperature probe	°C
oxygen content in water	dissolved oxygen probe	$mg\,dm^{-3}$

Many abiotic factors can be measured quickly and accurately using a range of sensors, which are advantageous for a number of reasons:

● Rapid changes can be detected.
● Human error in taking a reading is reduced.
● A high degree of precision can often be achieved.
● Data can be stored and tracked on a computer.

 Belt transect in a National Park

A group of students was asked to investigate the impact of human land use in a National Park on the species of plants which were found there. The students decided to carry out a belt transect of open countryside, across a main walking path, in mid-Dartmoor.

Part of the data the students collected is shown. Data was collected on the percentage cover of six species, and the maximum height of vegetation at each position was noted.

Species present	Quadrat position along line, from starting point / m									
	0	1	2	3	4	5	6	7	8	9
Grasses	25	5	10	85	85	80	75	40	45	0
Heathers	10	15	0	0	0	0	0	10	10	35
Mosses	5	0	0	0	0	0	0	0	0	10
Gorse	20	75	90	10	0	0	0	30	0	25
Bracken	30	0	0	0	0	0	0	0	40	20
Bare ground	5	0	0	5	15	20	25	20	5	0
Maximum vegetation height / cm	70	65	45	10	10	5	5	30	70	65

The students produced the following graph of this data:

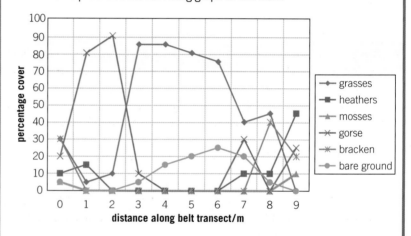

Summary questions

1 State which piece of equipment you would use to collect sample data on the following:
   a number of beetles on an oak tree trunk
   b number of moths in an area of woodland
   c pH of soil along a line transect
   d number of plant species present on a school playing field. (*4 marks*)

2 Describe the advantages of using a temperature probe over a thermometer. (*2 marks*)

3 Discuss the advantages and disadvantages of the different ways to measure species evenness. (*4 marks*)

Questions

1 Describe how you would use a transect to obtain the data which the students collected.

2 Explain why random sampling would not be appropriate to develop this data.

3 Plot a graph showing how the maximum vegetation height varied with distance along the belt transect.

4 Using the data gathered by the students, suggest how land use in this region affects the vegetation present.

5 A range of abiotic factors may also affect the species in the region studied. Suggest and explain how you would investigate the effect of one of these factors on the vegetation present.

6 Evaluate the quality of the data collected by the students.

11.4 Calculating biodiversity

Specification reference: 4.2.1

Ecologists, such as those working for the Environment Agency, often perform calculations using specific formulae to determine the biodiversity of an area. One such calculation is a measure of the species diversity. The diversity of the organisms present in an area is normally proportional to the stability of the ecosystem, so the greater the species diversity the greater the stability. The most stable communities have large numbers of fairly evenly distributed species, in good-sized populations.

Pollution often reduces biodiversity. As a result of harsh conditions, a few species tend to dominate. If corrective steps are taken to improve environmental conditions, biodiversity levels usually increase. Monitoring biodiversity is therefore a useful tool in successful conservation and environmental management.

How to calculate biodiversity

The simplest way to measure biodiversity is to count up the number of species present – the species richness. However, this measure does not take into account the number of individuals present. Therefore in a meadow containing two daisies and 1000 buttercups, the daisies have as much influence on the richness of the area as 1000 buttercups. A community dominated by one or two species is considered to be less diverse than one in which several different species have a similar abundance.

Simpson's Index of Diversity (D) is a better measure of biodiversity as it takes into account both species richness, and species evenness.

It is calculated using the formula:

$$D = 1 - \sum \left(\frac{n}{N}\right)^2$$

where:

Σ = sum of (total)

N = the total number of organisms of all species and

n = the total number of organisms of a particular species.

When using a technique such as Simpson's Index of Diversity, scientists normally have to estimate population size using a variety of sampling techniques, such as using a quadrat to estimate the population of a plant species in an area.

Simpson's Index of Diversity always results in a value between 0 and 1, where 0 represents no diversity and a value of 1 represents infinite diversity. The higher the value of Simpson's Index of Diversity, the more diverse the habitat.

Biodiversity values

What do low and high biodiversity values tell us about a habitat?

▼ **Table 1** *Typical habitat features for environments with low and high biodiversity*

Habitat features	Low biodiversity	High biodiversity
number of successful species	relatively few	a large number
nature of the environment	stressful and/or extreme with relatively few ecological niches	relatively benign/not stressful, with more ecological niches
adaptation of species to environment	relatively few species live in the habitat, often with very specific adaptations for the environment	many species live in the habitat, often with few specific adaptations to the environment
type of food webs	relatively simple	complex
effect of a change to the environment on ecosystem as a whole	major effects on the ecosystem	often relatively small effect

▲ **Figure 1** *A habitat with a low biodiversity value*

Although some habitats of low biodiversity are unable to support a large species diversity, those organisms that are present in the habitat can be highly adapted to the extreme environment of the habitat. These organisms may not survive elsewhere. It is therefore important to conserve some habitats with low biodiversity, as well as those with high biodiversity, in order to conserve rare species that may be too specialised to survive elsewhere.

  Worked example: Calculating Simpson's Index of Diversity

Individuals of the species living in a pond habitat were identified and counted (Table 2). They were sampled by sweeping a net through the pond.

▼ **Table 2** *Species identified in a pond*

Species	Number
water boatman	4
water strider	6
dragonfly larvae	3
mayfly larvae	8
caddisfly larvae	2

Calculate Simpson's Index of Diversity for this habitat.

Use the following formula:

$$D = 1 - \sum \left(\frac{n}{N} \right)^2$$

1 Calculate total number of organisms (N)

$N = 4 + 6 + 3 + 8 + 2 = 23$

2 Calculate $\left(\dfrac{n}{N}\right)^2$ for each organism. For example, for the water boatman:

n = number of organisms = 4

N = total number of organisms = 23

$\left(\dfrac{n}{N}\right)^2 = \left(\dfrac{4}{23}\right)^2 = 0.03$

3 Sum $\left(\dfrac{n}{N}\right)^2$ for all organisms

$\sum\left(\dfrac{n}{N}\right)^2 = 0.03 + 0.07 + 0.02 + 0.12 + 0.01 = 0.25$

4 Calculate Simpson's Index of Diversity

$D = 1 - \sum\left(\dfrac{n}{N}\right)^2 = 1 - 0.25 = 0.75$

This indicates that the pond habitat has a relatively high biodiversity.

▲ **Figure 2** *Ecologist studying biodiversity in a pond habitat*

▲ **Figure 3** *Pollution reduces biodiversity in a habitat. These trees have been killed by acid rain*

Summary questions

1 Describe the conditions likely to be found in an area of high biodiversity. *(3 marks)*

2 The organisms present in two pond habitats were sampled. The following values of Simpson's Index of Diversity were calculated:

Pond A: $D = 0.27$ Pond B: $D = 0.63$

 a Which pond habitat is the more biodiverse? *(1 mark)*

 b Which pond habitat is most likely to be polluted? Explain your answer. *(3 marks)*

3 A sample was taken of wildflowers growing in a meadow. Calculate Simpson's Index of Diversity for this habitat. *(4 marks)*

Species	Number
Bird's-foot trefoil	2
Crested dog's-tail	5
Meadow buttercup	9
Oxeye daisy	7
Rough hawkbit	2
Smaller cat's-tail	3

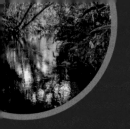

11.5 Calculating genetic biodiversity

Specification reference: 4.2.1

Maintaining genetic biodiversity is essential to the survival of a species. In isolated populations, such as those present within a captive breeding programme, genetic biodiversity is often reduced. This means that the individuals may suffer from a range of problems associated with in-breeding.

Scientists can calculate the genetic biodiversity of a population of a species (sometimes referred to as the gene pool) to monitor the health of the population and ensure its long-term survival.

The importance of genetic biodiversity

Within a species, *individuals* have very little variation within their DNA.

All members of the species share the same genes. However, they may have different versions of some of these genes. The different 'versions' of genes are called alleles. The differences in the alleles among individuals of a species creates genetic biodiversity within the species, or within a population of the species. The more alleles present in a population, the more genetically biodiverse the population.

Species that contain greater genetic biodiversity are likely to be able to adapt to changes in their environment, and hence are less likely to become extinct. This is because there are likely to be some organisms within the population that carry an advantageous allele, which enables them to survive in the altered conditions. For example, when a potentially fatal new disease is introduced to a population, all organisms will be killed unless individuals carry resistance to the disease. Those organisms are likely to survive the disease, and therefore be able to reproduce – leading to the survival of the species.

Factors that affect genetic biodiversity

For genetic biodiversity to increase, the number of possible alleles in a population must also increase. This can occur through:

- **mutation(s)** in the DNA of an organism, creating a new allele.
- interbreeding between different populations. When an individual migrates from one population and breeds with a member of another population, alleles are transferred between the two populations. This is known as **gene flow**.

In order for genetic biodiversity to decrease, the number of possible alleles in a population must also decrease. This can occur through:

- selective breeding (also known as artificial selection), where only a few individuals within a population are selected for their advantageous characteristics and bred. For example, the breeding of pedigree animals or of human food crops

- captive breeding programmes in zoos and conservation centres, where only a small number of *captive* individuals of a species are available for breeding. Often the wild population is endangered or extinct

- rare breeds, where selective breeding has been used historically to produce a breed of domestic animal or plant with characteristics which then become less popular or unfashionable, so the numbers of the breed fall catastrophically. When only a small number of individuals of a breed remain and are available for breeding, and all of these animals will have been selected for the specific breed traits, the genetic diversity of the remaining population will be low. This can cause serious problems when trying to restore numbers yet maintain breed characteristics, for example, a Gloucester Old Spot pig must have at least one spot on the body to be accepted into the registry of this rare breed

- artificial cloning (asexual reproduction), for example using cuttings to clone a farmed plant

- **natural selection.** As a result, species will evolve to contain primarily the alleles which code for advantageous characteristics. Over time, alleles coding for less advantageous characteristics will be lost from a population, or only remain in a few individuals.

- **genetic bottlenecks**, where few individuals within a population survive an event or change (e.g., disease, environmental change or habitat destruction), thus reducing the 'gene pool'. Only the alleles of the surviving members of the population are available to be passed on to offspring

- the **founder effect**, where a small number of individuals create a new colony, geographically isolated from the original. The gene pool for this new population is small.

- genetic drift, due to the random nature of alleles being passed on from parents to their offspring, the frequency of occurrence of an allele will vary. In some cases, the existence of a particular allele can disappear from a population altogether. Genetic drift is more pronounced in populations with a low genetic biodiversity.

▲ **Figure 1** *Gloucester Old Spot pig*

> ### Synoptic link
>
> An example of the founder effect is given in Topic 10.8, Changing population characteristics.

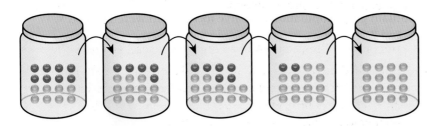

◀ **Figure 2** *The effect of genetic drift on genetic biodiversity can be shown using the 'marbles in a jar' analogy. At each generation, the random nature of alleles being passed on leads to a change in the frequency of the alleles present in the population. In this case, by the fifth generation the purple allele is removed from the population altogether*

Measuring genetic biodiversity

One way in which scientists quantify genetic biodiversity is by measuring polymorphism. Polymorphic genes have more than one

Summary questions

1 Describe how genetic biodiversity in a population can increase. *(2 marks)*

2 Explain why it is advantageous for a species to be genetically biodiverse. *(3 marks)*

3 A scientist was studying two species of *Drosophila* (flies). DNA was extracted from each species and 25 gene loci compared.
For species A, 12 of the loci studied were polymorphic.
For species B, 15 loci were polymorphic.

Use the data collected to explain which of the species was more genetically diverse. *(4 marks)*

allele. For example, different alleles exist for the immunoglobulin gene, which plays a role in determining human blood type – this is therefore defined as a polymorphic gene. The three alleles are:

- I^A – resulting in the production of antigen A
- I^B – resulting in the production of antigen B
- I^O – resulting in the production of neither antigen

Most genes are not polymorphic. These genes are said to be monomorphic – a single allele exists for this gene. This ensures that the basic structure of individuals within a species remains consistent. The proportion of genes that are polymorphic can be measured using the formula:

$$\text{proportion of polymorphic gene loci} = \frac{\text{number of polymorphic gene loci}}{\text{total number of loci}}$$

(The locus (plural – loci) of a gene refers to the position of the gene on a chromosome.)

The greater the proportion of polymorphic gene loci, the greater the genetic biodiversity within the population.

Worked example: Measuring genetic biodiversity

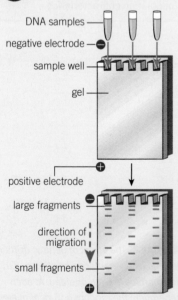

Gel electrophoresis is a technique used to separate fragments of DNA, based on their size. In this technique restriction enzymes are used to cut DNA into smaller pieces, which are then placed in a gel. The gel is placed between positive and negative electrodes, which cause the negatively charged DNA to move towards the positive side. The smaller the fragment of DNA, the further the movement through the gel. The pattern produced, known as a banding pattern, can be used to compare DNA samples from different individuals.
The following section of data was collected from the gel electrophoresis of five genetic loci within 20 individuals in an ibex (mountain goat) population. The five loci studied were labelled V, W, X, Y, and Z.

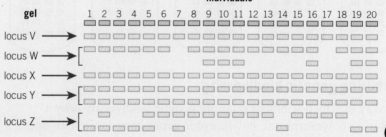

From this sample, the scientist could tell that:

- Loci W Y and Z were polymorphic
- Loci V and X were not polymorphic

Calculate the proportion of polymorphic gene loci for this mountain ibex population.

1   Count the number of polymorphic gene loci: 3 (W, Y and Z)

2   Count the total number of loci: 5 (V, W, X, Y and Z)

3   Calculate the proportion of polymorphic gene loci:

proportion of polymorphic gene loci

$$= \frac{\text{number of polymorphic gene loci}}{\text{total number of loci}}$$
$$= \frac{3}{5}$$
$$= 0.6$$

4   The proportion of polymorphic gene loci is often expressed as a percentage:

percentage of polymorphic gene loci

$$= \text{proportion of polymorphic gene loci} \times 100$$
$$= 0.6 \times 100$$
$$= 60\%$$

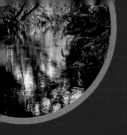

11.6 Factors affecting biodiversity

Specification reference: 4.2.1

Maintaining biodiversity is essential for preserving a balanced ecosystem for all organisms. As species are interconnected within an ecosystem, the removal of one species can have a profound effect on others. For example, it could lead to a loss of another species' food source or shelter.

As part of the human population you rely on biodiversity for many of the materials you need to survive, such as food, wood, and oxygen. However, humans are the leading cause of loss of biodiversity.

Human influence on biodiversity

The human population is growing at a dramatic rate. There are now over seven billion people living in the world, over double the number alive in the 1960s and over seven times more than in 1800. This increasing growth rate is linked to improvements in medicine, hygiene, housing, and infrastructure, which enable people to live for longer.

To create enough space for housing, industry, and farming to support the increasing population, humans are severely disrupting the ecology of many areas. The main problems are occurring as a result of:

● deforestation – the permanent removal of large areas of forest to provide wood for building and fuel (known as logging), and to create space for roads, building and agriculture.

● agriculture – an increasing amount of land has to be farmed in order to feed the growing population. This has resulted in large amounts of land being cleared and in many cases planted with a single crop (**monoculture**).

● climate change – there is much evidence that the release of carbon dioxide and other pollutants into the atmosphere from the burning of fossil fuels is increasing global temperatures.

Other forms of pollution result from industry and agriculture, such as the chemical pollution of waterways. The improper disposal of waste and packaging is a form of environmental pollution called littering.

▲ **Figure 1** *This photo shows a remnant of rainforest surrounded by farmland near Iguacu National Park in Brazil. Rainforests are being destroyed at a rapid pace. Almost 90% of West Africa's rainforest has been destroyed and the island of Madagascar has lost two thirds of its original rainforest since humans arrived 2000 years ago. Removal of tropical rainforests causes the greatest loss of global biodiversity – even though rainforests cover less than 10% of the Earth's surface, they contain approximately 80% of the world's documented species*

Deforestation

Deforestation can occur naturally, for example as a result of forest fires caused by lightning or extreme heat and dry weather. However, most deforestation now occurs deliberately as a result of human action. Some areas of forest have also been destroyed indirectly by humans through acid rain, which forms as a result of pollutants being released into the atmosphere.

Deforestation affects biodiversity in a number of ways. For example:

- It directly reduces the number of trees present in an area.

- If only a specific type of tree is felled, the species diversity is reduced. For example, rosewood is often extracted from rainforests (it is used in the manufacture of furniture and guitars), but less useable trees may be left intact.

- It reduces the number of animal species present in an area as it destroys their habitat, including their food source and home. This in turn reduces the number of other animal species that are present, by reducing or removing their food source.

- Animals are forced to migrate to other areas to ensure their survival. This may result in the biodiversity of neighbouring areas increasing.

In some areas forests are now being replaced. Although this helps to restore biodiversity, generally only a few commercially viable tree species are planted. Therefore, biodiversity is still significantly reduced from its original level.

▲ **Figure 2** *This land is being cleared to make space for housing. As well as reducing biodiversity, burning the trees increases carbon dioxide levels in the atmosphere*

Agriculture

In general, farmers will only grow a few different species of crop plants, or rear just a few species of animals. Farmers often select the species based on characteristics that give a high yield (high levels of production), for example, wheat that produces the most grain or dairy cows that produce the most milk. The selection of only a few species greatly reduces the biodiversity of the area.

In order to be economically viable, once the farmers have selected their desired species, a number of techniques are used to produce as many of the desired species as possible, maximising food production. Unfortunately many of these techniques lead to a reduction in biodiversity, for example:

- Deforestation – to increase the area of land available for growing crops or rearing animals.

- Removal of hedgerows – as a result of mechanisation, farmers remove hedgerows to enable them to use large machinery to help them plant, fertilise, and harvest crops. It also frees up extra land for crop growing. This reduces the number of plant species present in an area and destroys the habitat of animals such as blackbirds, hedgehogs, mice and many invertebrates.

- Use of chemicals such as pesticides and herbicides. Pesticides are used to kill pests that would eat the crops or live on the animals. This reduces species diversity directly as it destroys the pest species (normally insects), and indirectly by destroying the food source of other organisms.

- Herbicides are used to kill weeds. A weed is any plant growing in an area where it is not wanted. Weeds are destroyed as they compete with the cultivated plants for light, minerals, and

▲ **Figure 3** *This hedgerow supports a large diversity of species. As well as destroying the habitats of many organisms, removing hedges or trees causes soil erosion. Hedges act as natural windbreaks. Once they are removed, when the fields are left bare during winter the soil can be blown or washed away*

water. By destroying weeds, plant diversity is reduced directly, and animal diversity may also be reduced by the removal of an important food source.

- Monoculture – many farms specialise in the production of only one crop, with many acres of land being used for the growth of one species. This has an enormous local effect in lowering biodiversity as only one species of plant is present. As relatively few animal species will be supported by only one type of plant, this results in low overall biodiversity levels. The growth of vast oil palm plantations is one of the leading causes of rainforest deforestation, leading to a loss of habitat for critically endangered species like the rhino.

▲ **Figure 4** *This rice field covers a huge area of land. It is an example of monoculture – only rice is being grown*

Climate change

In 2007, the Intergovernmental Panel on Climate Change (IPCC) released a report summarising scientists' current understanding of climate change. The report took six years to produce and involved over 2500 scientific personnel in its production. Some of the key findings include the following:

- The warming trend over the last 50 years (about 0.13°C per decade) is nearly twice that for the previous 100 years.

- The average amount of water vapour in the atmosphere has increased since the 1980s over land and ocean. The increase is broadly consistent with the extra water vapour that warmer air can hold.

- Since 1961, the average temperature of the global ocean down to depths of 3 km has increased. The ocean has been absorbing more than 80% of the heat added to the climate system, causing seawater to expand and contributing to sea-level rise.

- The global average sea level rose by an average of 1.8 mm per year from 1961 to 2003. There is high confidence that the rate of observed sea level rise increased from the 19th to the 20th century.

- Average Arctic temperatures have increased at almost twice the global average rate in the past 100 years.

- Mountain glaciers and snow cover have declined on average in both hemispheres. Widespread decreases in glaciers and ice caps have contributed to sea-level rise.

- Long-term upward trends in the amount of precipitation have been observed over many regions from 1900 to 2005.

To enable our understanding of climate change to develop, significant quantities of data have been developed charting changes to the Earth's climate over time. This has required an enormous international co-operative effort over many years. It is only on the basis of reliable, irrefutable evidence that decisions of an international significance can take place. Decisions made now may have far-reaching consequences for the populations of individual countries or continents today, as well as far-reaching global implications for the future.

The need to produce reliable data for issues of this scale is paramount. Despite the weight of evidence for climate change, some scientists still believe that a causal link between human activity and climate change is yet to be established.

Global warming refers to a rise in the Earth's mean surface temperature. The Earth's climate has shown fluctuations in temperature throughout its history, so it is not possible to say for certain that humans are directly causing global warming. However, carbon dioxide levels in the atmosphere have significantly increased since the industrial revolution, trapping more thermal energy in the atmosphere. Therefore most scientists believe that human activities are contributing to global warming.

If global warming continues biodiversity will be affected. For example:

- The melting of the polar ice caps could lead to the extinction of the few plant and animal species living in these regions. Some species of animals present in the Arctic are migrating further and further north to find favourable conditions as their habitat shrinks. Increasing global temperatures would allow temperate plant and animal species to live further north than currently.

- Rising sea levels from melting ice caps and the thermal expansion of oceans could flood low-lying land, reducing the available terrestrial habitats. Saltwater would flow further up rivers, reducing the habitats of freshwater plants and animals living in the river and surrounding areas.

- Higher temperatures and less rainfall would result in some plant species failing to survive, leading to drought-resistant species (**xerophytes**, Figure 5) becoming more dominant. The loss of non-drought-resistant species of plants would lead to the loss of some animal species dependent on them as a food source. These would be replaced by other species that feed on the xerophytes.

- Insect life cycles and populations will change as they adapt to climate change. Insects are key pollinators of many plants, so if the range of an insect changes, it could affect the lives of the plants it leaves behind, causing extinction. And as insects carry many plant and animal pathogens, if tropical insects spread, this in turn could lead to the spread of tropical diseases towards the poles.

If climate change is slow, species may have time to adapt (for example by eating a different food source) or to migrate to new areas. This will lead to a loss of native species, but in turn other species may move into the area – so biodiversity would not necessarily be lost. The species mix would simply change.

▲ **Figure 5** *These Joshua trees (Yucca brevifolia) are an example of a xerophyte. A xerophyte is a species of plant that has adapted to survive in an environment with little water, such as a desert or an ice- or snow-covered region in the Alps or the Arctic*

Synoptic link

You learned about xerophyte adaptations in Topic 9.5, Plant adaptations to water availability.

▲ **Figure 6** *The small red-eyed damselfly (Erythromma viridulum). Changing species distribution provides evidence for climate change. Since 1980, 34 of the 37 British species of dragonfly and damselfly have expanded their range northwards by an average of 74 km. This is evidence that the UK's climate is growing warmer*

Loss of biodiversity in the UK

▲ **Figure 7** *Heathland*

▲ **Figure 8** *Woodland*

Scientists have estimated that the present worldwide rate of extinction is between 100 and 1000 times greater than at any other point in evolutionary history. This is primarily the result of the increase in the world human population. This has resulted in large areas of land being cleared worldwide, to meet the demand for food. Twelve to fifteen million hectares of forest are lost worldwide each year – the equivalent of 36 football fields per minute. These highly diverse habitats are replaced with agricultural land, which has far lower levels of biodiversity.

Conservation agencies have estimated the percentage of various habitats that have been lost in the UK since 1900. Their findings are summarised in Table 1.

1   State the key reason why the habitats stated in Table 1 have been lost over the past century.

2   Compare the proportion of chalk grassland that has been lost since 1900 with lowland mixed woodland.

3   There are approximately 1500 hectares of hay meadow in the UK. What was the equivalent figure in 1900?

4   Farmers can receive financial subsidies for farming areas of land in a traditional, sustainable manner. State and explain how this may affect species diversity in these areas.

▼ **Table 1**

Habitat	Habitat loss since 1900 (%)	Main reason for habitat loss
Hay meadow	95	Conversion to highly productive grass and silage
Chalk grassland	80	Conversion to highly productive grass and silage
Lowland fens and wetlands	50	Drainage and reclamation of land for agriculture
Limestone pavements in England	45	Removal for sale as rockery stone
Lowland heaths on acid soils	40	Conversion to grasslands and commercial forests
Lowland mixed woodland	40	Conversion to commercial conifer plantations and farmland
Hedgerows	30	To make larger fields to accommodate farm machinery

Summary questions

1   State and explain how the following factors reduce species diversity:
   a   monoculture   (*2 marks*)
   b   building of roads   (*2 marks*)
   c   use of pesticides.   (*2 marks*)

2   Explain why there is a reduction in species diversity when an area of forest is cleared to create additional land for the grazing of cattle.   (*2 marks*)

3   Suggest and explain ways in which climate change can affect the biodiversity in an area.   (*3 marks*)

11.7 Reasons for maintaining biodiversity

Specification reference: 4.2.1

Can you imagine what it would be like to live in a world where all the landscapes looked the same? Not only would it be visually (aesthetically) unappealing, but biodiversity levels would be low as only a small number of organisms would be supported by the habitats.

Reasons for maintaining biodiversity

It is important to maintain biodiversity for a number of reasons. These reasons can be broadly arranged into three groups – aesthetic, economic and ecological, although some reasons fall across more than one category.

Aesthetic reasons

- The presence of different plants and animals in our environment enriches our lives. For example, you might like to relax on a beach, walk in your local woodland or park or visit a rainforest.

- The natural world provides inspiration for people such as musicians and writers, who in turn provide pleasure for many others through music and books.

- Studies have shown that patients recover more rapidly from stress and injury when they are supported by plants and a relatively natural environment.

Economic reasons

If biodiversity in an ecosystem is maintained, levels of long-term productivity are higher.

- Soil erosion and desertification may occur as a result of deforestation. These reduce a country's ability to grow crops and feed its people, which can lead to resource- and economic-dependence on other nations.

- It is important to conserve all organisms that we use to make things. Non-sustainable removal of resources, such as hardwood timber, will eventually lead to the collapse of industry in an area. Once all or enough of the raw material has been lost, it does not become economically viable to continue the industry. Note that even when 'sustainable' methods are used – for example replanting forest areas – the new areas will not be as biodiverse as the established habitats they replace.

- Large-scale habitat and biodiversity losses mean that species with potential economic importance may become extinct before they are even discovered. For example, undiscovered species in tropical rainforests may be chemically or medically useful. A number of marine species use a chemical-based defence mechanism. These are rich potential sources of new and economically important medicines.

Learning outcomes

Demonstrate knowledge, understanding, and application of:

→ the ecological, economic and aesthetic reasons for maintaining biodiversity.

▲ **Figure 1** *Most people would agree that natural, biodiverse areas are more attractive than acres of cultivated crops and concrete. Visiting these areas provides space for relaxation and exercise, both essential for a healthy life. Protecting these landscapes is therefore essential for human well-being*

- Continuous monoculture results in soil depletion – a reduction in the diversity of soil nutrients. It happens because the crop takes the same nutrients out of the soil year after year and is then harvested, not left for the nutrients to be recycled. This depletion of soil nutrients makes the ecosystem more fragile. The crops it can support will be weaker, increasing vulnerability to opportunistic insects, plant competitors, and microorganisms. The farmer will become increasingly dependent on expensive pesticides, herbicides, and fertilisers in order to maintain productivity.

- High biodiversity provides protection against abiotic stresses (including extreme weather and natural disasters) and disease. When biodiversity is not maintained, a change in conditions or a disease can destroy entire crops. The Irish potato famine of the 1840s was a direct consequence of the reliance on only two varieties of potato. When a new disease spread to the area (the oomycete *Phytophthora infestans*), neither species contained alleles for genetic resistance, so the entire crop was destroyed. This led to widespread famine and the deaths of around 1 million people.

- Areas rich in biodiversity provide a pleasing, attractive environment that people can enjoy. Highly biodiverse areas can promote tourism in the region, with its associated economic advantages.

- The greater the diversity in an ecosystem, the greater the potential for the manufacture of different products in the future. These products may be beneficial to humans. For example, it may make food production more financially viable or provide cures or treatment for disease.

- Plant varieties are needed for cross breeding, which can lead to better characteristics such as disease resistance or increased yield. The wild relatives of cultivated crop plants provide an invaluable reservoir of genetic material to aid the production of new varieties of crops. Also, through genetic engineering, scientists aim to use genes from wild plants and animals to make crop plants and animals more efficient, thus reducing the land required to feed more people. If these wild varieties are lost, the crop plants may themselves also become more vulnerable to extinction. This is also important ecologically.

Ecological reasons

- All organisms are interdependent on others for their survival. The removal of one species may have a significant effect on others, for example a food source or a place to live may be lost. For example, decomposers break down dead plant and animal remains, releasing nutrients into the soil, which plants later use for healthy growth. Plants rely on bees for pollination – this is important for both wild plant species and commercially produced crops. Fruit farmers use bees to pollinate their crops; a decrease in the wild bee population would decrease crop yields.

- Some species play a key role in maintaining the structure of an ecological community. These are known as **keystone species**. They have a disproportionately large effect on their environment relative to their abundance (in terms of their biomass or productivity). They affect many other organisms in an ecosystem and help to determine the species richness and evenness in the community. When a keystone species is removed the habitat is drastically changed. All other species are affected and some may disappear altogether. It is therefore essential to protect keystone species to maintain biodiversity. (See the Application for examples of keystone species.)

Human activity versus biodiversity

We have discussed the negative impact humans have on biodiversity, such as deforestation and clearing land for monoculture. However, human activity also plays an important role in increasing biodiversity. In many countries, including the UK, the natural habitat is created by human intervention and the management of land. For example, farming, grazing, planting of hedges, meadows, and forest management have changed the landscapes, the habitats and the ecology over thousands of years. Even the wildest of habitats, such as Dartmoor and the Scottish mountains, are a result of farmers and landowners managing the ecosystems.

One example is sheep grazing on downlands. This enables rare species like the Glanville fritillary (an orange patterned butterfly) to survive. By maintaining the grass at low levels it allows the plantains that the caterpillars feed on to thrive and therefore maintains biodiversity. Research has also shown that after annual controlled burning of gorse and heather in the New Forest (an area of lowland heath), biodiversity soars. If left to its own devices, bracken and pioneer tree species such as pine and silver birch would start to dominate. Areas of lowland heath worldwide are now rarer than rainforest and provide habitats for rare UK bird and reptile species such as the nightjar and sand lizard.

Keystone species

Sea stars, American alligators and prairie dogs are all examples of key stone species:

- Like many keystone species, sea stars are predators. They maintain a balanced ecosystem by limiting the population of other species. Sea stars eat mussels and sea urchins, which have no other natural predators. If the sea star is removed from the ecosystem, the mussels undergo a population explosion, reducing the number of other species present in an area (such as barnacles and limpets) as they compete for space and other resources. Similarly, if sea urchins are not eaten, their growing population crowds coral reefs, preventing other species from occupying the same area.

▲ **Figure 2** *Sea stars (*Pisaster ochraceus*) feeding on mussels. Sea stars are keystone species in the intertidal zones of the Pacific ocean*

- Alligators make burrows for nesting and to stay warm. When they abandon their burrow, fresh water fills the space, which is used by other species during the dry season for breeding and drinking. Alligators are predators, which also contributes to the maintenance of biodiversity in these habitats.

- It is estimated that up to 200 species rely on prairie-dog colonies, primarily due to their tunnelling activities. Prairie colonies provide a food source and burrows for other animals such as snakes. Their tunnelling aerates the soil, which, combined with their droppings, leads to a redistribution of nutrients. It also channels rainwater into the water table. These processes help to maintain a biodiverse range of plant life in the region. So essential is the prairie dog to its habitat that its loss would lead to a change in the ecosystem itself.

▲ **Figure 3** *American alligators (Aligator mississippiensis) are keystone species in the Everglade wetlands*

▲ **Figure 4** *Prairie dogs (Cynomys spp.) are a keystone species for the prairies – their existence adds to a diversity of life*

1   Define the term 'keystone species'.
2   Explain why keystone species are often predators.
3   The purple coneflower is a plant species found on the North American prairies. Extracts from the plant have anti-bacterial properties, which have been used for centuries to treat fevers and infections. Explain how a reduction in the population of prairie dogs could affect the number of purple coneflowers in this habitat.

Summary questions

1   State the differences between aesthetic, economic, and ecological arguments for maintaining biodiversity.   (*1 mark*)

2   ⚙ Suggest two ethical reasons why we should maintain biodiversity.   (*3 marks*)

3   The Irish potato famine of the 1840s had a devastating effect on the population.
   a   Explain how a lack of agricultural biodiversity led to this disaster.   (*2 marks*)
   b   Suggest and explain how a similar famine could be prevented from occurring in the future.   (*2 marks*)

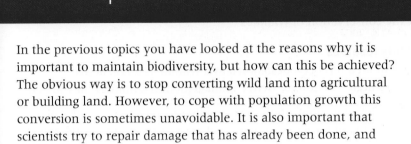

In the previous topics you have looked at the reasons why it is important to maintain biodiversity, but how can this be achieved? The obvious way is to stop converting wild land into agricultural or building land. However, to cope with population growth this conversion is sometimes unavoidable. It is also important that scientists try to repair damage that has already been done, and increase biodiversity.

Maintaining biodiversity

Conservation is the name given to the preservation and careful management of the environment and of natural resources. By conserving the natural habitat in an area, organisms' chances of survival are maintained, allowing them to reproduce. As a consequence species and genetic diversity can be safeguarded.

There are many different ways in which scientists try to conserve biodiversity. They can be divided into two main categories:

- *in situ* **conservation** – within the natural habitat
- *ex situ* **conservation** – out of the natural habitat.

Scientists are currently trying to conserve a number of species to prevent their extinction. Species are classified, for the purposes of conservation, according to their abundance in the wild:

- extinct – no organisms of the species exist anywhere in the world
- extinct in the wild – organisms of the species only exist in captivity
- endangered – a species that is in danger of extinction
- vulnerable – a species that is considered likely to become endangered in the near future.

Non-threatened and categories of least concern follow below. Many conservation techniques focus on increasing the numbers of organisms from species that are classified as endangered.

Scientists also promote the practice of **sustainable development** – economic development that meets the needs of people today, without limiting the ability of future generations to meet their needs.

In situ conservation

In situ conservation takes place inside an organism's natural habitat. This maintains not only the genetic diversity of species, but also the evolutionary adaptations that enable a species to adapt continually to changing environmental conditions, such as changes in pest populations or climate. By allowing the endangered species to interact with other species, it also preserves the interdependent relationships present in ·

Learning outcomes

Demonstrate knowledge, understanding, and application of:

→ *in situ* and *ex situ* methods of maintaining biodiversity

→ international and local conservation agreements made to protect species and habitats.

▲ **Figure 1** *The giant panda (*Ailuropoda melanoleuca*) is an example of an endangered species. Its numbers have been severely reduced by loss of habitat and poaching. Current estimates suggest there are less than 2000 giant pandas living in the wild*

▲ **Figure 2** *This is Wistman's Wood on Dartmoor. It is a SSSI – a Site of Special Scientific Interest. The UK has over 4000 conservation areas where habitats are protected, covering around 8% of the nation's land*

▲ **Figure 3** *This white rhinoceros is having its horn removed in Umhlametsi Private Nature Reserve, South Africa as an anti-poaching measure. The horn is highly desirable for its use in ornaments and some traditional medicines*

▲ **Figure 4** *In Britain, the rhododendron has taken over large areas, virtually eliminating some native plants. Invasive plants and animals are the second greatest threat to biodiversity after habitat loss*

a habitat, therefore interlinked species may also be preserved. *In situ* conservation is generally cheaper than *ex situ* conservation.

Marine (saltwater), aquatic (freshwater) and terrestrial (land) nature reserves are examples of areas that have been specifically designated for the conservation of wildlife.

Wildlife reserves

Once an area has been designated as a wildlife reserve, active management is required. Active management techniques may include:

- controlled grazing – only allowing livestock to graze a particular area of land for a certain period of time to allow species time to recover, or keeping a controlled number of animals in a habitat to maintain it (see below)

- restricting human access – for example, not allowing people to visit a beach during the seal reproductive season, or by providing paths which must be followed to prevent plants being trampled

- controlling poaching – this includes creating defences to prevent access, issuing fines, or more drastic steps such as the removal of rhino horns

- feeding animals – this technique can help to ensure more organisms survive to reproductive age

- reintroduction of species – adding species to areas that have become locally extinct, or whose numbers have decreased significantly

- culling or removal of invasive species – an invasive species is an organism that is not native to an area and has negative effects on the economy, environment, or health. These organisms compete with native species for resources

- halting succession – **succession** is a natural process in which early colonising species are replaced over time until a stable mature population is achieved. For example, as a result of natural succession any piece of land left alone for long enough in the UK will develop into woodland. The only way to protect some habitats such as heath-, down- or moorland from becoming woodland is through controlled grazing. In different parts of the country ponies, deer, sheep, and cows eat tree seedlings as they appear, preventing succession from heathland to scrubland to woodland. This is an important role played by humans in maintaining some of our most beautiful habitats for future generations.

Marine conservation zones

Marine conservation zones are less well established than terrestrial ones. Lundy Island is currently the only statutory marine reserve in England, but there are many other protected areas.

Marine reserves are vital in preserving species-rich areas such as coral reefs, which are being devastated by non-sustainable fishing methods. The purpose of the marine reserve is not to prevent fisherman from visiting the entire area, but to create areas of refuge within which populations can build up and repopulate adjacent areas. Large areas of sea are required for marine reserves as the target species often move large distances, or breed in geographically different areas.

Ex situ conservation

Ex situ conservation involves the removal of organisms from their natural habitat. It is normally used in addition to *in situ* measures, ensuring the survival of a species.

Botanic gardens

Plant species can be grown successfully in botanic gardens. Here the species are actively managed to provide them with the best resources to grow, such as the provision of soil nutrients, sufficient watering, and the removal or prevention of pests.

There are roughly 1500 botanic gardens worldwide, holding 35 000 plant species. Although this is a significant number (more than 10% of the world's flora), the majority of species are not conserved. Many wild relatives of selectively bred crop species are under-represented amongst the conserved species. These wild species are a potential source of genes, conferring resistance to diseases, pests, and parasites.

Seed banks

A **seed bank** is an example of a gene bank – a store of genetic material. Seeds are carefully stored so that new plants may be grown in the future. They are dried and stored at temperatures of $-20\,°C$ to maintain their viability, by slowing down the rate at which they lose their ability to germinate. Almost all temperate seeds, and many tropical seeds, can be stored in this way. Scientists expect that they will remain viable for centuries, providing a back-up against the extinction of wild plants. The Svalbard 'Doomsday Vault' in Norway stores seeds in the permafrost and already houses around 800 000 species. It will eventually have 3 million different types of seeds and aims to provide a back-up against the extinction of plants in the wild by storing seeds for future reintroduction and research, for breeding and for genetic engineering in the future.

Seed banks don't work for all plants. Some seeds die when dried and frozen, and sadly the seeds of most tropical rainforest trees fall into this category.

Captive breeding programmes

Captive breeding programmes produce offspring of species in a human-controlled environment. These are often run and managed by zoos and aquatic centres. For example, The National Marine Aquarium in South West England is playing an important role in the conservation of sea horse species. Several species are now solely represented by animals in captivity.

Scientists working on captive breeding programmes aim to create a stable, healthy population of a species, and then gradually reintroduce the species back into its natural habitat. The Arabian Oryx is an example of a species that was extinct in the wild before its reintroduction.

▲ **Figure 5** *The Millennium Seed Bank Project at Kew Gardens contains over a billion seeds from over 34 000 species in underground frozen vaults. It is the world's largest collection of seeds and aims to provide a back-up against the extinction of plants in the wild by storing seeds for future use*

▲ **Figure 6** *These critically endangered Western lowland gorillas (*<u>Gorilla</u> <u>gorilla</u> <u>gorilla</u>*) are part of a New York Zoo captive breeding programme. Although far more invertebrates than vertebrates face extinction, most captive breeding programmes focus on vertebrates as people find it easier to relate to, and have sympathy with vertebrates. This generates financial support for their conservation and extends public education to wider issues*

Captive breeding programmes provide the animals with shelter, an abundant supply of nutritious food, an absence of predators and veterinary treatment. Suitable breeding partners or semen (which can be used to artificially inseminate females) can be imported from other zoos if not available within the zoo's own population.

Maintaining genetic diversity within a captive breeding population can be difficult. As only a small number of breeding partners are available, problems related to inbreeding can occur. To overcome this, an international catalogue is maintained, detailing genealogical data on individuals. Mating can thus be arranged to ensure that genetic diversity is maximised. Techniques such as artificial insemination, embryo transfer and long-term cryogenic storage of embryos allow new genetic lines to be introduced without having to transport the adults to new locations, and do not require the animals' cooperation.

Some organisms born in captivity may not be suitable for release in the wild. These are some of the reasons:

- Diseases – there may be a loss of resistance to local diseases in captive-bred populations. Also, new diseases might exist in the wild, to which captive animals have yet to develop resistance.

- Behaviour – some behaviour is innate, but much has to be learned through copying or experience. In an early case of reintroduction, a number of monkeys starved because they had no concept of having to search for food – they had become domesticated. Now food is hidden in cages, rather than just supplied, so that the animals learn to look for it.

- Genetic races – the genetic make-up of captive animals can become so different from the original population that the two populations cannot interbreed.

- Habitat – in many cases the natural habitat must first be restored to allow captive populations to be reintroduced. If only a small suitable habitat exists it is likely that there are already as many individuals as the habitat can support. The introduction of new individuals can lead to stress and tension as individuals fight for limited territory and resources such as food.

Conservation agreements

To conserve biodiversity successfully, local, and international cooperation is required to ensure habitats and individual species are preserved. Animals do not respect a country's boundaries. Therefore, to increase the chances of a species' survival, cross-border protections should be offered.

International Union for the Conservation of Nature

Intergovernmental organisations, such as the International Union for the Conservation of Nature (IUCN), assist in securing agreements between nations. At least once a year the IUCN publishes the Red List, detailing the current conservation status of threatened animals. Countries can then work together to conserve these species.

The IUCN was also involved in the establishment of the Convention on International Trade in Endangered Species (CITES). This treaty regulates the international trade of wild plant and animal specimens and their products. As the trade in wild animals and plants crosses borders between countries, the effort to regulate it requires international cooperation to safeguard certain species from over-exploitation. Today, more than 35 000 species of animals and plants are protected by this treaty.

The Rio Convention

In 1992, an historic meeting of 172 nations was held in Rio de Janeiro, which became known as the Earth Summit. The summit resulted in some new agreements between nations in the Rio Convention:

- The Convention on Biological Diversity (CBD) requires countries to develop national strategies for sustainable development, thus ensuring the maintenance of biodiversity.

- The United Nations Framework Convention on Climate Change (UNFCCC) is an agreement between nations to take steps to stabilise greenhouse gas concentrations within the atmosphere.

- The United Nations Convention to Combat Desertification (UNCCD) aims to prevent the transformation of fertile land into desert and reduce the effects of drought through programmes of international cooperation.

Each convention contributes to maintaining biodiversity. They are intrinsically linked, operating in many ecosystems and addressing interdependent issues.

Countryside stewardship scheme

Many conservation schemes are set up at a more local level. An example is the Countryside Stewardship Scheme in England. The scheme, which operated from 1991–2014, offered governmental payments to farmers and other land managers to enhance and conserve the English landscape. Its general aim was to make conservation a part of normal farming and land management practice. Specific aims of the scheme included:

- sustaining the beauty and diversity of the landscape
- improving, extending and creating wildlife habitats
- restoring neglected land and conserving archaeological and historic features
- improving opportunities for countryside enjoyment.

This scheme has now been replaced by the Environmental Stewardship Scheme, which operates similarly.

▲ **Figure 7** *A number of non-governmental organisations exist whose aim is to promote conservation and sustainability. One example is the World Wide Fund for Nature (WWF)*

Summary questions

1   Describe three methods of *ex situ* conservation.   (*3 marks*)

2   Discuss the advantages and disadvantages of captive breeding programmes.
    (*4 marks*)

3   State and explain four techniques used in the active management of wildlife reserves.   (*4 marks*)

4   Explain why local and international agreements can help to preserve biodiversity.
    (*4 marks*)

Practice questions

1 Which of the following can be used to measure the biodiversity of an area?

   A Student's *t*-test

   B Simpson's Index

   C Correlation coefficient

   D Standard deviation   (*1 mark*)

2 Which of the following statements are true with respect to what you would expect to find in an area of high biodiversity?

   Statement 1: A large number of successful species

   Statement 2: Complex food chains

   Statement 3: An extreme environment

   A 1, 2 and 3 are correct

   B Only 1 and 2 are correct

   C Only 2 and 3 are correct

   D Only 1 is correct   (*1 mark*)

3 The table below contains a number of terms ecologists use when studying biodiversity. Complete the table using the appropriate term or description.   (*5 marks*)

Term	Description
a.	The variety of living organisms present in an area.
Genetic biodiversity	b.
c.	A sample produced without bias; each individual has an equal likelihood of selection.
Abiotic factor	d.
e.	A sampling technique where different areas within a habitat are identified and sampled separately; for example, by using a line transect.

4 When studying biodiversity, scientists often take samples of a habitat.

   a (i) State what is meant by the term sampling.   (*1 mark*)

      (ii) State two reasons why sampling is carried out.   (*1 mark*)

   b Describe the difference between random and non-random sampling, stating one advantage of each technique.   (*3 marks*)

   c Explain how you could sample an area of grassland, to study the organisms present.   (*6 marks*)

5 a The black poplar tree was once a common tree throughout southern Britain. Its numbers have decreased by 94% since 1942 and it is in danger of becoming extinct in the wild. There are thought to be approximately 2500 black poplars surviving in Britain today. Use the information above to calculate the original number of black poplar trees in 1942. Show your working.   (*2 marks*)

   b Species such as the black poplar contribute to biodiversity in the UK.

      Suggest three reasons why the conservation of the black poplar is important.   (*3 marks*)

   c Botanic gardens are important in the conservation of plant species.

      (i)   State why the conservation of a species in a botanic garden is described as *ex situ*.   (*1 mark*)

      (ii) Many botanic gardens use seed banks as a method of plant conservation.

            Outline the advantages of using a seed bank, as opposed to adult plants, in order to conserve an endangered plant species.   (*4 marks*)

      (iii) Suggest why it is important to ensure that, for each species, the seeds in a seed bank have been collected from several different sites in the wild.
            (*3 marks*) OCR F212 2012

6 On a biology field trip, a pair of students collected some data about plant species in an area of ash woodland. Their results are shown in the table.

species	Number of individuals (*n*)	*n*/*N*	(*n*/*N*)^2
Dog's mercury	40		
Wild strawberry	13	0.13	0.0169
Common avens	43		
Wood sorrel	4		
	N =		
			$\sum (n/N)^2 =$
			$1 - [\sum (n/N)^2] =$

   a (i) Use the information in the table to work out Simpson's index of diversity (*D*) for the area of woodland sampled using the formula: $D = 1 - (\sum (n/N)^2)$
            Copy and complete the table.

(ii) Simpson's index of diversity takes into account both species richness and species evenness. In a school exercise book a student wrote the following definitions:

Species richness is a measure of the amount of species in an area

Species evenness shows how many individuals there are of a species in an area.

The teacher did not award a mark for either of these statements. Suggest how each statement could be improved. (*2 marks*)

(iii) if the value for Simpson's Index of Diversity is high, this indicates that the biodiversity of the habitat is high.

Outline the **implications** for a habitat if the Simpson's Index of Diversity is **low**. (*2 marks*)

b when collecting data on the field trip, the students placed quadrats in 15 locations and calculated a mean number of plants for each species.

Suggest two **other** steps they could have taken to ensure that their value for Simpson's Index of Diversity was as accurate as possible. (*2 marks*)

OCR F212/01 2013

7 Scientists have identified approximately 1.8 million different species. The number of species that actually exist is likely to be significantly higher than 1.8 million.

a Suggest two reasons why the number of species identified is likely to be lower than the actual number of species present on Earth. (*2 marks*)

b Many organisations, such as the International Union for the Conservation of Nature (IUCN), gather annual data about the number of species that are known to exist and to what extent they are considered to be endangered.

Figure 1 shows the total number of species assessed by the IUCN over a 10 year period and the number of those species assessed that are considered to

be threatened with extinction.

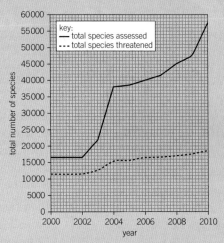

(i) Using the graph, **compare** the changes in the total number of species assessed with the changes in the total number of threatened species over the 10 year period. (*3 marks*)

(ii) Using the graph, calculate the percentage of species assessed that were threatened with extinction in **2010**. Show your working. Give your answer to the **nearest whole number**. (*2 marks*)

(iii) Suggest explanations for the shape of the two curves between **2005** and **2010**. (*2 marks*)

c A study of the biodiversity of an area considers not only the total number of species but also the relative number of individuals within each species.

State **one** relative factor that could be taken into account when describing the biodiversity of an area. (*1 mark*)

d In any attempt to protect global diversity, cooperation between countries is important.

Two examples of such international cooperation are:

• Convention on International trade in Endangered Species (CITES)

• Rio Convention on Biological Diversity.

Other than the conservation of biodiversity, state **two** aims for each of these conventions. (*4 marks*)

OCR F212/01 2013

Communicable diseases are caused by infective organisms known as **pathogens**. Pathogens include bacteria, viruses, fungi, and protoctista. Each has particular characteristics that affect the way they are spread and the ways we can attempt to prevent or cure the diseases they cause.

A communicable disease can be passed from one organism to another. In animals they are most commonly spread from one individual of a species to another, but they can also be spread between species. Communicable diseases in plants are spread directly from plant to plant. **Vectors**, which carry pathogens from one organism to another, are involved in the spread of a number of important plant and animal diseases. Common vectors include water and insects.

Globally, around 13 million people a year die as a result of communicable diseases. That is 23% of all deaths – non-communicable diseases cause around 68% of deaths, and injuries cause the rest. Communicable diseases are also a major problem in domestic and wild animals, and in the plants on which life on Earth depends.

Types of pathogens
Bacteria

There are probably more bacteria than any other type of organism. A small proportion of these bacteria are pathogens, causing communicable diseases.

Bacteria are prokaryotes, so they have a cell structure that is very different from the eukaryotic organisms they infect. They do not have a membrane-bound nucleus or organelles.

Bacteria can be classified in two main ways:

- By their basic shapes – they may be rod shaped (bacilli), spherical (cocci), comma shaped (vibrios), spiralled (spirilla), and corkscrew (spirochaetes).

bacillus (rod)   chain of bacilli (known as streptobacilli)   coccus (spherical)   pair of cocci   chain of cocci (streptococci)   cluster of cocci (staphylococci)   vibrio (comma)   spirillum (spiral)   spirochaete (corkscrew)

▲ **Figure 1** *The main types of bacteria*

- By their cell walls – the two main types of bacterial cell walls have different structures and react differently with a process called Gram staining. Following staining **Gram positive bacteria** look purple-blue under the light microscope, for example methicillin-resistant *Staphylococcus aureus* (MRSA). **Gram negative bacteria** appear red, for example the gut bacteria *Escherichia coli* (*E.coli*). This is useful because the type of cell wall affects how bacteria react to different **antibiotics** (a compound that kills or inhibits the growth of bacteria).

Synoptic link

You learnt about the structure of proteins and nucleic acids in Topic 3.1 Biological elements.

Viruses

Viruses are non-living infectious agents. At 0.02–0.3 µm in diameter, they are around 50 times smaller in length than the average bacterium. The basic structure of a virus is some genetic material (DNA or RNA) surrounded by protein. Viruses invade living cells, where the genetic material of the virus takes over the biochemistry of the host cell to make more viruses. Viruses reproduce rapidly and evolve by developing adaptations to their host, which makes them very successful pathogens. All naturally occurring viruses are pathogenic. They cause disease in every other type of organism. There are even viruses that attack bacteria, known as **bacteriophages**. They take over the bacterial cells and use them to replicate, destroying the bacteria at the same time. People now use bacteriophages both to identify and treat some diseases, and they are very important in scientific research. Medical scientists consider viruses to be the ultimate **parasites**.

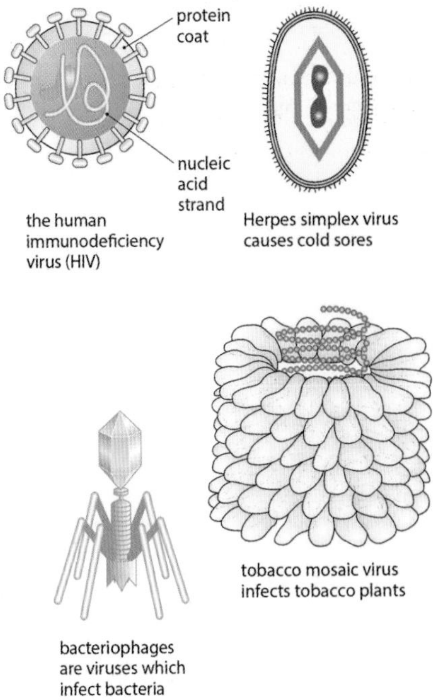

Protoctista (protista)

The protoctista (now widely known as protista) are a group of eukaryotic organisms with a wide variety of feeding methods. They include single-celled organisms and cells grouped into colonies. A small percentage of protoctista act as pathogens, causing devastating communicable diseases in both animals and plants. The protists which cause disease are parasitic – they use people or animals as their host organism. Pathogenic protists may need a vector to transfer them to their hosts – malaria and sleeping sickness are examples – or they may enter the body directly through polluted water – amoebic dysentery and *Giardia* are examples of these.

▲ **Figure 2** *The structure of viruses*

protein coat

nucleic acid strand

the human immunodeficiency virus (HIV)

Herpes simplex virus causes cold sores

tobacco mosaic virus infects tobacco plants

bacteriophages are viruses which infect bacteria

Fungi

Fungal diseases are not a major problem in animals, but they can cause devastation in plants. Fungi are eukaryotic organisms that are often multicellular, although the yeasts which cause human diseases such as thrush are single-celled. Fungi cannot photosynthesise and they digest their food extracellularly before absorbing the nutrients. Many fungi are saprophytes which means they feed on dead and decaying matter. However some fungi are parasitic, feeding on living plants and animals. These are the pathogenic fungi which cause communicable diseases. Because fungal infections often affect the leaves of plants, they stop them photosynthesising and so can quickly kill the plant. When fungi reproduce they produce millions of tiny spores which can spread huge distances, this adaptation

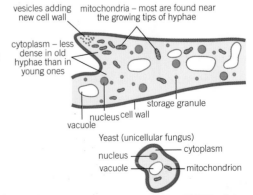

Fungal hyphae (filamentous fungus)

vesicles adding new cell wall

mitochondria – most are found near the growing tips of hyphae

cytoplasm – less dense in old hyphae than in young ones

storage granule

nucleus cell wall

vacuole

Yeast (unicellular fungus)

nucleus

cytoplasm

vacuole

mitochondrion

▲ **Figure 3** *The structure of fungi*

means they can spread rapidly and widely through crop plants. Fungal diseases of plants cause hardship and even starvation in many countries around the world.

Pathogens – modes of action

Damaging the host tissues directly

Many types of pathogen damage the tissues of their host organism. It is this damage, combined with the way in which the body of the host responds to the damage, that causes the symptoms of disease. Different types of pathogens attack and damage the host tissues in different ways:

- Viruses take over the cell metabolism. The viral genetic material gets into the host cell and is inserted into the host DNA. The virus then uses the host cell to make new viruses which then burst out of the cell, destroying it and then spread to infect other cells

- Some protoctista also take over cells and break them open as the new generation emerge, but they do not take over the genetic material of the cell. They simply digest and use the cell contents as they reproduce. Proctists which cause malaria are an example of this.

- Fungi digest living cells and destroy them. This combined with the response of the body to the damage caused by the fungus gives the symptoms of disease.

Producing toxins which damage host tissues:

- Most bacteria produce toxins that poison or damage the host cells in some way, causing disease. Some bacterial toxins damage the host cells by breaking down the cell membranes, some damage or inactivate enzymes and some interfere with the host cell genetic material so the cells cannot divide. These toxins are a by-product of the normal functioning of the bacteria

- Some fungi produce toxins which affect the host cells and cause disease.

1. attachment of virus to host cell

2. insertion of viral nucleic acid

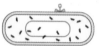

3. replication of viral nucleic acid

4. synthesis of viral protein

5. assembly of virus particles

6. lysis of host cell

▲ **Figure 4** *This virus is a bacteriophage and it destroys the bacterial cell as the new viruses burst out. Viruses have the same effect in multicellular eukaryotic organisms such as animals and plants*

Summary questions

1. **a** Explain what is meant by the term 'communicable disease'. *(2 marks)*
   b Produce a pie chart to show the main causes of death worldwide each year. *(4 marks)*

2. Make a table to compare bacteria, viruses, protoctista, and fungi. *(4 marks)*

3. If approximately 13 million people die of communicable diseases each year, give an approximate figure for the numbers who die of non-communicable disease and explain how you arrived at your answer. *(4 marks)*

4. Give four ways in which pathogens can attack the cells of their host organism and cause disease. *(6 marks)*

5. **a** Explain the difference between the way viruses and protists cause disease. *(4 marks)*
   b Suggest why viruses are described as the ultimate parasite. *(2 marks)*

Plant diseases

Plant diseases threaten people, because when crop plants fail, people suffer. They may starve, economies may struggle and jobs are lost. Plant diseases threaten ecosystems too – entire species can be threatened.

Learning outcomes

Demonstrate knowledge, understanding, and application of:

→ the different types of pathogens that can cause communicable diseases in plants and animals.

The threat to English oak trees

Oak woodlands are a traditional part of the British countryside. These mighty trees (*Quercus robur*) can live for centuries and are home to up to 284 species of insects. As many as 324 different lichens have been identified on a single tree. Future generations, however may not see these trees in the countryside – they are under threat from a new disease. Acute oak decline first appeared in the UK the 1980s, having spread from continental Europe. It causes dark fluid to ooze from the bark, with a rapid decline in the tree and often death.

Scientists still do not know exactly what the cause of this deadly tree disease is – which makes it very difficult to understand how to prevent it spreading. Some of the evidence so far includes:

- The discovery of previously unknown bacteria in the tree which may play a role in the disease

- Evidence of oak jewel beetle activity in infected trees – they may be important in disease development. For example, they may act as a vector or their presence may just be a coincidence

A massive research project is currently underway involving DNA analysis of the microorganisms on infected and healthy oak trees, along with a careful study of the behaviour of the oak jewel beetle. For now, the advice is to try and avoid spreading the disease by careful hygiene procedures for both people who work in oak woodlands and their machinery.

1   Suggest two pieces of evidence which would help to show whether oak jewel beetles are vectors of Acute oak decline or not.

Plant diseases are caused by a range of pathogens. They include:

Disease	Effect on plants
Ring rot – a bacterial disease of potatoes, tomatoes, and aubergines caused by the Gram positive bacterium *Clavibacter michiganensis*. It damages leaves, tubers and fruit. It can destroy up to 80% of the crop and there is no cure. Once bacterial ring rot infects a field it cannot be used to grow potatoes again for at least two years.	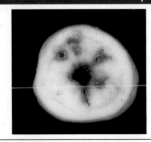 ◀ **Figure 1** *Bacterial ring rot infects a field so it cannot be used to grow potatoes for at least two years*

Disease	Effect on plants
Tobacco mosaic virus (TMV) – a virus that infects tobacco plants and around 150 other species including tomatoes, peppers, cucumbers, petunias and delphiniums. It damages leaves, flowers and fruit, stunting growth and reducing yields, and can lead to an almost total crop loss. Resistant crop strains are available but there is no cure.	  ▲ **Figure 2** *TMV and its effects*
Potato blight (tomato blight, late blight) – caused by the fungus-like protoctist oomycete *Phytophthora infestans*. The hyphae penetrate host cells, destroying leaves, tubers and fruit, causing millions of pounds worth of crop damage each year. There is no cure but resistant strains, careful management and chemical treatments can reduce infection risk.	 ◀**Figure 3** *Tomato blight caused by the protoctista* <u>Phytophthora</u> <u>infestans</u>
Black sigatoka – a banana disease caused by the fungus *Mycosphaerella fijiensis*, which attacks and destroys the leaves. The hyphae penetrate and digest the cells, turning the leaves black. If plants are infected it can cause a 50% reduction in yield. Resistant strains are being developed – good husbandry and **fungicide** (a chemical that kills fungi) treatment can control the spread of the disease but there is no cure.	 ▲ **Figure 4** *Banana leaf infected by Black sigatoka*

➕ Banana diseases and food security ⚙️

Food security is one of the biggest issues globally. In an ideal world everyone would consistently have a balanced diet provided in a sustainable way. One of the main problems, however, for many people is getting enough to eat. If plant diseases threaten staple crops such as  rice, maize, cassava, and bananas then they threaten food security and the survival of the population.

- Bananas are grown in over 130 countries where they are important both as a food crop and economically as a cash crop. They are the 4th most important crop in the developing world after rice, wheat, and maize

- In East Africa bananas (known as plantains) are the staple food for around 50% of the population. People eat around 400 kg of bananas per year

- 90% of the bananas cultivated are produced on small farms and eaten locally. In recent years, as a result of Black Sigatoka there has been a 40% fall in banana yields

- Around 10% of bananas are produced on big plantations for Western supermarkets. These are all from the same clone of a variety called Cavendish so they are genetically very similar. Black Sigatoka is invading these plantations too.

1 There are around 150 million people in East Africa. How many bananas need to be cultivated to feed the 50% of the population for whom bananas are their staple diet?
2 Suggest the effect on Black Sigatoka on the population of East Africa
3 What problems are likely to affect the control of the disease on
  a   small local farms
  b   large plantations of Cavendish bananas

Animal diseases

The diseases that affect animals – and in particular human beings – have a profound effect on human health and wellbeing – and on national economies. Communicable diseases range from mild to fatal. Examples include:

Tuberculosis (TB)

A bacterial disease of humans, cows, pigs, badgers, and deer commonly caused by *Mycobacterium tuberculosis* and *M. bovis*. TB damages and destroys lung tissue and suppresses the immune system, so the body is less able to fight off other diseases. Worldwide in 2012 around 8.6 million people had TB of which 1.3 million died. The global rise of HIV/AIDS has had a big impact on the numbers of people also suffering from diseases such as TB, because people affected by HIV/AIDS are much more likely to develop TB infections. In people TB is both curable (by antibiotics) and preventable (by improving living standards and vaccination).

 ## TB, cows, and badgers

TB affects animal populations. In 2013, almost 33 000 UK cattle were destroyed because they were infected with bovine TB. There is clear evidence that TB is passed from wild animals, such as badgers or possums, to cattle, and vice versa. This presents a problem as cattle can be tested and culled, but it is very difficult to prevent them becoming re-infected from wildlife, particularly when they are out at pasture.

Scientists are still unsure how this wildlife infection can best be controlled. One method is to cull the wildlife source – in countries where this has been done, TB rates in cattle have fallen substantially, however it must be carried out carefully and thoroughly or it can lead to greater disease spread as animals are dispersed.

Some people, however, feel culling is not an acceptable approach and vaccination of either cattle or the wild animals is a better route. The test for TB cannot currently distinguish between an infected animal and a vaccinated animal, so current EU law bans cattle vaccines. Research is continuing on an improved version of both the vaccine and test. Vaccinating a population of wild animals is not an easy task, and it is as yet an unproven method to control the spread of disease.

The problem of TB in animals will not be solved easily, but the research and the debate continues.

1 Why do you think it is so difficult to prevent re-infection of cattle when they are out grazing in the fields?
2 Suggest why EU law bans the vaccination of cattle against TB
3 Why do you think that people are against culling badgers, when thousands of cattle infected with TB are slaughtered each year?
4 What might be the difficulties of vaccinating a wild population of animals?
5 Investigate the impact of TB on infected cows and badgers

- Bacterial meningitis – a bacterial infection (commonly *Streptococcus pneumoniae* or *Neisseria meningitidis*) of the meninges of the brain (protective membranes on the surface of the brain), which can spread into the rest of the body causing septicaemia (blood poisoning) and rapid death. It mainly affects very young children and teenagers aged 15–19. They have different symptoms but in both, a blotchy red/purple rash that does not disappear when a glass is pressed against it is a symptom of septicaemia and immediate medical treatment is needed. About 10% of people infected will die. Up to 25% of those who recover have some permanent damage. Antibiotics will cure the disease if delivered early. Vaccines can protect against some forms of bacterial meningitis.

- HIV/AIDS (acquired immunodeficiency syndrome) – caused by HIV (human immunodeficiency virus), which targets T helper cells in the immune system of the body (see Topic 12.6, The specific immune system). It gradually destroys the immune system so affected people are open to other infections, such as TB and pneumonia, as well as some types of cancer. HIV/AIDS can affect humans and some non-human primates. HIV is a retrovirus with RNA as its genetic material. It contains the enzyme reverse transcriptase, which transcribes the RNA to a single strand of DNA to produce a single strand of DNA in the host cell. This DNA interacts with the genetic material of the host cell. The virus is passed from one person to another in bodily fluids, most commonly through unprotected sex, shared needles, contaminated blood products and from mothers to their babies during pregnancy, birth or breast feeding. In 2012 around 35 million people worldwide were living with HIV infection and about 1.6 million died of the disease. There is as yet no vaccine and no cure, but anti-retroviral drugs slow the progress of the disease to give many years of healthy life. Girls and women are at particularly high risk of HIV/AIDS in many countries. Traditional practices such as female genital mutilation (FGM) increase the infection rate – if the same equipment is used multiple times then this can spread the infection, in addition, women who have undergone FGM are also more vulnerable to infection during intercourse. Sub-Saharan Africa is the region worst affected by HIV/AIDS, with 25 million people living with HIV/AIDS – around 70% of the global total. This disease has massive social and economic consequences as well as the personal impact to each person infected.

- Influenza (flu) – a viral infection (*Orthomyxoviridae* spp.) of the ciliated epithelial cells in the gas exchange system. It kills them, leaving the airways open to secondary infection. Flu can be fatal, especially to young children, old people and people with chronic illnesses. Many of these deaths are from severe secondary bacterial infections such as pneumonia on top of the original viral infection. Flu affects mammals, including humans and pigs, and birds, including chickens. There are three main strains – A, B and C. Strain

Synoptic link

You will learn more about TB in Topic 12.3, The transmission of communicable diseases and Topic 12.7, Preventing and treating disease.

Synoptic link

You learnt about ciliated epithelial cells of the gaseous exchange system in Topics 6.4, Cell specialisation and levels of organisation.

A viruses are the most virulent and they are classified further by the proteins on their surfaces, for example A(H1N1) and A(H3N3). Flu viruses mutate regularly. The change is usually quite small, so having flu one year leaves you with some immunity for the next. Every so often, however there is a major change in the surface antigens and this heralds a flu epidemic or pandemic as there are no antibodies available. Vulnerable groups are given a flu vaccine annually to protect against ever changing strains. There is no cure.

Zoonotic Influenza

A disease which people can catch from animals is known as a zoonosis. Influenza, for example attacks a range of animals including birds and pigs. Sometimes the virus which causes bird flu or swine (pig) flu mutates and becomes capable of infecting people. These new strains can be particularly serious, because few people have any natural immunity to them.

In March 2009 60% of the population of a small town in Mexico became infected with a new disease and two babies died. Some of those infected tested positive for H1N1, a form of flu usually found in pigs, rather than the usual human flu strains.

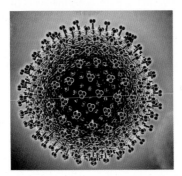

▲ The H1N1 virus which caused a flu pandemic in 2009

The outbreak spread to the US, where using DNA analysis techniques, the virus was identified as a new mutant strain of the H1N1 swine flu virus, which had not been seen before in either pigs or people. Three months after it first appeared people were infected with H1N1 flu in 62 countries, and some of them were dying. None of the available flu vaccines were any use against this zoonotic virus.

Within five months almost 3000 people around the world had died.

Fortunately, only six months after swine flu H1N1 was first identified, scientists produced an effective vaccine. In spite of this, recent analyses of the data suggest between 200 000–300 000 people died as a result of H1N1 infection in the 2009 outbreak – and up to 80% of those deaths were in people aged 65 and younger. In a normal seasonal flu outbreak, only around 10% of deaths occur in people who are under 65.

H1N1 is now part of the normal seasonal flu vaccine and scientists remain on the lookout for the next mutation which may enable the flu virus to pass from pigs or birds to people, known as a species jump.

1 H1N1 is a virus not a bacterium. Why does this make it so much more dangerous?
2 Explain how modern DNA technology helps in the case of a zoonotic disease outbreak such as this
3 a How did the age profile of the people who died from H1N1 differ from the normal pattern of flu-related deaths?
   b Suggest reasons for this difference.

- Malaria – caused by the protoctista *Plasmodium* and spread by the bites of infected *Anopheles* mosquitoes (the vector – Topic 12.3). The *Plasmodium* parasite has a complex life cycle with two hosts – mosquitoes and people. They reproduce inside the female mosquito. The female needs to take two blood meals to provide her with protein before she lays her eggs – and this is when *Plasmodium* is passed on to people. It invades the red blood cells, liver, and even the brain. Around 200 million people are reported to have malaria

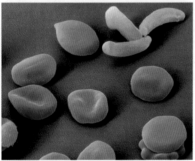

▲ **Figure 5** *Top: Only female* <u>Anopheles</u> *mosquitoes (top) spread malaria – they need two blood meals before they lay their eggs. When infected females feed they transmit the parasite to the human host – bottom: scanning electron microscope showing red blood cells and* <u>Plasmodium falciparum</u> *protozoa. × 500 magnification.*

each year, and over 600 000 die. The disease recurs, making people weak and vulnerable to other infections. There is no vaccine against malaria and limited cures, but preventative measures can very effective. The key is to control the vector. *Anopheles* mosquitoes can be destroyed by insecticides and by removing the standing water where they breed. Simple measures such as mosquito nets, window and door screens and long sleeved clothing can prevent them biting people and spreading the disease.

● Ring worm – a fungal disease affecting mammals including cattle, dogs, cats and humans. Different fungi infect different species – in cattle, ring worm is usually caused by *Trichophyton verrucosum*. It causes grey-white, crusty, infectious, circular areas of skin. It is not damaging but looks unsightly and may be itchy. Antifungal creams are an effective cure.

● Athlete's foot – a human fungal disease caused by *Tinia pedia*, a form of human ring worm that grows on and digests the warm, moist skin between the toes. It causes cracking and scaling, which is itchy and may become sore. Antifungal creams are an effective cure.

➕ Identifying pathogens

When an outbreak of a disease occurs in plants or animals, the key to successful control or cure is to identify the pathogens involved. Our ability to do this has increased along with our understanding of the causes of disease and developments in technology:

● Traditionally pathogens were cultured in the laboratory and identified using a microscope.

● Monoclonal antibodies can be used now to identify pathogenic organisms in both plants and animals.

● DNA sequencing technology means pathogens can be identified precisely, down to a single mutation (see Zoonotic influenza).

Case study: In a transplant ward, four patients developed infections caused by methicillin resistant *Staphylococcus aureus* (MRSA). If the bacterium was being transmitted to patients by a member of staff, this was a serious outbreak. But DNA sequencing at the Sanger Institute gave rapid results – new technology means a bacterial genome can be sequenced in less than 24 hours. Researchers showed that each of the patients had a different strain of MRSA. The cases were not linked, so it was not a hospital-based outbreak requiring staff to be screened or treated.

1 Why is it important to identify the pathogen causing a communicable disease?
2 Suggest the benefits and limitations of culturing pathogens and using a light microscope to identify them.
3 Find out as much as you can about the use of monoclonal antibodies in the detection of plant diseases.

Summary questions

1 Make a table to summarise all the bacterial, viral, protoctist, and fungal diseases described in this section, including the main organisms that are affected. *(6 marks)*

2 Compare and contrast a bacterial disease of plants and of animals. *(6 marks)*

3 Suggest three ways in which animal diseases and three ways in which plant diseases may be spread from one organism to the next. *(6 marks)*

12.3 The transmission of communicable diseases

Specification reference: 4.1.1

For the pathogens that cause communicable diseases to be successful, they have to be transmissible. So how are pathogenic bacteria, viruses, protoctista, and fungi transmitted from one host to another?

Transmission of pathogens between animals

Understanding how diseases are transmitted from one individual to another allows us to work out ways to reduce or prevent it happening. There are two main types of transmission – direct transmission and indirect transmission.

Direct transmission

Here the pathogen is transferred directly from one individual to another by:

Direct contact (contagious diseases):

- kissing or any contact with the body fluids of another person, for example, bacterial meningitis and many sexually transmitted diseases
- direct skin-to-skin contact, for example, ring worm, athlete's foot
- microorganisms from faeces transmitted on the hands, for example, diarrhoeal diseases.

Inoculation:

- through a break in the skin, for example, during sex (HIV/AIDS)
- from an animal bite, for example, rabies
- through a puncture wound or through sharing needles, e.g. septicaemia.

Ingestion:

- taking in contaminated food or drink, or transferring pathogens to the mouth from the hands, for example, amoebic dysentery, diarrhoeal diseases.

Indirect transmission

This is where the pathogen travels from one individual to another indirectly.

Fomites:

- inanimate objects such as bedding, socks, or cosmetics can transfer pathogens, for example, athlete's foot, gas gangrene and *Staphylococcus infections*.

Droplet infection (inhalation):

- Minute droplets of saliva and mucus are expelled from your mouth as you talk, cough or sneeze. If these droplets contain pathogens, when healthy individuals breathe the droplets in they may become infected, for example, influenza, tuberculosis.

Vectors:

- A vector transmits communicable pathogens from one host to another. Vectors are often but not always animals, for example,

Preventing the spread of communicable diseases in humans

Key factors in reducing the spread of communicable diseases in humans include:

- hand washing – regular hand washing is the single most effective way of preventing the spread of many communicable diseases

- improvements in living and working conditions, for example, reducing overcrowding, ensuring good nutrition

- disposal of both bodily and household waste effectively.

1 Explain why hand washing is so effective at preventing disease transmission.

2 Why does improving living standards have such an impact on disease transmission?

3 Effective management of household waste, leaving no empty containers around and keeping drains clear, substantially reduces the incidence of malaria in an area. Explain why this happens.

4 Research and report on other low-tech ways in which the incidence of malaria can be prevented.

mosquitoes transmit malaria, rat fleas transmit bubonic plague, dogs, foxes and bats transmit rabies.

- Water can also act as a vector of disease, for example, diarrhoeal diseases.

Transmission between animals and humans

Some communicable diseases can be passed from animals to people, for example the bird flu strain H1N1 and brucellosis, which is passed from sheep to people. Minimising close contact with animals and washing hands thoroughly following any such contact can reduce infection rates. People can also act as vectors of some animal diseases, sometimes with fatal results, for example foot-and-mouth disease.

Factors affecting the transmission of communicable diseases in animals

The probability of catching a communicable disease is increased by a number of factors:

- overcrowded living and working conditions
- poor nutrition
- a compromised immune system, including (in humans) having HIV/AIDS or needing immunosuppressant drugs after transplant surgery
- (in humans) poor disposal of waste, providing breeding sites for vectors
- climate change – this can introduce new vectors and new diseases, for example increased temperatures promote the spread of malaria as the vector mosquito species is able to survive over a wider area
- Culture and infrastructure – in many countries traditional medical practises can increase transmission
- Socioeconomic factors – for example, a lack of trained health workers and insufficient public warning when there is an outbreak of disease can also affect transmission rates.

Transmission of pathogens between plants

Plants do not move around, cough or sneeze, yet diseases spread rapidly through plant communities, plant pollen and seed, for example move widely. Plants also have a less well developed immune system than humans.

Direct transmission

This involves direct contact of a healthy plant with any part of a diseased plant. Examples are ring rot, tobacco mosaic virus (TMV), tomato and potato blight, and black sigatoka.

Indirect transmission

Soil contamination

Infected plants often leave pathogens (bacteria or viruses) or reproductive spores from protoctista or fungi in the soil. These can infect the next crop. Examples are black sigatoka spores, ring rot bacteria, spores of *P. infestans* and TMV. Some pathogens (often as spores) can survive the composting process so the infection cycle can be completed when contaminated compost is used.

Vectors

- Wind – bacteria, viruses and fungal or oomycete spores may be carried on the wind, e.g. Black sigatoka blown between Caribbean islands, *P. infestans* sporangia form spores which are carried by the wind to other potato crops/tomato plants.

- Water – spores swim in the surface film of water on leaves; raindrop splashes carry pathogens and spores, etc. Examples are spores of *P. infestans* (potato blight) which swim over films of water on the leaves.

- Animals – insects and birds carry pathogens and spores from one plant to another as they feed. Insects such as aphids inoculate pathogens directly into plant tissues.

- Humans – pathogens and spores are transmitted by hands, clothing, fomites, farming practices and by transporting plants and crops around the world. For example, TMV survives for years in tobacco products, ring rot survives on farm machinery, potato sacks, etc.

Factors affecting the transmission of communicable diseases in plants

A number of factors are responsible:

- planting varieties of crops that are susceptible to disease
- over-crowding increases the likelihood of contact
- poor mineral nutrition reduces resistance of plants
- damp, warm conditions increase the survival and spread of pathogens and spores
- climate change – increased rainfall and wind promote the spread of diseases; changing conditions allow animal vectors to spread to new areas; drier conditions may reduce the spread of disease.

▲ **Figure 1** *Disease can spread rapidly through monocultures such as these crop plants*

Summary questions

1   Explain the difference between direct and indirect transmission of communicable pathogens.   (*2 marks*)

2   Compare and contrast *direct* transmission of animal and plant diseases.   (*4 marks*)

3   Compare and contrast *indirect* transmission of animal and plant diseases.   (*4 marks*)

4   Suggest different approaches to control the spread of malaria.   (*5 marks*)

Preventing the spread of communicable diseases in plants

Key factors in reducing the spread of communicable diseases in plants:

- Leave plenty of room between plants to minimise the spread of pathogens.
- Clear fields as thoroughly as possible – remove all traces of plants from the soil at harvesting.
- Rotate crops – the spores or bacteria will eventually die if they do not have access to the host plant.
- Follow strict hygiene practices – measures such as washing hands, washing boots, sterilising storage sacks, washing down machinery, etc.
- Control insect vectors.

1   Discuss ways of controlling communicable plant diseases with reference to the plant disease triangle in Figure 2.

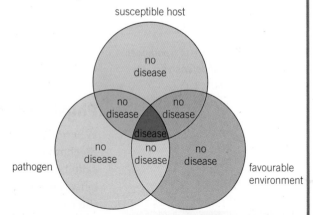

▲ **Figure 2** *Disease in plants is the result of the interaction between a susceptible host, a pathogen and a favourable environment. Modifying any of these factors can reduce the incidence of disease*

12.4 Plant defences against pathogens

Specification reference: 4.1.1

Synoptic link

You learnt about cell division and meristems in plants in Topic 6.2, Mitosis and Topic 6.3, Meiosis.

Plants have evolved a number of ways to defend themselves against the pathogens that cause communicable diseases. The waxy cuticle of plant leaves, the bark on trees, and the cellulose cell walls of individual plant cells act as barriers, which prevent pathogens getting in. Unlike animals, plants do not heal diseased tissue – they seal it off and sacrifice it. Because they are continually growing at the meristems, they can then replace the damaged parts.

Recognising an attack

Plants are not passive – they respond rapidly to pathogen attacks. Receptors in the cells respond to molecules from the pathogens, or to chemicals produced when the plant cell wall is attacked. This stimulates the release of signalling molecules that appear to switch on genes in the nucleus. This in turn triggers cellular responses, which include producing defensive chemicals, sending alarm signals to unaffected cells to trigger their defences, and physically strengthening the cell walls.

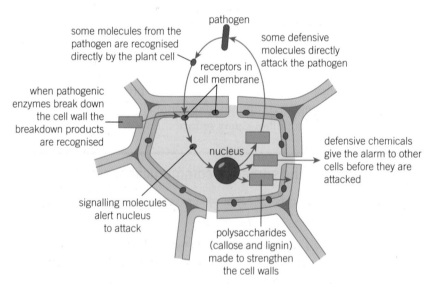

▶ **Figure 1** *The presence of a pathogen stimulates a series of defensive strategies in the plant*

Physical defences

When plants are attacked by pathogens they rapidly set up extra mechanical defences. They produce high levels of a polysaccharide called **callose**, which contains β-1,3 linkages and β-1,6 linkages between the glucose monomers. Scientists still do not fully understand the roles played by callose in the defence mechanisms of the plant but current research suggests that:

Synoptic link

You learnt about α and β linkages in polysaccharides in Topic 3.3, Carbohydrates.

- within minutes of an initial attack, callose is synthesised and deposited between the cell walls and the cell membrane in cells next to the infected cells. These callose papillae act as barriers, preventing the pathogens entering the plant cells around the site of infection

- large amounts of callose continue to be deposited in cell walls after the initial infection. Lignin is added, making the mechanical barrier to invasion even thicker and stronger
- callose blocks sieve plates in the phloem, sealing off the infected part and preventing the spread of pathogens
- callose is deposited in the plasmodesmata between infected cells and their neighbours, sealing them off from the healthy cells and helping to prevent the pathogen spreading.

Chemical defences

Many plants produce powerful chemicals that either repel the insect vectors of disease or kill invading pathogens. Some of these chemicals are so powerful that we extract and use them or synthesise them to help us control insects, fungi and bacteria. Some have strong flavours and are used as herbs and spices. Examples of plant defensive chemicals include:

- insect repellents – for example, pine resin and citronella from lemon grass
- insecticides – for example, pyrethrins – these are made by chrysanthemums and act as insect neurotoxins; and caffeine – toxic to insects and fungi
- antibacterial compounds including antibiotics – for example, phenols – antiseptics made in many different plants; antibacterial gossypol produced by cotton; defensins – plant proteins that disrupt bacterial and fungal cell membranes; lysosomes – organelles containing enzymes that break down bacterial cell walls
- antifungal compounds – for example, phenols – antifungals made in many different plants; antifungal gossypol produced by cotton; caffeine – toxic to fungi and insects; saponins – chemicals in many plant cell membranes that interfere with fungal cell membranes; **chitinases** – enzymes that break down the chitin in fungal cell walls
- anti-oomycetes – for example, **glucanases** – enzymes made by some plants that break down glucans; polymers found in the cell walls of oomycetes (e.g., *P.infestans*)
- general toxins – some plants make chemicals that can be broken down to form cyanide compounds when the plant cell is attacked. Cyanide is toxic to most living things.

▲ **Figure 2** *Castor oil beans (shown here) produce ricin, which is used as a defensive chemical. Just 0.2 milligrams would be fatal if ingested*

Summary questions

1 Describe how the plant response to a pathogen attack is triggered.
(*3 marks*)

2 Make a table or diagram to summarise plant defences against pathogens.
(*4 marks*)

3 Investigate the evidence for one of the roles of callose in plant defence responses.
(*6 marks*)

12.5 Non-specific animal defences against pathogens

Specification reference: 4.1.1

Learning outcomes

Demonstrate knowledge, understanding, and application of:

→ the primary non-specific defences against pathogens in animals

→ the structure and mode of action of phagocytes

→ how to examine and draw cells observed in blood smears.

Synoptic link

You learnt about the gas exchange system in Topic 7.2, Mammalian gaseous exchange system.

Mammals (for example humans) have two lines of defence against invasion by pathogens. The primary non-specific defences against pathogens are always present or activated very rapidly. This system defends against all pathogens in the same way. Mammals have a specific immune response, which is specific to each pathogen but is slower to respond (as discussed in Topic 12.6).

Non-specific defences – keeping pathogens out

The body has a number of barriers to the entry of pathogens:

- The skin covers the body and prevents the entry of pathogens. It has a skin flora of healthy microorganisms that outcompete pathogens for space on the body surface. The skin also produces sebum, an oily substance that inhibits the growth of pathogens.

- Many of the body tracts, including the airways of the gas exchange system, are lined by **mucous membranes** that secrete sticky mucus. This traps microorganisms and contains lysozymes, which destroy bacterial and fungal cell walls. Mucus also contains phagocytes, which remove remaining pathogens.

- Lysozymes in tears and urine, and the acid in the stomach, also help to prevent pathogens getting into our bodies.

We also have expulsive reflexes. Coughs and sneezes eject pathogen-laden mucus from the gas exchange system, while vomiting and diarrhoea expel the contents of the gut along with any infective pathogens.

Blood clotting and wound repair

If you cut yourself, the skin is breached and pathogens can enter the body. The blood clots rapidly to seal the wound. When platelets come into contact with collagen in skin or the wall of the damaged blood vessel, they adhere and begin secreting several substances. The most important are:

- thromboplastin, an enzyme that triggers a cascade of reactions resulting in the formation of a blood clot (or thrombus, Figure 1)

- serotonin, which makes the smooth muscle in the walls of the blood vessels contract, so they narrow and reduce the supply of blood to the area.

The clot dries out, forming a hard, tough scab that keeps pathogens out. This is the first stage of wound repair. Epidermal cells below the scab start to grow, sealing the wound permanently, while damaged blood vessels regrow. Collagen fibres are deposited to give the new

Synoptic link

You learnt about the enzymatic-control of the blood clotting cascade in Topic 4.4, Cofactors, coenzymes, and prosthetic groups.

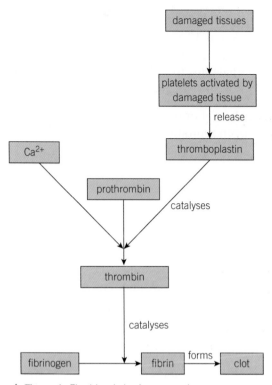

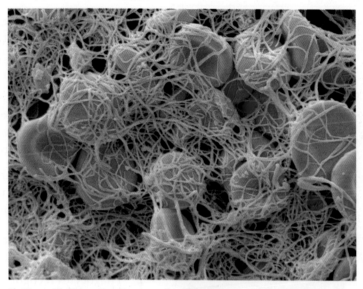

▲ **Figure 1** *The blood clotting cascade – once a clot forms, blood cannot leak out of the body and pathogens cannot get in*

▲ **Figure 2** *Scanning electron micrograph showing red blood cells in a fibrin mesh. × 800 magnification*

tissue strength. Once the new epidermis reaches normal thickness, the scab sloughs off and the wound is healed.

Inflammatory response

The inflammatory response is a localised response to pathogens (or damage or irritants) resulting in **inflammation** at the site of a wound. Inflammation is characterised by pain, heat, redness, and swelling of tissue.

Mast cells are activated in damaged tissue and release chemicals called **histamines** and **cytokines**.

- Histamines make the blood vessels dilate, causing localised heat and redness. The raised temperature helps prevent pathogens reproducing.

- Histamines make blood vessel walls more leaky so blood plasma is forced out, once forced out of the blood it is known as tissue fluid. Tissue fluid causes swelling (oedema) and pain.

- Cytokines attract white blood cells (phagocytes) to the site. They dispose of pathogens by **phagocytosis**.

If an infection is widespread, the inflammatory response can cause a whole-body rash.

Non-specific defences – getting rid of pathogens

If the pathogens get into the body, the next lines of defence are adaptations to prevent them growing or to destroy them.

Synoptic link

You learnt about phagocytosis as a form of bulk transport in Topic 5.4, Active transport.

Synoptic link

You will learn more about the control of body temperature in Chapter 15, Homeostasis.

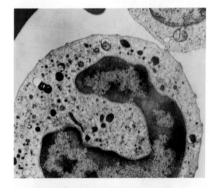

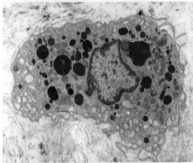

▲ **Figure 3** *Common phagocytes in the body: 70% of white blood cells are neutrophils (top) with their multi-lobed nuclei and rapid action against pathogens. Macrophages (bottom), with their simpler round nuclei, make up 4% of the white blood cells and are involved in both the non-specific defence system and the specific immune system. Coloured transmission electron micrographs, × 2000 magnification*

Fevers

Normal body temperature of around 37 °C is maintained by the hypothalamus in your brain. When a pathogen invades your body, cytokines stimulate your hypothalamus to reset the thermostat and your temperature goes up. This is a useful adaptation because:

- most pathogens reproduce best at or below 37 °C. Higher temperatures inhibit pathogen reproduction
- the specific immune system works faster at higher temperatures.

Phagocytosis

Phagocytes are specialised white cells that engulf and destroy pathogens. There are two main types of phagocytes – neutrophils and macrophages (Figure 2).

Phagocytes build up at the site of an infection and attack pathogens. Sometimes you can see pus in a spot, cut or wound. Pus consists of dead neutrophils and pathogens.

The stages of phagocytosis

1. Pathogens produce chemicals that attract phagocytes.
2. Phagocytes recognise non-human proteins on the pathogen. This is a response not to a specific type of pathogen, but simply a cell or organism that is non-self.
3. The phagocyte engulfs the pathogen and encloses it in a vacuole called a **phagosome.**
4. The phagosome combines with a lysosome to form a **phagolysosome.**
5. Enzymes from the lysosome digest and destroy the pathogen.

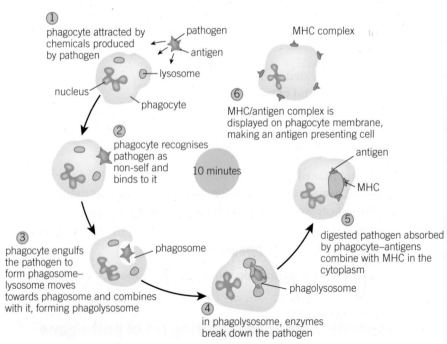

▲ **Figure 4** *Phagocytosis – a key process in both the non-specific and specific defence systems of the body*

It usually takes a human neutrophil under 10 minutes to engulf and destroy a bacterium. Macrophages take longer but they undergo a more complex process. When a macrophage has digested a pathogen, it combines antigens from the pathogen surface membrane with special glycoproteins in the cytoplasm called the **major histocompatibility complex** (**MHC**). The MHC complex moves these pathogen antigens to the macrophage's own surface membrane, becoming an **antigen-presenting cell** (APC). These antigens now stimulate other cells involved in the specific immune system response (see Topic 11.6, Factors affecting biodiversity).

Counting blood cells

In your previous studies you learnt how to examine microscope slides and draw the cells you saw. You also learned how to count the number of cells in a given area of a slide. Both of these skills are very important when looking at blood smears, made by spreading a single drop of blood very thinly across a slide. They are often stained to show up the nuclei of the lymphocytes, making them easier to identify. Identifying the numbers of different types of lymphocytes in a blood smear indicates if a non-specific or specific immune response is taking place.

Helpful chemicals

Phagocytes that have engulfed a pathogen produce chemicals called **cytokines.** Cytokines act as cell-signalling molecules, informing other phagocytes that the body is under attack and stimulating them to move to the site of infection or inflammation. Cytokines can also increase body temperature and stimulate the specific immune system.

Opsonins are chemicals that bind to pathogens and 'tag' them so they can be more easily recognised by phagocytes. Phagocytes have receptors on their cell membranes that bind to common opsonins, and the phagocyte then engulfs the pathogen. There are a number of different opsonins, but antibodies such as immunoglobulin G (IgG) and immunoglobulin M (IgM) have the strongest effect.

Synoptic link

You learnt about smear slides in Topic 2.1, Microscopy. It may help to look back at Topics 6.4, The organisation and specialisation of cells, and 6.5, Stem cells, to remind yourself of the different types of cells in the blood.

Summary questions

1 Make a table to show the main adaptations of the body that prevent the entry of pathogens. *(4 marks)*

2 A woman gets a bad scratch from a bramble. The scratch gets very red and hot and the next day it contains pus. Explain what is happening. *(4 marks)*

3 a Describe the process of phagocytosis. *(6 marks)*
   b Explain how cytokines and opsonins make the process of phagocytosis more effective than it would be without them. *(5 marks)*

12.6 The specific immune system

Specification reference: 4.1.1

All cells have molecules called **antigens** on their surfaces. The body recognises the difference between *self* antigens on your own cells and *non-self* antigens on the cells of pathogens. Some toxins also act as antigens. Antigens trigger an immune response, which involves the production of polypeptides called **antibodies**.

The **specific immune system** (also known as active or acquired immunity) is slower than the non-specific responses – it can take up to 14 days to respond effectively to a pathogen invasion. However, the immune memory cells mean it reacts very quickly to a second invasion by the same pathogen.

Learning outcomes

Demonstrate knowledge, understanding, and application of:

→ modes of action of B and T lymphocytes in the specific immune response

→ the structure and functions of antibodies

→ actions of opsonins, agglutinins and anti-toxins

→ the primary and secondary immune responses

→ autoimmune diseases.

Antibodies

Antibodies are Y-shaped glycoproteins called **immunoglobulins**, which bind to a specific antigen on the pathogen or toxin that has triggered the immune response. There are millions of different antibodies, and there is a specific antibody for each antigen.

Antibodies are made up of two identical long polypeptide chains called the heavy chains and two much shorter identical chains called the light chains (Figure 1). The chains are held together by disulfide bridges and there are also disulfide bridges within the polypeptide chains holding them in shape.

Antibodies bind to antigens with a protein-based 'lock-and-key' mechanism similar to the complementarity between the active site of an enzyme and its substrate. The binding site is an area of 110 amino acids on both the heavy and the light chains, known as the variable region. It is a different shape on each antibody and gives the antibody its specificity. The rest of the antibody molecule is always the same, so it is called the constant region.

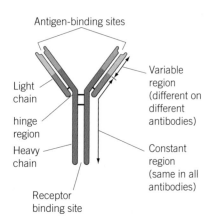

▲ **Figure 1** *All antibodies have the same basic structure*

When an antibody binds to an antigen it forms an **antigen–antibody complex.**

The hinge region of the antibody provides the molecule with flexibility, allowing it to bind two seperate antigens, one at each of its antigen-binding sites.

How antibodies defend the body

1   The antibody of the antigen–antibody complex acts as an opsonin so the complex is easily engulfed and digested by phagocytes.

2   Most pathogens can no longer effectively invade the host cells once they are part of an antigen–antibody complex.

3   Antibodies act as **agglutinins** causing pathogens carrying antigen–antibody complexes to clump together. This helps prevent them spreading through the body and makes it easier for phagocytes to engulf a number of pathogens at the same time.

Synoptic link

You learnt about protein structure and glycoproteins in Topic 3.5, Lipids and Topic 3.6, Structure of proteins, and about the active sites of enzymes in Topic 4.1 Enzyme action.

Lymphocytes and the immune response

The specific immune system is based on white blood cells called **lymphocytes**. **B lymphocytes** mature in the **B**one marrow, while **T lymphocytes** mature in the **T**hymus gland.

The main types of T lymphocytes:

- **T helper cells** – these have CD4 receptors on their cell-surface membranes, which bind to the surface antigens on APCs (Topic 12.5). They produce **interleukins,** which are a type of cytokine (cell-signalling molecule). The interleukins made by the T helper cells stimulate the activity of B cells, which increases antibody production, stimulates production of other types of T cells and attracts and stimulates macrophages to ingest pathogens with antigen–antibody complexes.

- **T killer cells** – these destroy the pathogen carrying the antigen. They produce a chemical called **perforin,** which kills the pathogen by making holes in the cell membrane so it is freely permeable.

- **T memory cells** – these live for a long time and are part of the **immunological memory**. If they meet an antigen a second time, they divide rapidly to form a huge number of clones of T killer cells that destroy the pathogen.

- **T regulator cells** – these cells suppress the immune system, acting to control and regulate it. They stop the immune response once a pathogen has been eliminated, and make sure the body recognises self antigens and does not set up an **autoimmune response**. Interleukins are important in this control.

The main types of B lymphocytes:

- **Plasma cells** – these produce antibodies to a particular antigen and release them into the circulation. An active plasma cell only lives for a few days but produces around 2000 antibodies per second while it is alive and active.

- **B effector cells** – these divide to form the plasma cell clones.

- **B memory cells** – these live for a very long time and provide the immunological memory. They are programmed to remember a specific antigen and enable the body to make a very rapid response when a pathogen carrying that antigen is encountered again.

Cell-mediated immunity

In cell-mediated immunity, T lymphocytes respond to the cells of an organism that have been changed in some way, for example by a virus infection, by antigen processing or by mutation (for example cancer cells) and to cells from transplanted tissue. The cell-mediated response is particularly important against viruses and early cancers.

1 In the non-specific defence system, macrophages engulf and digest pathogens in phagocytosis. They process the antigens from the surface of the pathogen to form antigen-presenting cells (APCs).

2   The receptors on some of the T helper cells fit the antigens. These T helper cells become activated and produce interleukins, which stimulate more T cells to divide rapidly by mitosis. They form clones of identical activated T helper cells that all carry the right antigen to bind to a particular pathogen.

3   The cloned T cells may:

- develop into T memory cells, which give a rapid response if this pathogen invades the body again
- produce interleukins that stimulate phagocytosis
- produce interleukins that stimulate B cells to divide
- stimulate the development of a clone of T killer cells that are specific for the presented antigen and then destroy infected cells.

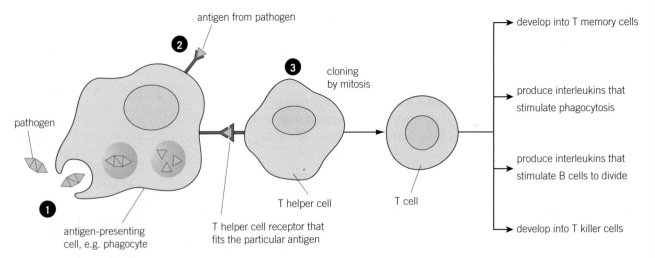

▲ **Figure 2** *Model of cell-mediated immunity*

Humoral immunity

In humoral immunity the body responds to antigens found outside the cells, for example bacteria and fungi, and to APCs. The humoral immune system produces antibodies that are soluble in the blood and tissue fluid and are not attached to cells.

B lymphocytes have antibodies on their cell-surface membrane (immunoglobulin M or IgM) and there are millions of different types of B lymphocytes, each with different antibodies. When a pathogen enters the body it will carry specific antigens, or produce toxins that act as antigens. A B cell with the complementary antibodies will bind to the antigens on the pathogen, or to the free antigens. The B cell engulfs and processes the antigens to become an APC (see Figure 3).

1   Activated T helper cells bind to the B cell APC. This is **clonal selection** – the point at which the B cell with the correct antibody to overcome a particular antigen is selected for cloning.

2   Interleukins produced by the activated T helper cells activate the B cells.

3   The activated B cell divides by mitosis to give clones of plasma cells and B memory cells. This is **clonal expansion.**

4 Cloned plasma cells produce antibodies that fit the antigens on the surface of the pathogen, bind to the antigens and disable them, or act as opsonins or agglutinins. This is the **primary immune response** and it can take days or even weeks to become fully effective against a particular pathogen. This is why we get ill – the symptoms are the result of the way our body reacts when the pathogens are dividing freely, before the primary immune response is fully operational.

5 Some cloned B cells develop into B memory cells. If the body is infected by the same pathogen again, the B memory cells divide rapidly to form plasma cell clones. These produce the right antibody and wipe out the pathogen very quickly, before it can cause the symptoms of disease. This is the **secondary immune response.**

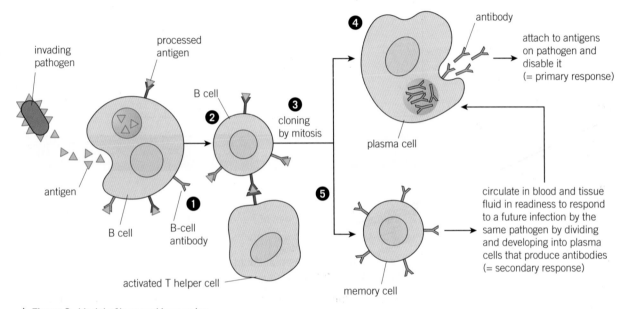

▲ **Figure 3** *Model of humoral immunity*

Autoimmune diseases

Sometimes the immune system stops recognising 'self' cells and starts to attack healthy body tissue. This is termed an **autoimmune disease**. Scientists still do not understand fully why this happens. There appears to be a genetic tendency in some families, sometimes the immune system responds abnormally to a mild pathogen or normal body microorganisms and in some cases the T regulator cells do not work effectively.

There are around 80 different autoimmune diseases that can cause chronic inflammation or the complete breakdown and destruction of healthy tissue. Immunosuppressant drugs, which prevent the immune system working, may be used as treatments but they deprive the body of its natural defences against communicable diseases.

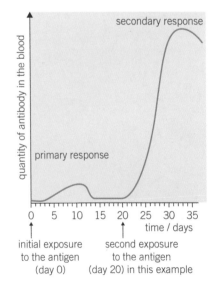

▲ **Figure 4** *Primary and secondary immune responses*

Synoptic link

Look back to the Application on gene therapy in Topic 6.5, Stem cells, for an example of an autoimmune disease.

▼ **Table 1** *Some common autoimmune diseases*

Autoimmune disease	Body part affected	Treatment
Type 1 diabetes	• the insulin-secreting cells of the pancreas	• insulin injections • pancreas transplants • immunosuppressant drugs
Rheumatoid arthritis	• joints—especially in the hands, wrists, ankles and feet	• no cure • anti-inflammatory drugs • steroids • immunosuppressants • pain relief
Lupus	• often affects skin and joints and causes fatigue • can attack any organ in the body including kidneys, liver, lungs or brain	• no cure • anti-inflammatory drugs • steroids • immunosuppressants • various others

Summary questions

1 What are antibodies and how do they work? *(2 marks)*

2 Compare the main types of T and B lymphocytes – what are their similarities and differences? *(6 marks)*

3 Discuss the problems that could arise from treating an autoimmune disease with immunosuppressant drugs. *(3 marks)*

4 The humoral immune system deals well with bacterial and fungal infections but the cell-mediated immune system is more effective at tackling viral infections. Explain the biology behind this statement. *(6 marks)*

12.7 Preventing and treating disease

Specification reference: 4.1.1

Non-communicable diseases cannot be passed from one person to another. They include heart disease, most types of cancer and many diseases of the nervous, endocrine and digestive systems. Communicable diseases are caused by pathogens and can be passed from person to person.

When you come into contact with a foreign antigen, you need some form of immunity to prevent you getting the disease. There are several ways of achieving this immunity.

Natural immunity

Some forms of immunity occur naturally in the body:

- When you meet a pathogen for the first time, your immune system is activated and antibodies are formed, which results in the destruction of the antigen (Topic 12.6, The specific immune system). The immune system produces T and B memory cells so if you meet a pathogen for a second time, your immune system recognises the antigens and can immediately destroy the pathogen, before it causes disease symptoms. This is known as **natural active immunity**. It is known as active because the body has itself acted to produce antibodies and/or memory cells.

- The immune system of a new-born baby is not mature and it cannot make antibodies for the first couple of months. A system has evolved to protect the baby for those first few months of life. Some antibodies cross the placenta from the mother to her fetus while the baby is in the uterus, so it has some immunity to disease at birth. The first milk a mammalian mother makes is called **colostrum**, which is very high in antibodies. The infant gut allows these glycoproteins to pass into the bloodstream without being digested. So within a few days of birth, a breast-fed baby will have the same level of antibody protection against disease as the mother. This is **natural passive immunity** and it lasts until the immune system of the baby begins to make its own antibodies. The antibodies the baby receives from the mother are likely to be relevant to pathogens in its environment, where the mother acquired them.

Artificial immunity

Some diseases can kill people before their immune system makes the antibodies they need. Medical science can give us immunity to some of these life-threatening diseases without any contact with live pathogens.

Artificial passive immunity

For certain potentially fatal diseases, antibodies are formed in one individual (often an animal), extracted and then injected into the bloodstream of another individual. This **artificial passive immunity** gives temporary immunity – it doesn't last long but it can be lifesaving. For example, tetanus is caused by a toxin released by the bacterium *Clostridium tetani*, found in the soil and animal faeces. It causes the muscles to go into spasm so you cannot swallow or breathe. People who might be infected with tetanus (for example after a contaminated cut) will be injected with tetanus antibodies extracted from the blood of horses, preventing the development of the disease but not providing long-term immunity. Rabies is another fatal disease that is treated with a series of injections that give artificial passive immunity.

Artificial active immunity – the principles of vaccination

In **artificial active immunity** the immune system of the body is stimulated to make its own antibodies to a safe form of an antigen (a **vaccine**), which is injected into the bloodstream (vaccination). The antigen is not usually the normal live pathogen, as this could cause the disease and have fatal results. The main steps are as follows:

1 The pathogen is made safe in one of a number of ways so that the antigens are intact but there is no risk of infection. Vaccines may contain:

   ● killed or inactivated bacteria and viruses, for example, whooping cough (pertussis)

   ● attenuated (weakened) strains of live bacteria or viruses, for example, rubella, BCG against TB, polio (vaccine taken orally)

   ● toxin molecules that have been altered and detoxified, for example, diphtheria, tetanus

   ● isolated antigens extracted from the pathogen, for example, the influenza vaccine

   ● genetically engineered antigens, for example, the hepatitis B vaccine.

2 Small amounts of the safe antigen, known as the vaccine, are injected into the blood.

3 The primary immune response is triggered by the foreign antigens and your body produces antibodies and memory cells as if you were infected with a live pathogen.

4 If you come into contact with a live pathogen, the secondary immune response is triggered and you destroy the pathogen rapidly before you suffer symptoms of the disease.

The artificial active immunity provided by vaccines may last a year, a few years or a lifetime. Sometimes boosters (repeat vaccinations) are needed to increase the time you are immune to a disease.

Vaccines and the prevention of epidemics

Vaccines are used to give long-term immunity to many diseases. However, they are also used to help prevent epidemics. An **epidemic**

▲ **Figure 2** *In the UK babies are given vaccinations against a range of diseases including pertussis, diphtheria, tetanus, polio, meningitis, measles, mumps, and rubella. As a result, the number of children who die from preventable infections is now very low*

is when a communicable disease spreads rapidly to a lot of people at a local or national level. A **pandemic** is when the same disease spreads rapidly across a number of countries and continents.

At the beginning of an epidemic, mass vaccination can prevent the spread of the pathogen into the wider population. When vaccines are being deployed to prevent epidemics, they often have to be changed regularly to remain effective.

When a significant number of people in the population have been vaccinated, this gives protection to those who do not have immunity. This is known as herd immunity, as there is minimal opportunity for an outbreak to occur.

Case Study: Influenza

Flu is a disease that has caused epidemics at intervals throughout history. The virus that causes strain A influenza mutates regularly, so the antigens on the surface change too. Some forms of this virus can cross the species barrier from animals such as birds or pigs to people. Although people develop resistance to one strain of flu, the next year the antigens on the surface of the virus may have changed so much that the immune system does not recognise it and many of the same people become ill again.

Every year in the UK older people and anyone who has a compromised immune system are given a flu vaccine. Every year the mixture of flu antigens in the vaccine is different, reflecting the forms of the virus that the World

Health Organisation (WHO) predicts will be most common and most likely to cause serious disease.

If a flu epidemic begins, more people are vaccinated to control infection rates. Because people travel freely and frequently, an epidemic can spread rapidly from one country to another. People across the world need to be vaccinated to stop the spread of disease and stock piles of vaccines are in place in case this becomes necessary.

SARS was a new flu-like disease that appeared in 2002 and spread from birds to people. It spread rapidly across countries as people travelled around. However, in spite of a lack of vaccine, careful management of cases by isolation meant the outbreak was contained and quickly closed down with relatively few deaths (Figure 3).

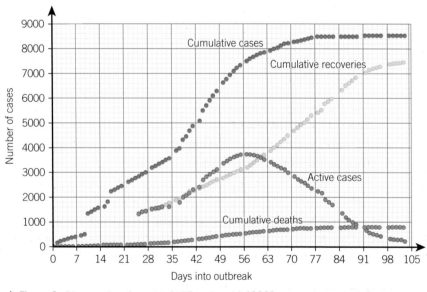

▲ **Figure 3** *Disease data from the SARS outbreak of 2002.*

1 Suggest three ways in which an outbreak of 'flu' or 'flu-like illness' might be contained.

Some communicable diseases that cause problems at a global level cannot yet be prevented by vaccination. Examples include:

- malaria – *Plasmodium*, the protoctist that causes malaria. It is very evasive – it spends time inside the erythrocytes so it is protected by self antigens from the immune system, and within an infected individual its antigens reshuffle

- HIV, the human immunodeficiency virus that causes AIDS. It enters the macrophages and T helper cells, so it has disabled the immune system itself.

So far scientists have been unable to develop a vaccine for these diseases, which between them affect millions of people globally every year.

Medicines and the management of disease

Medicines can be used to treat communicable and non-communicable diseases. Medicines can be used to treat symptoms and cure them, making people feel better. Common medicines include painkillers, anti-inflammatories and anti-acid medicines (which reduce indigestion).

Medicines that cure people include chemotherapy against some cancers, antibiotics that kill bacteria, and antifungals that kill fungal pathogens.

Sources of medicines

Penicillin was the first widely used, effective, safe antibiotic capable of curing bacterial diseases. It comes from a mould, *Penicillium chrysogenum*, famously discovered by Alexander Fleming in 1928, when he found it growing on his *Staphylococcus* spp. cultures. Fleming saw what the mould did to his bacteria but could not extract enough to test its potential. It needed Howard Florey and Ernst Chain to develop an industrial process for making the new drug, which has since saved millions of lives around the world.

The medicines we use today come from a wide range of sources. Scientists design drugs using complex computer programmes. They can build up 3-dimensional models of key molecules in the body, and of pathogens and their antigen systems. This allows models of potential drug molecules to be built up which are targeted at particular areas of a pathogen. Computers are also used to search through enormous libraries of chemicals, to isolate any with a potentially useful action against a specific group of feature of a pathogen, or against the mutated cells in a cancer.

Analysis of the genomes of pathogens and genes which have been linked to cancer enable scientists to target their novel drugs to attack any vulnerabilities. However, many of the drugs most commonly used in medicine are still either derived from, or based on, bioactive compounds discovered in plants, microorganisms or other forms of life. Table 2 lists some examples.

▼ **Table 2** *Some common medicinal drugs, derived from living organisms*

Drug	Source	Action
Penicillin	commercial extraction originally from mould growing on melons	antibiotic – the first effective treatment against many common bacterial diseases
Docetaxel/paclitaxel	derived originally from yew trees	treatment of breast cancer
Aspirin (acetylsalicylic acid)	based on compounds from sallow (willow) bark	painkiller, anti-coagulant, anti-pyretic (reduces fever) and anti-inflammatory
Prialt	derived from the venom of a cone snail from the oceans around Australia	new pain-killing drug 1000 times more effective than morphine
Vancomycin	derived from a soil fungus	one of our most powerful antibiotics
Digoxin	based on digitoxin, originally extracted from foxgloves	powerful heart drug used to treat atrial fibrillation and heart failure

In the 21ˢᵗ century biodiversity is rapidly being lost around the world, including the destruction of rain forests, the loss of coral reefs and loss of habitat for natural ecosystems in countries all around the world. This is at least partly due to human activities. Scientists have not yet explored and identified and analysed a fraction of life on Earth. One of many reasons why it is so important to maintain biodiversity is to make sure we do not destroy a plant, animal or microorganism which could give us the key to a life-saving drug.

Drug design for the future
Pharmacogenetics

Personalised medicine – a combination of drugs that work with your individual combination of genetics and disease – is the direction in which medicine is going. The human genome can be analysed relatively rapidly and cheaply, giving a growing understanding of the genetic basis of many diseases. The science of interweaving knowledge of drug actions with personal genetic material is known as **pharmacogenomics**. We already know that genotypes and drugs interact. For example, in approximately 30% of all breast cancers there is a mutation in the HER2 gene. The activity of this gene can be shut down by specific drugs – trastuzumab (known an Herceptin) and lapatinib. By analsying breast tumours and treating those which have this mutation with the relevant drugs, doctors can reduce the deaths from HER2 breast cancer by up to 50%. In future, this type of treatment, where clinicians looks at the genome of their patients and the genome of the invading pathogen before deciding how to treat them, will become increasingly common.

▲ **Figure 4** *Mould, a cone snail and foxgloves – this diverse range of organisms provide the origins of some important medicines*

Synthetic biology

Another major step forward in drug development is synthetic biology. Using the techniques of genetic engineering, we can develop populations of bacteria to produce much needed drugs that would otherwise be too rare, too expensive or just not available. Synthetic biology enables the use of bacteria as biological factories. Mammals have also been genetically modified to produce much needed therapeutic proteins in their milk. This re-engineering of biological systems for new purposes has great potential in medicine. Nanotechnology is another strand of synthetic biology, where tiny, non-natural particles are used for biological purposes – for example, to deliver drugs to very specific sites within the cells of pathogens or tumours.

The antibiotic dilemma

At the beginning of the 20th century, 36% of all deaths – and 52% of all childhood deaths – were from communicable diseases. Antibiotics interfere with the metabolism of the bacteria without affecting the metabolism of the human cells – this is called **selective toxicity**. They gave doctors, for the first time, medicines that were effective against bacteria, so antibiotics were understandably, widely used. By the start of this century, the numbers of children dying per year had fallen dramatically and, of those remaining few deaths, only about 7% were due to communicable diseases.

There are many different types of bacteria and a range of antibiotics is used against them, including streptomycin, amoxicillin (very like penicillin), cephalosporins, tetracyclines, sulfonamides, polymixines, ampicillin, and vancomycin. In 2014 up to 1 in 6 of all prescriptions were still for antibiotics. They are often used for relatively minor infections where the immune system of the patient would deal with the infection with no serious difficulty.

Unfortunately, antibiotics are becoming less effective in the treatment of bacterial diseases. Bacteria are becoming resistant to more and more antibiotics. This trend started with penicillin – now there are microorganisms that are resistant to all of the antibiotics we have.

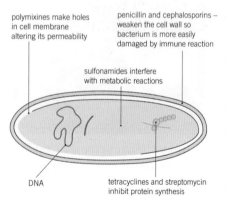

polymixines make holes in cell membrane altering its permeability

penicillin and cephalosporins – weaken the cell wall so bacterium is more easily damaged by immune reaction

sulfonamides interfere with metabolic reactions

DNA

tetracyclines and streptomycin inhibit protein synthesis

▲ **Figure 5** *Some of the different ways in which antibiotic drugs damage bacteria*

The development of antibiotic resistance

There is an evolutionary race between scientists and bacteria. An antibiotic works because a bacterium has a binding site for the drug, and a metabolic pathway that is affected by the drug. If a random mutation during bacterial reproduction produces a bacterium that is not affected by the antibiotic, that is the one which is best fitted to survive and reproduce, passing on the antibiotic resistance mutation to the daughter cells. Bacteria reproduce very rapidly, so once a mutation occurs it does not take long to grow a big population of **antibiotic-resistant bacteria**.

In a few decades we have reached a stage where increasing numbers of bacterial pathogens are resistant to most or all of our antibiotics.

In some countries, including the US, farmers routinely add antibiotics to animal feed prophylactically to prevent animals losing condition due to infections, and reducing business profits. There are concerns that such routine exposure to the antibiotics accelerates natural selection of antibiotic-resistant strains of both human and animal pathogens. However, in the UK it is illegal to give animals routine antibiotics this way. Evidence suggests that it is the over subscription of antibiotics to people which is the prime cause of the rise in antibiotic resistance.

MRSA and *C. difficile*

Antibiotic-resistant bacteria are a particular problem in hospitals and care homes for older people, where antibiotics are often needed and used. **MRSA (methicillin-resistant *Staphylococcus aureus*)** and *Clostridium difficile* (*C. difficile*) have been high-profile examples of antibiotic-resistant bacteria. They are summarised in Table 3.

Figure 6 ▶
Development of resistance in bacteria to ciprofloxacin, an antibiotic used for treating a wide variety of infections

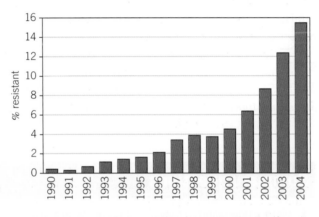

▼ Table 3

MRSA	C. difficile
• bacterium carried by up to 30% of the population on their skin or in their nose • in the body it can cause boils, abscesses and potentially fatal septicaemia • was treated effectively with methicillin, a penicillin-like antibiotic but mutation has produced methicillin-resistant strains	• bacterium in the guts of about 5% of the population • produces toxins that damage the lining of the intestines, leading to diarrhoea, bleeding and even death • not a problem for healthy person *but* when commonly-used antibiotics kill off much of the 'helpful' gut bacteria it survives, reproduces and takes hold rapidly

Antibiotic-resistant infections can be reduced in the long term by measures including:

- minimising the use of antibiotics, and ensuring that every course of antibiotics is completed to reduce the risk of resistant individuals surviving and developing into a resistant strain population

- good hygiene in hospitals, care homes and in general – this has a major impact on the spread of all infections, including antibiotic-resistant strains.

Solving the problem

The development of antibiotic-resistant bacteria is one of the biggest health problems of our time – there is a fear that we may return to the days when bacterial infections killed thousands of people each year in the UK alone. Scientists are working on developing new antibiotics using computer modelling and looking at possible sources in a wide variety of places, including soil microorganisms, crocodile blood, fish slime, honey and the deepest abysses of the oceans. But at the moment, bacterial resistance is building faster than new antibiotics can be found. In 2014 it was announced that a new Lottery-funded prize of £10 million named Longitude will be reserved for anyone who can come up with a cost-effective, accurate and easy-to-use test for bacterial infections so that doctors all over the world can use the right antibiotics at the right time, and only when they are needed.

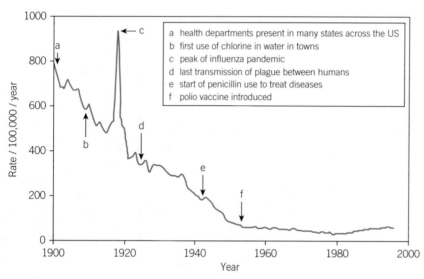

a health departments present in many states across the US
b first use of chlorine in water in towns
c peak of influenza pandemic
d last transmission of plague between humans
e start of penicillin use to treat diseases
f polio vaccine introduced

▲ **Figure 7** *Graph showing crude death rate per year per 100 000 in the US from 1900–1996*

Summary questions

1   Make a flow diagram to explain how artificial active immunity is stimulated when a vaccination is given.   (*4 marks*)

2   Suggest why some children still die of communicable diseases in the UK even in the 21st century.   (*4 marks*)

3   Using the data from Figure 3.
   a When did the peak of the SARS outbreak occur?   (*1 mark*)
   b How many people were infected during the outbreak of the disease   (*1 mark*)
   c How many people died in the outbreak   (*1 mark*)
   d Approximately what percentage of the people affected by the SARS virus recovered?   (*3 marks*)

4   Explain how vaccinations may be used:
   a to prevent rabies after a person has been bitten by a rabid dog, fox or bat   (*2 marks*)
   b to control epidemics.   (*4 marks*)

5   Using data from Figure 7, discuss the factors that affect mortality from communicable diseases and comment on the effects of using antibiotics to treat bacterial infections on mortality since penicillin was first introduced.   (*6 marks*)

Practice questions

1 Doctors regularly receive advice with regard to the prescribing of antibiotics.

   Which of the following statements is/are likely to form part of this advice.

   Statement 1: avoid the use of broad spectrum antibiotics

   Statement 2: only prescribe antibiotics when patients have a bacterial infection

   Statement 3: use targeted antibiotics whenever possible

   A 1, 2 and 3 are correct

   B Only 1 and 2 are correct

   C Only 2 and 3 are correct

   D Only 1 is correct   (*1 mark*)

2 a Define the term pathogen.   (*2 marks*)

   b Describe the physical barriers of plants which act as a defence against infection from pathogens.

                                          (*3 marks*)

   c Explain why plants cannot acquire immunity.   (*4 marks*)

3 The 2014 Ebola virus outbreak was more correctly called an epidemic. It is often fatal and there is no known cure or vaccine available.

   The diagram below shows the progression of the disease in the first few months of the outbreak.

   a (i) Describe the trends shown in the graph.
                                          (*4 marks*)

      (ii) Suggest the reasons for the difference in the trends.   (*3 marks*)

   b Describe the structure of a virus and explain why viruses are not considered to be living organisms.   (*4 marks*)

   c Ebola outbreaks usually happen in remote areas of developing countries. They are quite rare and usually on a relatively small scale.

      Discuss why there are no effective drugs or vaccines available to treat Ebola.   (*5 marks*)

4 The graph below shows the change in incidence of TB in patients and their HIV status, if known.

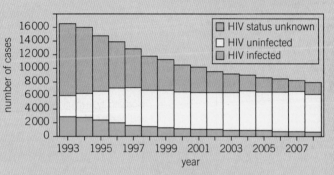

   a Describe and explain the changes from 1993 to 2008.   (*5 marks*)

   b Describe what is meant by the term opportunistic infection.   (*3 marks*)

   c Explain the differences in the way TB and HIV are transmitted.   (*4 marks*)

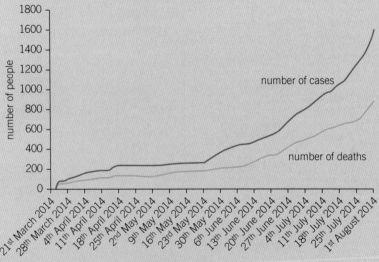

Module 4 summary

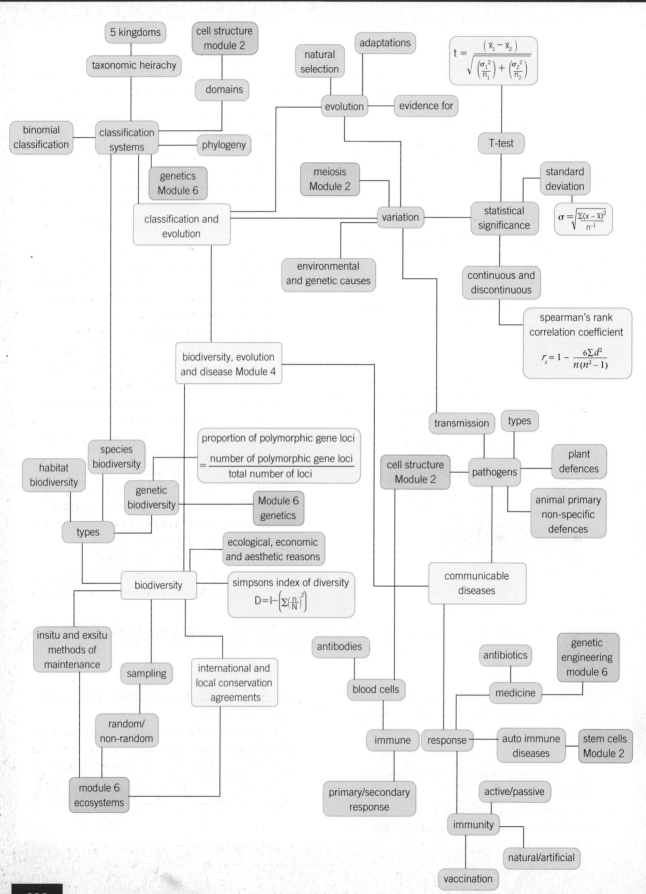

Application

Doctors at a Cambridge hospital were puzzled. A 50 year old Chinese male patient reported seizures, headaches, a changed sense of smell and memory flashbacks, however, tests for communicable diseases such as TB, HIV, syphilis and Lyme disease were all negative. MRI scans showed an abnormal region in his brain, however a biopsy showed only inflammation. Over four years, the symptoms changed and scans showed the abnormal region move across the patient's brain until finally in 2014 doctors operated. They were amazed to remove a 1 cm long tapeworm which was the cause of the symptoms. The patient recovered completely once the worm was removed!

This type of tapeworm was new to the UK. A tissue sample was sent to the Wellcome Trust Sanger Institute for genome sequencing and identified as *Spirometra erinaceieuropae*, a tapeworm species found in China, Japan, South Korea and Thailand. Known human infections are rare – only 300 in the last 60 years. The normal lifecycle of this tapeworm is: Eggs hatch in water → Infects tiny crustaceans → eaten by and infects reptiles e.g., snakes and amphibians e.g., frogs → infects

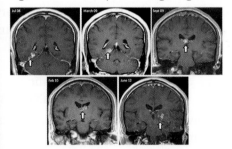

▲ **Figure 1** *The movement of a parasitic tapeworm through a human brain over time*

carnivores e.g., cats, dogs which eat intermediate hosts → proglottids passed out in faeces to start the cycle again.

Sequencing also showed the tapeworm genome was approximately 10 times bigger than any other sequenced tapeworm species. This gives the potential for it to express many different proteins, and therefore invade a wide variety of host animals in its lifecycle. It also showed that the species is not sensitive to a common anti-tapeworm drug, but might be sensitive to an alternative treatment.

1 a What is a communicable disease?
   b The patient described above was tested for TB and HIV/AIDS. What are the similarities and differences between these two diseases and the tapeworm infestation affecting the patient?
   c What is inflammation? Suggest a role for inflammation in the symptoms experienced by the patient during the infestation by the tapeworm?
2 a What methods are available to classify an organism such as *Spirometra erinaceieuropae*, found in an unexpected place?
   b What is natural selection and how is it involved in the process of evolution?
   c The genome of *Spirometra erinaceieuropae* is 10 times larger than any other species of tapeworm sequenced and almost a third of the size of the human genome. How might this large genome be a successful adaptation for this parasite.
   d Using the identification of *Spirometra erinaceieuropae* as one example, discuss the role of DNA sequencing in both classification and medicine.

Extension

The parasite *Spirometra erinaceieuropae* remained in the brain of a patient for over 4 years. It was not destroyed by the specific immune system.

1 Summarise how you would expect the specific immune system to respond to an invasion by another organism.
2 Investigate and describe the adaptations which enable a tapeworm such as *Spirometra*

erinaceieuropae to avoid destruction by the immune system of the host.
3 Identify one biochemical response by a parasitic worm and produce a clear explanation of the way it interacts with the host cells to reduce or moderate the host immune response. Suggest potential uses for these types of responses in the pharmaceutical industry.

Paper 1 practice questions

1   Which of the following statements is/are correct with respect to phylogenetic trees?

Statement 1: Phylogenetic trees depict the evolutionary relationships among groups of organisms

Statement 2: Species and their most recent common ancestor form a clade within a phylogenetic tree

Statement 3: Phylogenetic trees produced more recently show the relationship between clades and taxonomic groups

   A   1, 2 and 3 are correct

   B   Only 1 and 2 are correct

   C   Only 2 and 3 are correct

   D   Only 1 is correct   *(1 mark)*

2   a   Define the following terms:

   cardiac output   *(2 marks)*

   stroke volume   *(2 marks)*

   heart rate   *(1 mark)*

   b   Copy and arrange them in the correct order in the formula below:

   .... × .... = ...   *(1 mark)*

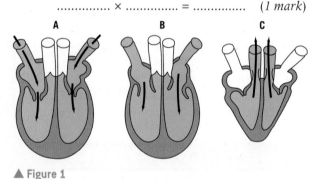

▲ Figure 1

   c   Describe, using the diagrams in Figure 1, what is happening at each stage in the cardiac cycle shown.   *(6 marks)*

3   Xylem vessels have two walls, a primary wall made of cellulose and a secondary wall composed of lignin. Lignin, a strong inflexible polymer, is not usually deposited uniformly but laid down in rings or spirals.

   a   Describe the role of lignin in xylem vessels.   *(4 marks)*

   b   Suggest the benefit to a plant of way that lignin is deposited in the walls of xylem vessels.   *(2 marks)*

4   Consider the statement:

   'evolution can be summarised as change over time'

   a   Discuss why this statement is an over simplification as a description of evolution.   *(4 marks)*

   b   Explain, using the finch populations observed by Darwin in the Galapagos islands the process of disruptive selection.   *(6 marks)*

5   a   Describe the difference between conservation and preservation.   *(3 marks)*

   b   Describe the difference between 'in situ' and 'ex situ' conservation.   *(4 marks)*

6   Fill in the missing words.

   T memory cells live for a long time and are part of the................. .... If they meet an .... for a second time, they undergo .... to form a large .... of T .... cells that destroy the pathogen.   *(5 marks)*

7   Figure 2 shows diagrams of four cells that have been placed in different solutions.

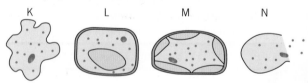

▲ Figure 2

   a   In the table write the letter **K**, **L**, **M** or **N** next to the description that best matches the diagram. One has been done for you.

description	letter
an animal cell that has been placed in distilled water	
an animal cell that has been placed in a concentrated sugar solution	
a plant cell that has been placed in distilled water	
a plant cell that has been placed in concentrated sugar solution	**M**

(3 marks)

b Explain, using the term **water potential**, what has happened to cell **M**. *(3 marks)*

c Small non-polar substances enter cells in different ways to large or polar substance.

Outline the ways in which the substances below, can enter a cell through the plasma (cell surface) membrane.

Small, non-polar substances

Large substances

Polar substances *(5 marks)*

OCR F211 2009

8 a The structure of cell membranes can be described as 'proteins floating in a sea of lipids'. This membrane structure allows certain substances to pass through freely whereas other substances cannot.

State the term used to describe a membrane through which some substances can pass freely but others cannot. *(1 mark)*

b Copy and complete the following paragraph about cell membranes through which some substances can pass freely but others cannot.

The model of cell membrane structure is called the ………. ……….. model. Phospholipid bilayers with specific membrane proteins account for the ability of the membrane to allow both passive and …………… transport mechanisms. Ions and most polar molecules are insoluble in the phospholipid bilayer. However, the bilayer allows diffusion of most non-polar molecules such as …………………. .

Protein channels, which may be gated, and …………… proteins enable the cell to control the movement of most polar substances. *(4 marks)*

c One function of membranes that is not mentioned in (b) is cell signalling.

(i) State what is meant by *cell signalling*. *(1 mark)*

(ii) Explain how cell surface membranes contribute to the process of cell signalling. In your answer you should use appropriate technical terms, spelled correctly. *(4 marks)*

OCR F211/01 2013

9 Various measurements of lung function are used to help diagnose lung disease and to monitor its treatment.

a State what is meant by the following terms:

(i) vital capacity; *(1 mark)*

(ii) forced expiratory volume 1 (FEV1). *(1 mark)*

One measure of lung function is:

$$\text{Percentage lung function} = \frac{\text{FEV 1}}{100} \times 100$$

This is particularly useful in identifying possible obstructive disorders of the airways and lungs, such as asthma or chronic obstructive pulmonary disease (COPD).

- Asthma is a condition that responds to, and can be controlled by, the use of bronchodilators.
  These are drugs that dilate the airways and improve airflow.

- COPD lasts for a long period of time and is caused by progressive and permanent damage to the lung tissue.

When the value calculated for the percentage lung function is less than or equal to 70%, this indicates an obstructive disorder. A 'normal' value is approximately 80%.

Table 1 shows data relating to three patients, **C**, **D** and **E**, before and after treatment with a bronchodilator drug.

patient	age (years)	before treatment vital capacity (dm³)	FEV1 (dm³)	percentage lung function	after treatment vital capacity (dm³)	FEV1 (dm³)	percentage lung function
C	18	5.5	3.8	69	5.6	4.5	
D	45	5.3	3.6	68	5.5	4.0	73
E	78	3.8	2.2	58	3.8	2.2	58

b (i) Calculate the percentage lung function for patient **C** after treatment with the bronchodilator drug.

Show your working and give your answer in percentage **to the nearest whole number**.

(2 marks)

(ii) Using the information in the table and your answer to **(c)(i)**, indicate with a tick (✓) in the table below a diagnosis for each patient.

patient	Diagnosis Asthma	COPD
C		
D		
E		

(3 marks)

OCR F221 2009

10 a Complete the following paragraph about cells by using the most appropriate term(s).

Cell that are not specialised but still have the ability to divide are called .... cells. Such cells can be found in the .... of the long bones of mammals. These cells can .... into other types of cell, such as erythrocytes that carry oxygen in the blood. In plants, ... tissues also contains cells that are not specialised. *(4 marks)*

b Sponges are simple eukaryotic multicellular organisms that live underwater on the surface of rocks. Sponges have a cellular level of organisation. This means they have no tissues.

Each cell type is specialised to perform a particular function. One type of cell found in a sponge is a collar cell. Collar cells are held in positions on the inner surface of the body of the sponge. Figure 3 is a diagram showing a vertical section through the body of a sponge and an enlarged drawing of a collar cell.

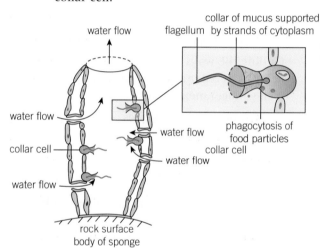

▲ Figure 3

(i) Suggest **one** function of the flagellum in the collar cell. *(1 mark)*

(ii) Suggest **one** possible role for the collar of mucus in the cell *(1 mark)*

c In more advanced organisms, cells are organised into tissues consisting of one or more types of specialised cells.

Describe how cells are organised into tissues, using **xylem** and **phloem** as examples *(4 marks)*

OCR F211 2012

11 A student investigating how different concentrations of sucrose solution affect the size of animal cells obtained the results shown in **Table 1**.

Concentration of sucrose solution / mol dm^{-3}					
	0.05	0.10	0.20	0.40	0.80
diameter of cell 1 / μm	8.4	7.2	6.6	5.7	2.3
diameter of cell 2 / μm	7.8	7.3	6.8	5.7	2.5
diameter of cell 3 / μm	8.1	7.4	7.0	5.5	2.4
mean diameter / μm	8.1	7.3	6.8		2.4
mean change in diameter / μm	+1.1	+0.3	−0.2		−4.6

a The original mean diameter of the cells was 7.0 μm. Copy and complete **Table 1**.
(*2 marks*)

b Plot a graph using the results obtained in the table. (*4 marks*)

c Describe and explain the trend shown by the graph. (*4 marks*)

d Suggest the type of cells the student used in the investigation. (*1 mark*)

12 A student investigating rate diffusion carried out the following procedure.

1 prepared a petri-dish containing a layer of agar

2 cut a 1cm well in the centre of the agar

3 placed 10 drops of a coloured solution in the well

4 measured the distance travelled by the colored solution from the edge of the well every 15 minutes.

The results obtained by the student are shown in **Table 2**.

time / min	distance diffused from well by coloured solution / mm
0	0
15	14
30	22
45	26
60	28
75	29

a Plot a graph using the results obtained in the table. (*4 marks*)

b Calculate, using the graph, the rate of diffusion of the solution between 10 minutes and 20 minutes. (*4 marks*)

c Describe and explain the trend shown by the graph. (*4 marks*)

The smallest units on the ruler used to measure the distances diffused by the coloured solution were 1mm.

d (i) State the uncertainty of the measurements obtained by the student. (*1 mark*)

(ii) Calculate the percentage error for the measurement taken at 45 minutes. (*4 marks*)

(iii) Suggest how the student could have improved the precision of the measurements. (*1 mark*)

(iv) Suggest how the student could have improved the reliability of the investigation. (*2 marks*)

e Evaluate the validity of the investigation. (*4 marks*)

Paper 2 practice questions

1 The diagram shows a normal ECG trace.

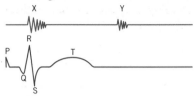

a (i) State which valves are closing at points X and Y. *(2 marks)*

(ii) Describe the function of the atrio-ventricular valves in the heart. *(2 marks)*

(iii) Explain why the right ventricular wall of the heart is less muscular than the left ventricular wall. *(5 marks)*

Figure 2 shows an ECG trace during a heart attack (a) and during fibrillation (b).

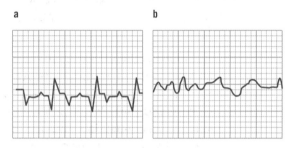

b Describe the differences between the ECG traces shown during a heart attack and fibrillation. *(3 marks)*

c *'A cardiac arrest occurs when there is a problem with the electrical activity in the heart, whereas a heart attack happens when there is a problem with the plumbing'.*
Discuss what you understand by this statement. *(6 marks)*

2 Variation is a fundamental characteristic of living organisms. Variation can be influenced by both genetics and the environment. The graphs below represent the two main types of variation.

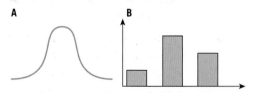

a Name the two types of variation shown in the graphs. *(2 marks)*

b Explain why evolution would not be possible without variation. *(4 marks)*

c Describe how variation arises. *(4 marks)*

d As humans, we have an anthropocentric view of evolution. This means we measure the success of a species by our own standards. Discuss whether our view of evolution is right. *(6 marks)*

3 a Explain, using the term **surface area to volume ratio**, why large, active organisms need a specialised surface for gaseous exchange. *(2 marks)*

b The table describes some of the key features of the mammalian gas exchange system.

Copy and complete the table by explaining how each feature improves the efficiency of gaseous exchange. The first one has been completed for you.

Feature of gas exchange system	How feature improves efficiency of gaseous exchange system
Many alveoli	This increases the surface across which oxygen and carbon dioxide can diffuse
The epithelium of the alveoli is very thin	
There are capillaries running over the surface of the alveoli	
The lungs are surrounded by the diaphragm and intercostal muscles	

(3 marks)

c Outline how the diaphragm **and** intercostal muscles cause **inspiration**. *(4 marks)*

d The graph shows the trace from a spirometer recorded from a 16-year old student.

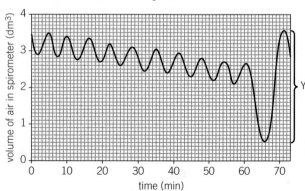

(i) Label on the trace, using the letter X, a point that indicates when the student was inhaling. (*1 mark*)

(ii) At the end of the trace the student measured his vital capacity. This is indicated by the letter Y.
State the vital capacity of the student.
(*1 mark*) OCR F211 2009

4 The diagram shows part of a DNA molecule.

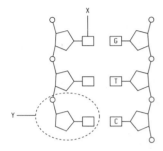

(i) Name the parts of the molecule represented by the letters **X** and **Y**.
(*2 marks*)

(ii) Copy and complete the diagram in Figure 2 by drawing hydrogen bonds to connect the two strands. (*2 marks*)

(iii) State two ways in which a diagram of part of an RNA molecule would appear different from the DNA molecule shown in Figure 2. (*2 marks*)

b DNA replication takes place during interphase of the cell cycle. It occurs by a semi-conservative mechanism.

(i) Explain why DNA replication is considered to be semi-conservative.
(*2 marks*)

(ii) Explain why complemenatry base-pairing is important in DNA replication. (*2 marks*)

c In 1958, two scientists, Meselson and Stahl, conducted an investigation into DNA replication.

- Bacteria were grown in a food source that contained only the 'heavy' isotope of nitrogen, ^{15}N After many generations, the bacterial DNA contained only the 'heavy' form of nitrogen.

- Some of the bacteria were then transferred to another food source containing only the normal, 'lighter' form of nitrogen, ^{14}N.

- DNA was extracted from the bacteria and centrifuged. (When a solution is centrifuged, the heavier, more dense moecules tend to settle nearer the bottom of the tube.)

Some results from the experiment are shown in the diagram.

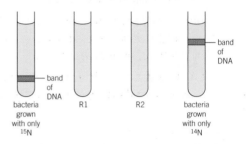

(i) In the figure, the tube labelled **R1** represents the results for DNA obtained from bacteria that had been **transferred** from the ^{15}N to the ^{14}N food source and left long enough for their DNA to replicate **once** only.

Copy the figure and draw **one** band on tube **R1** in the position you would expect the DNA to appear **after** centrifuging. (*1 mark*)

(ii) In the figure, the tube labelled **R2** represents the results for DNA obtained from bacteria that had been **transferred** from the ^{15}N to the ^{14}N food source and left long enough for their DNA to replicate **twice**.

On your drawing draw **two** bands on tube **R2** in the positions you would expect the DNA to appear **after** centrifuging. (*1 mark*)

d The technique of centrifugation used by Meselson and Stahl involves:

- Mixing the DNA sample with concentrated sugar solution

- Placing the mixture of DNA and sugar solution in test tubes

- Spinning the test tubes at a very high speed.

Suggest **three** precautions that Meselson and Stahl would have taken in order to ensure that the centrifugation part of their investigation produced valid results.

(*3 marks*) OCR F212/01 2013

5 The graph shows oxygen dissociation curves for both myoglobin and haemoglobin.

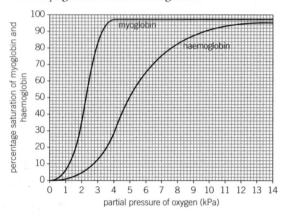

Calculate the decrease in percentage saturation of both myoglobin and haemoglobin between 4 kPa and 2 kPa partial pressure of oxygen.

(*1 mark*) OCR F224 2010

6 a breast cancer mortality in the UK has decreased in all age groups since the 1990's.

The graph shows the mortality rates from breast cancer in females in the UK between 1988 and 2008.

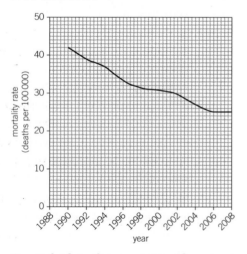

(i) Calculate the percentage decrease in mortality rate between 1990 and 2007.

Show your working and **give your answer to the nearest whole number**.

(*2 marks*)

(ii) Suggest three reasons for this reduction in mortality rate from breast cancer

(*3 marks*) OCR F222 2011

7 a Describe the process of phagocytosis.

(*5 marks*)

b Explain what is meant by the term pathogen. (*2 marks*)

c Explain why antibiotics are not prescribed to treat influenza. (*3 marks*)

d Suggest why antibiotics might still be prescribed to someone with influenza.

(*3 marks*)

e Outline the different ways in which pathogens are made safe to use in vaccines.

(*4 marks*)

f The diagram below shows a diagram of antibody.

Describe how an antibody, such as the one in the diagram above, has the ideal structure to carry out its role in an immune response. (*6 marks*) OCR F224 2011

8 Advances in technology particularly in biochemistry have provided evidence that casts doubt on the way in which some organisms have been grouped in the five kingdom classification system.

DNA sequencing suggests that some organisms are more closely related to organisms belonging to another kingdom than other members of their own kingdom.

a Explain the meaning of the following terms:

(i) Phylogeny (*2 marks*)

(ii) Hierarchy (*2 marks*)

(iii) Taxonomy (*2 marks*)

(iv) Cladistics (*2 marks*)

b Outline the differences between kingdom and domain classification systems. (*5 marks*)

c Explain the meaning of the term 'descent with modification'. (*6 marks*)

9 A student carried out an investigation into the effect of the enzyme inhibitor, lead nitrate, on the activity of the enzyme amylase.

The results that were obtained are shown in **Table 1**

percentage concentration of lead nitrate solution	transmission of light / arbitrary units			mean transmission of light / arbitrary units
	first run	second run	third run	
0	84	87	82	84.3
0.2	55	53	52	53.3
0.4	36	37	36	36.3
0.6	27	24	27	26.0
0.8	22	20	21	21.0
1.0	20	21	18	

a **(i)** Outline a procedure that the student could have followed to obtain the results. *(4 marks)*

(ii) Copy and complete **Table 1** *(1 mark)*

b **(i)** The third run for 1.0% solution of lead nitrate initially produced a reading of 70 arbitrary units.

The student discarded this result and repeated this run. Explain why. *(2 marks)*

(ii) Plot a graph using the results in **Table 1**. *(3 marks)*

c **(i)** Describe and explain the shape of the graph. *(4 marks)*

(ii) State an appropriate conclusion for these results. *(1 mark)*

The student stated in their conclusion:

'the enzyme amylase was inhibited at all concentrations of lead nitrate'

d Discuss, whether or not, the student was correct in making this statement. *(4 marks)*

e Evaluate the validity of this investigation. *(4 marks)*

f Suggest how this investigation could be improved, *(1 mark)*

10 A student carried out an investigation into the effect of pH on the activity of the enzyme amylase.

a Suggest a hypothesis that the student made before starting the investigation. *(2 marks)*

b **(i)** State the independent and dependent variables in this investigation. *(2 marks)*

(ii) State two variables that should be controlled in this investigation. *(2 marks)*

The student obtained the results shown in the table.

pH		5	6	7	8	9
Time taken for starch to be broken down/min	First run	11	7	3	4	10
	Second run	10	6	4	5	9
	Third run	8	7	3	6	10

c Draw a table to present the raw data correctly. Calculate and include the mean values in the table that you produce. *(4 marks)*

d Describe how the student increased the reliability of the investigation. *(2 marks)*

e Discuss whether the results obtained by the student support your hypothesis. *(4 marks)*

Glossary

abiotic factors non-living conditions in a habitat.

activation energy the energy required to initiate a reaction.

active site area of an enzyme with a shape complementary to a specific substrate, allowing the enzyme to bind a substrate with specificity.

active transport movement of particles across a plasma membrane against a concentration gradient. Energy is required.

adenosine diphosphate (ADP) a nucleotide composed of a nitrogenous base (adenine), a pentose sugar and two phosphate groups. Formed by the hydrolysis of ATP, releasing a phosphate ion and energy.

adenosine triphosphate (ATP) a nucleotide composed of a nitrogenous base (adenine), a pentose sugar and three phosphate groups. The universal energy currency for cells.

agglutinins chemicals (antibodies) that cause pathogens to clump together so they are easier for phagocytes to engulf and digest.

alleles different versions of the same gene.

amino acids monomer used to build polypeptides and thus proteins.

anabolism (anabolic) reactions of metabolism that construct molecules from smaller units. These reactions require energy from the hydrolysis of ATP.

analogous structures structures that have adapted to perform the same function but have a different origin.

anaphase third stage of mitosis when chromatids are separated to opposite poles of the cell.

antibiotic-resistant bacteria bacteria that undergo mutation to become resistant to an antibiotic and then survive to increase in number.

antibiotics a chemical or compound that kills or inhibits the growth of bacteria.

antibodies Y-shaped glycoproteins made by B cells of the immune system in response to the presence of an antigen.

antigen identifying chemical on the surface of a cell that triggers an immune response.

antigen–antibody complex the complex formed when an antibody binds to an antigen.

antigen-presenting cell (APC) a cell that displays foreign antigens complexed with major histocompatibility complexes on their surfaces.

antisense strand the strand of DNA that runs 3' to 5' and is complementary to the sense strand. It acts as a template strand during transcription.

anti-toxins chemicals (antibodies) that bind to toxins produced by pathogens so they no longer have an effect.

apoplast the cell walls and intercellular spaces of plant cells.

apoplast route movement of substances through the cell walls and cell spaces by diffusion and into cytoplasm by active transport.

arrhythmia an abnormal rhythm of the heart.

artefacts objects or structures seen through a microscope that have been created during the processing of the specimen.

artificial active immunity immunity which results from exposure to a safe form of a pathogen, for example, by vaccination.

artificial passive immunity immunity which results from the administration of antibodies from another animal against a dangerous pathogen.

asexual reproduction the production of genetically identical offspring from a single parent.

assimilates the products of photosynthesis that are transported around a plant, e.g., sucrose.

atrial fibrillation an abnormal rhythm of the **heart** when the **atria** beat very fast and incompletely.

atrio-ventricular node (AVN) stimulates the ventricles to contract after imposing a slight delay to ensure atrial contraction is complete.

autoimmune disease a condition or illness resulting from an autoimmune response.

autoimmune response response when the immune system acts against its own cells and destroys healthy tissue in the body.

autotrophic organisms that acquire nutrients by photosynthesis.

B effector cells B lymphocytes that divide to form plasma cell clones.

B lymphocytes (B cells) lymphocytes which mature in the bone marrow and that are involved in the production of antibodies.

B memory cells B lymphocytes that live a long time and provide immunological memory of the antibody needed against a specific antigen.

belt transect two parallel lines are marked along the ground and samples are taken of the area at specified points.

Benedict's reagent an alkaline solution of copper(II)sulfate used in the chemical tests for reducing sugars and non-reducing sugars. A brick-red precipitate indicates a positive result.

beta pleated sheet sheet-like secondary structure of proteins.

binomial nomenclature the scientific naming of a species with a Latin name made of two parts – the first indicating the genus and the second the species.

biodiversity the variety of living organisms present in an area.

biuret test the chemical test for proteins; peptide bonds form violet coloured complexes with copper ions in alkaline solutions.

Bohr effect the effect of carbon dioxide concentration on the uptake and release of oxygen by haemoglobin.

bradycardia a slow heart rhythm of below 60 beats per minute.

breathing rate the number of breaths (inhalation and exhalation) taken per minute.

bulk transport a form of active transport where large molecules or whole bacterial cells are moved into or out of a cell by endocytosis or exocytosis.

bundle of his conducting tissue composed of purkyne fibres that passes through the septum of the heart

callose a polysaccharide containing β 1-3 linkages and β 1-6 linkages between the glucose monomers that is important in the plant response to infection.

carbaminohaemoglobin the compound formed when carbon dioxide combines with haemoglobin.

carbohydrates organic polymers composed of the elements carbon, hydrogen and oxygen in the ratio $C_x(H_2O)_y$. Also known as saccharides or sugars.

carbonic anhydrase enzyme which catalyses the reversible reaction between carbon dioxide and water to form carbonic acid.

cardiac cycle the events of a single heartbeat, composed of diastole and systole.

carrier proteins membrane proteins that play a part in the transport of substances through a membrane.

cartilage strong, flexible connective tissue found in many areas of the bodies of humans and other animals.

catabolism (catabolic) reactions of metabolism that break molecules down into smaller units. These reactions release energy.

catalase an enzyme that catalyses the breakdown of hydrogen peroxide.

cell cycle the highly ordered sequence of events that takes place in a cell, resulting in division of the nucleus and the formation of two genetically identical daughter cells.

cell signalling a complex system of intercellular communication.

cellulose a polysacchardie formed from beta glucose molecules where alternate beta glucose molecules are turned upside down. It is unable to coil or form branches but makes hydrogen bonds with other cellulose molecules to produce strong and insoluble fibres. Major component of plant cell walls.

cell wall a strong but flexible layer that surrounds some cell-types.

centrioles component of the cytoskeleton of most eukaryotic cells, composed of microtubules.

centromere region at which two chromatids are held together.

channel proteins membrane proteins that provide a hydrophilic channel through a membrane.

checkpoints control mechanisms of the cell cycle.

chiasmata sections of DNA, which became entangled during crossing over, break and rejoin during anaphase 1 of meiosis sometimes resulting in an exchange of DNA between bivalent chromosomes, forming recombinant chromatids and providing genetic variation.

chloride shift the movement of chloride ions into the red blood cells as hydrogen ions move out to maintain the electrochemical equilibrium.

chloroplasts organelles that are responsible for photosynthesis in plant cells. Contain chlorophyll pigments, which are the site of the light reactions of photosynthesis.

chromatids two identical copies of DNA (a chromosome) held together at a centromere.

chromatin uncondensed DNA in a complex with histones.

chromosomes structures of condensed and coiled DNA in the form of chromatin. Chromosomes become visible under the light microscope when cells are preparing to divide.

circulatory system the transport system of an animal.

clonal expansion the mass proliferation of antibody-producing cells by clonal selection.

clonal selection the theory that exposure to a specific antigen selectively stimulates the proliferation of the cell with the appropriate antibody to form numerous clones of these specific antibody-forming cells (clonal expansion).

closed circulatory system a circulatory system where the blood is enclosed in blood vessels and does not come into direct contact with the cells of the body beyond the blood vessels.

Clostridium difficile (C. difficile) a species of Gram positive bacteria that is resistant to most antibiotics.

codon a three-base sequence of DNA or RNA that codes for an amino acid.

cofactors non-protein components necessary for the effective functioning of an enzyme.

cohesion-tension theory the best current model explaining the movement of water through a plant during transpiration.

communicable diseases diseases that can be passed from one organism to another, of the same or different species.

community all the populations of living organisms in a particular habitat.

companion cells the active cells found next to sieve tube elements that supply the phloem vessels with all of their metabolic needs.

competitive inhibitor an inhibitor that competes with substrate to bind to active site on an enzyme.

complementary base pairing specific hydrogen bonding between nucleic acid bases. Adenine (A) binds to thymine (T) or uracil (U) and cytosine (C) binds to guanine (G).

compound light microscope a light microscope which uses two lenses to magnify an object; the objective lens, which is placed near to the specimen and an eyepiece lens, through which the specimen is viewed.

condensation reaction a reaction between two molecules resulting in the formation of a larger molecule and the release of a water molecule. The opposite reaction to a hydrolysis reaction.

continuous variation a characteristic that can take any value within a range, e.g. height.

contrast staining or treating specific cell components so they are visible compared to untreated components.

convergent evolution organisms evolve similarities because the organisms adapt to similar environments or other selection pressures.

correlation coefficient statistical test used to consider the relationship between two sets of data.

countercurrent exchange system a system for exchanging materials or heat when the two different components flow in opposite directions past each other.

counterstain application of second stain with a contrasting colour to sample for microscopy.

crossing over see chiasmata.

cytokines cell-signalling molecules produced by mast cells in damaged tissues that attract phagocytes to the site of infection or inflammation.

cytokinesis cell division stage in the mitotic phase of the cell cycle that results in the production of two identical daughter cells.

cytolysis the bursting of an animal cell caused by increasing hydrostatic pressure as water enters by osmosis.

cytoplasm internal fluid of cells, composed of cytosol (water, salts and organic molecules), organelles and cytoskeleton.

cytoskeleton a network of fibres in the cytoplasm of a eukaryotic cell.

denatured (denaturation) change in tertiary structure of a protein or enzyme, resulting in loss of normal function.

deoxyribonucleic acid (DNA) the molecule responsible for the storage of genetic information.

diastole the stage of the cardiac cycle in which the heart relaxes and the atria and then the ventricles fill with blood.

dicotyledonous plants (dicots) plants that produce seeds containing two cotyledons, which act as food stores for the developing embryo and form the first leaves when the seed germinates.

differential staining using specific stains to distinguish different types of cell.

differentiation the process of a cell becoming differentiated. Involves the selective expression of genes in a cell's genome.

diploid normal chromosome number; two chromosomes of each type – one inherited from each parent.

disaccharide a molecule comprising two monosaccharides, joined together by a glycosidic bond.

discontinuous variation a characteristic that can only result in certain discrete values, for example, blood type.

divergent evolution species diverge over time into two different species, resulting in a new species becoming less like the original one.

DNA helicase enzyme that catalyses the unwinding and separating of strands in DNA replication.

DNA polymerase enzyme that catalyses the formation of phosphodiester bonds between adjacent nucleotides in DNA replication.

DNA replication the semi-conservative process of the production of identical copies of DNA molecules.

double circulatory system a circulatory system where the blood travels twice through the heart for each complete circulation of the body. In the first circulation blood is pumped by the heart to the lungs. In the second circulation oxygenated blood is pumped by the heart to the brain and body to supply cells with oxygen.

ectopic heartbeat extra heartbeats that are out of the normal rhythm.

elastic recoil the ability to return to original shape and size following stretching. Particularly of the alveoli of the lungs and of the arteries.

electrocardiogram (ECG) a technique for measuring tiny changes in the electrical conductivity of the skin that result from the electrical activity of the heart. This produces a trace which can be used to analyse the health of the heart.

electron microscopy microscopy using a microscope that employs a beam of electrons to illuminate the specimen. As electrons have a much smaller wavelength than light they produce images with higher resolutions than light microscopes.

emulsion test laboratory test for lipids using ethanol; a white emulsion indicates the presence of a lipid.

endocytosis the bulk transport of materials into cells via invagination of the cell-surface membrane forming a vesicle.

endosymbiosis the widely-accepted theoretical process by which eukaryotic cells evolved from prokaryotic cells.

end-product inhibition the product of a reaction inhibits the enzyme required for the reaction.

enzyme–product complex complex formed as a result of an enzyme-catalysed reaction, when a substrate is converted to a product or products while bound to the active site of an enzyme.

enzymes biological catalysts that interact with substrate molecules to facilitate chemical reactions. Usually globular proteins.

enzyme-substrate complex complex formed when a substrate is bound to the active site of an enzyme.

epidemic when a communicable disease spreads rapidly to a lot of people at a local or national level.

eukaryotes multicellular eukaryotic organisms like animals, plants and fungi and single-celled protoctista.

eukaryotic cells cells with a nucleus and other membrane-bound organelles.

ex situ **conservation** conservation methods out of the natural habitat.

exchange surfaces surfaces over which materials are exchanged from one area to another.

exocytosis the bulk transport of materials out of cells. Vesicles containing the material fuse with the cell-surface membrane and the contents are released to the outside of the cell.

exoskeleton an external skeleton of some organisms, e.g. insects.

expiratory reserve volume the extra amount of air that can be forced out of the lungs over and above the normal exhalation (tidal volume).

facilitated diffusion diffusion across a plasma membrane through protein channels.

fatty acids long chain carboxylic acids used in the formation of triglycerides.

fibrous proteins long, insoluble, structural proteins.

fluid-mosaic model model of the structure of a cell membrane in which phospholipids within the phospholipid bilayer are free to move and proteins of various shapes and sizes are embedded in various positions.

fossils the remains or impression of a prehistoric plant or animal preserved in rock.

founder effect when a few individuals of a species colonise a new area, their offspring initially experience a loss in genetic variation, and rare alleles can become much more common in the population.

Fungi biological kingdom containing yeasts, moulds and mushrooms.

gametes haploid sex cells produced by meiosis in organisms that reproduce sexually.

gaseous exchange system the complex systems in which the respiratory gases oxygen and carbon dioxide are exchanged in an organism.

gene a section of DNA that contains the complete sequence of bases (codons) to code for a protein.

gene flow when alleles are transferred from one population to another by interbreeding.

genetic bottleneck when large numbers of a population die prior to reproducing, leading to reduced genetic biodiversity within the population.

genetic code the sequences of bases in DNA are the 'instructions' for the sequences of amino acids in the production of proteins.

genetic variation a variety of different combinations of alleles in a population.

gills the gaseous exchange organs of fish, comprised of gill plates, gill filaments and gill lamellae.

globular proteins spherical, water-soluble proteins.

glucose a monosaccharide with the chemical formula $C_6H_{12}O_6$. One of the main products of photosynthesis in plants.

glycerol alcohol found in triglycerides.

glycogen a branched polysaccharide formed from alpha glucose molecules. A chemical energy store in animal cells.

glycolipids cell-surface membrane lipids with attached carbohydrate molecules of varying lengths and shapes.

glycoproteins extrinsic membrane proteins with attached carbohydrate molecules of varying lengths and shapes.

glycosidic bond a covalent bond between two monosaccharides.

goblet cells differentiated cells specialised to secrete mucus.

Golgi apparatus organelle in most eukaryotic cells formed from an interconnected network of flattened, membrane-enclosed sacs, or cisternae. Play a role in modifying and packaging proteins into vesicles.

Gram negative bacteria bacteria with cell walls that stain red with Gram stain.

Gram positive bacteria bacteria with cell walls that stain purple-blue with Gram stain.

granum (plural grana) a structure inside chloroplasts composed of a stack of several thylakoids. Contains chlorophyll pigments, where light reactions occur during photosynthesis.

guard cells cells that can open and close the stomatal pores, controlling gaseous exchange and water loss in plants.

habitat biodiversity the number of different habitats found within an area.

haemoglobin the red, oxygen-carrying pigment of red blood cells.

haemoglobinic acid the compound formed when haemoglobin accepts free hydrogen ions in its role as a buffer in the blood.

haemolymph the transport medium or 'blood' in insects

haploid half the normal chromosome number; one chromosome of each type.

heterotrophic organisms that acquire nutrients by the ingestion of other organisms.

hexose monosaccharide a monosaccharide composed of six carbons.

histamines chemicals produced by mast cells in damaged tissues that make the blood vessels dilate (causing redness and heat) and the blood vessel walls leaky (causing swelling and pain).

histones proteins that form a complex with DNA called chromatin.

homologous chromosomes matching pair of chromosomes, one inherited from each parent.

homologous structure a structure which appears superficially different but has the same underlying structure.

hydrolysis reaction the breakdown of a molecule into two smaller molecules requiring the addition of a water molecule. The opposite reaction to a condensation reaction.

hydrophilic the physical property of a molecule that is attracted to water.

hydrophobic the physical property of a molecule that is repelled by water.

hydrophytes plants with adaptations that enable them to survive in very wet habitats or submerged or at the surface of water.

hydrostatic pressure the pressure created by water in an enclosed system.

immune response a biological response that protects the body by recognising and responding to antigens and by destroying substances carrying non-self antigens.

immunoglobulins Y-shaped glycoproteins that form antibodies.

in situ **conservation** conservation methods within the natural habitat.

independent assortment the arrangement of each homologous chromosome pair (bivalent) in metaphase 1 and metaphase 2 of meiosis is independent of each other and results in genetic variation.

induced-fit hypothesis modified lock and key explanation for enzyme action; the active site of the enzyme is modified in shape by binding to the substrate.

inflammation biological response of vascular tissues to pathogens, damaged cells, or irritants, resulting in pain, heat, redness and swelling.

inhibitor a factor that prevents or reduces the rate of an enzyme-catalysed reaction.

inspiratory reserve volume the maximum volume of air that can be breathed in over and above a normal inhalation (tidal volume).

insulin a globular protein hormone involved in the regulation of blood glucose concentration.

intercostal muscles the muscles between the ribs that pull the ribs upwards during inhalation (internal intercostal muscles) and downwards during forced exhalation (external intercostal muscles)

interleukins a type of cytokine produced by T helper cells.

interphase growth period of the cell cycle, between cell divisions (mitotic phase). Consists of stages G_1, S and G_2.

interspecific variation the differences between organisms of different species.

intraspecific variation the differences between organisms of the same species.

iodine test a chemical test for the presence of starch using a potassium iodide solution. A colour change to purple/black indicates a positive result.

ion an atom or molecule with an overall electric charge because the total number of electrons is not equal to the total number of protons. See anion and cation.

ionic bond a chemical bond that involves the donating of an electron from one atom to another, forming positive and negative ions held together by the attraction of the opposite charges.

keystone species species which are essential for maintaining biodiversity – they have a disproportionately large effect on their environment relative to their abundance.

kingdom the second biggest and broadest taxonomic group.

lactose a disaccharide made up of a galactose and glucose monosaccharide.

laser scanning confocal microscope a microscope that employs a beam of fluorescence and a pin-hole aperture to produce an image with a very high resolution.

light microscope an instrument that uses visible light and glass lenses to enable the user to see objects magnified many times.

line transect a line is marked along the ground and samples are taken at specified points.

lipids non-polar macromolecules containing the elements carbon, hydrogen and oxygen. Commonly known as fats (solid at room temperature) and oils (liquid at room temperature).

lung surfactant chemical mixture containing phospholipids and both hydrophilic and hydrophobic proteins, which coats the surfaces of the alveoli and prevents them collapsing after every breath.

lymph modified tissue fluid that is collected in the lymph system.

lymphocytes white blood cells that make up the specific immune system.

lysosomes specialised vesicles containing hydrolytic enzymes for the breakdown of waste materials within a cell.

macromolecules large complex molecules with a large molecular weight.

maltose two glucose molecules linked by a 1, 4 glycosidic bond.

mass transport system a transport system where substances are transported in a mass of fluid.

meiosis form of cell division where the nucleus divides twice (meiosis I and meiosis II) resulting in a halving of the chromosome number and producing four haploid cells from one diploid cell.

membrane a selectively-permeable barrier surrounding all cells and forming compartments within eukaryotic cells.

membrane proteins protein components of cell-surface membranes.

meristematic tissue (meristems) tissue found at regions of growth in plants. Contains stem cells.

messenger (m)RNA short strand of RNA produced by transcription from the DNA template strand. It has a base sequence complementary to the DNA from which it is transcribed, except it has uracil (U) in place of thymine (T).

metaphase second stage of mitosis when chromosomes line up at the metaphase plate.

mitochondrial DNA DNA present within the matrix of mitochondria.

mitosis nuclear division stage in the mitotic phase of the cell cycle.

mitotic phase period of cell division of the cell cycle. Consists of the stages mitosis and cytokinesis.

monoculture the cultivation of a single crop in a given area.

monomers individual molecules that make up a polymer.

monosaccharide a single sugar molecule.

mRNA see messenger (m)RNA.

MRSA (methicillin-resistant *Staphylococcus aureus*) a mutated strain of the bacterium *Staphylococcus aureus* that is resistant to the antibiotic, methicillin.

mucous membranes membranous linings of body tracts that secrete a sticky mucus.

multipotent a stem cell that can only differentiate into a range of cell types within a certain type of tissue.

mutation A change in the genetic material which may affect the phenotype of the organism.

myogenic muscle which has its own intrinsic rhythm.

natural active immunity immunity which results from the response of the body to the invasion of a pathogen.

natural passive immunity the immunity given to an infant mammal by the mother through the placenta and the colostrum.

natural selection the process by which organisms best suited to their environment survive and reproduce, passing on their characteristics to their offspring through their genes.

non-competitive inhibitor an inhibitor that binds to an enzyme at an allosteric site.

non-random sampling an alternative sampling method to random sampling, where the sample is not chosen at random. It can be opportunistic, stratified or systematic.

normal distribution a distribution of continuous data where the mean, median, and mode have the same value, there is symmetry around the mean with most data points being close to the mean and fewer data points further away from the mean. When plotted produces a bell-shaped or normal distribution curve.

nucleic acids large polymers formed from nucleotides. Contain the elements carbon, hydrogen, nitrogen , phosphorus, and oxygen.

nucleotides the monomers used to form nucleic acids. Made up of a pentose monosaccharide, a phosphate group and a nitrogenous base.

oncotic pressure the tendency of water to move into the blood by osmosis as a result of the plasma proteins.

open circulatory system a circulatory system with a heart but few vessels to contain the transport medium.

operculum the bony flap covering the gills of bony fish. Part of the mechanism that maintains a constant flow of water over the gas exchange surfaces.

opportunistic sampling sampling using the organisms that are conveniently available. The weakest form of sampling as it may not be representative of the population.

opsonins chemicals that bind to pathogens and tag them so they are recognised more easily by phagocytes, e.g. antibodies.

organelle membrane-bound compartments with varying functions inside eukaryotic cells.

osmosis diffusion of water through a partially permeable membrane down a water potential gradient. A passive process.

oxygen dissociation curve graph showing the relationship between oxygen and haemoglobin at different partial pressures of oxygen.

oxygenated blood blood that has passed through the gas exchange organs (e.g. lungs) and is high in oxygen.

pandemic when a communicable disease spreads rapidly to a lot of people across a number of countries.

partially permeable membrane that allows some substances to cross but not others.

passive transport transport that is a passive process (does not require energy) and does not use energy from cellular respiration.

pathogens microorganisms that cause disease.

penicillin the first widely used, safe antibiotic, derived from a mould, *Penicillium notatum*.

pentose monosaccharide a monosaccharide composed of five carbons.

peptide bond bond formed between two amino acids.

peptides chains of two or more amino acid molecules.

phagocytosis process by which white blood cells called phagocytes recognise non-self cells, engulf them digest them within a vesicle called a phagolysosome.

phagosome the vesicle in which a *pathogen* or damaged cell is engulfed by a phagocyte.

phloem plant transport tissue that carries the products of photosynthesis (assimilates) to all cells of the plant.

phosphodiester bonds covalent bonds formed between the phosphate group of one nucleotide and the hydroxyl (OH) group of another.

phospholipid bilayer arrangement of phospholipids found in cell membranes; the hydrophilic phosphate heads form both the inner and outer surface of a membrane, sandwiching the fatty acid tails to form a hydrophobic core.

phospholipids modified triglycerides, where one fatty acid has been replaced with a phosphate group.

phylogeny the evolutionary relationships between organisms.

pinocytosis endocytosis of liquid materials.

plasma the main component of blood, a yellow fluid containing many dissolved substances and carrying the blood cells.

plasma cells B lymphocytes that produce about 2000 antibodies to a particular antigen every second and release them into the circulation.

plasma membrane all the membranes of cells, which have the same basic structure described by the fluid-mosaic model.

pluripotent a stem cell that can differentiate into any type of cell, but not form a whole organism.

polymers long-chain molecules composed of linked (bonded) multiple individual molecules (monomers) in a repeating pattern.

polypeptide chains of three or more amino acids.

polysaccharide a polymer made up of many sugar monomers (monosaccharides).

primary immune response the relatively slow production of a small number of the correct antibodies the first time a pathogen is encountered.

prokaryotes single-celled prokaryotic organisms from the kingdom Prokaryotae.

prokaryotic cells cells with no membrane-bound nucleus or organelles.

prophase first stage of mitosis when chromatin condenses to form visible chromosomes and the nuclear envelope breaks down.

prosthetic group non-protein component of a conjugated protein.

proteases enzymes that catalyse the breakdown of proteins and peptides into amino acids.

proteins one or more polypeptides arranged as a complex macromolecule.

Protista biological kingdom containing unicellular eukaryotes.

purines double-ringed, nitrogenous bases that form part of a nucleotide.

Purkyne fibres tissue that conducts the wave of excitation to the apex of the heart.

pyrimidines single-ringed, nitrogenous bases that form part of a nucleotide.

quaternary structure the association of two or more protein subunits.

random sampling sampling where each individual in the population has an equal likelihood of selection.

receptors extrinsic glycoproteins that bind chemical signals, triggering a response by the cell.

recombinant chromatids chromatids with a combination of DNA from both homologous chromosomes, formed by crossing over and chiasmata in meiosis.

reducing sugars saccharides (sugars) that donate electrons resulting in the reduction (gain of electrons) of another molecule.

reduction division cell division resulting in the production of haploid cells from a diploid cell; meiosis.

residual volume the volume of air that is left in the lungs after forced exhalation. It cannot be measured directly

resolution the shortest distance between two objects that are still seen as separate objects.

R-groups variable groups on amino acids.

ribonucleic acid (RNA) molecules involved in the copying and transfer of genetic information from DNA. Polynucleotides consisting of a ribose sugar and one of four bases; uracil (U), cytosine (C), adenine (A), and guanine (G).

ribose the pentose monosaccharide present in RNA molecules.

ribosomal (r)RNA form of RNA that makes up the ribosome.

RNA polymerase enzyme that catalyses the formation of phosphodiester bonds between adjacent RNA nucleotides.

root hair cells cells found just behind the growing tip of a plant root that have long hair-like extensions that greatly increase the surface area available for the absorption of water and minerals from the soil.

root pressure the active pumping of minerals into the xylem by root cells that produces a movement of water into the xylem by osmosis.

saprophytic organisms that acquire nutrients by absorption – mainly of decaying material.

scanning electron microscopy an electron microscope in which a beam of electrons is sent across the surface of a specimen and the reflected electrons are focused to produce a three-dimensional image of the specimen surface.

secondary immune response the relatively fast production of very large quantities of the correct antibodies the second time a pathogen is encountered as a result of immunological memory – the second stage of a specific immune response.

seed bank a store of genetic material from plants in the form of seeds.

selection pressure factors that affect an organism's chance of survival or reproductive success.

selective toxicity the ability to interfere with the metabolism of a pathogen without affecting the cells of the host.

selectively permeable plasma membrane with protein channels that allows specific substances to cross only.

semi-conservative replication DNA replication results in one old strand and one new strand present in each daughter DNA molecule.

sense strand the strand of DNA that runs 5' to 3' and contains the genetic code for a protein.

sieve plates areas between the cells of the phloem where the walls become perforated giving many gaps and a sieve-like appearance that allows the phloem contents to flow through.

sieve tube elements the main cells of the phloem that have a greatly reduced living content and sieve plates between the cells.

Simpson's Index of Diversity (D) a measure of biodiversity between 0 and 1 that takes into account both species richness and species evenness.

single circulatory system a circulatory system where the blood flows through the heart and is pumped out to travel all around the body before returning to the heart.

sinks (in plants) regions of a plant that require assimilates to supply their metabolic needs, e.g. roots, fruits.

sino-atrial node (SAN) region of the heart that initiates a wave of excitation that triggers the contraction of the heart.

smooth endoplasmic reticulum endoplasmic reticulum lacking ribosomes; the site of lipid and carbohydrate synthesis, and storage.

sources (in plants) regions of a plant that produce assimilates (e.g. glucose) by photosynthesis or from storage materials, e.g. leaves, storage organs.

Spearman's rank correlation coefficient a specific type of correlation test that compares the ranked orders of two datasets in order to consider their relationships.

specialised having particular structure to serve a specific function.

species the smallest and most specific taxonomic group.

specific immunity also known as active immunity or acquired immunity – the immune system 'remembers' an antigen after an initial response leading to an enhanced response to subsequent encounters.

spiracles small openings along the thorax and abdomen of an insect that open and close to control the amount of air moving in and out of the gas exchange system and the level of water loss from the exchange surfaces.

stage graticule a slide with a scale in micrometres (µm) etched into it. Used to measure the size of a sample under a light microscope.

stains (staining) dyes used in microscopy sample preparation to increase contrast or identify specific components.

starch a polysaccharide formed from alpha glucose molecules either joined to form amylose or amylopectin.

stem cells undifferentiated cells with the potential to differentiate into any of the specialised cell types of the organism.

stomata pores in the surface of a leaf or stem that may be opened and closed by guard cells.

stratified sampling sampling where populations are divided into sub-groups (strata) based on a particular characteristic. A random sample is then taken from each of these strata proportional to its size.

stroma fluid interior of chloroplasts.

Student's t test statistical test used to compare the means of data values of two populations.

substrate a substance used, or acted on, by another process or substance. For example a reactant in an enzyme-catalysed reaction.

succession the progressive replacement of one dominant type of species or community by another in an ecosystem, until a stable climax community is established

sucrose a disaccharide made up of a fructose and glucose monosaccharides.

sustainable development economic development that meets the needs of people today, without limiting the ability of future generations to meet their needs.

symplast the continuous cytoplasm of living plant cells connected through the plasmodesmata.

symplast route phloem loading through the cytoplasm of the cells via plasmodesmata by diffusion (passive).

systematic sampling different areas of a habitat are identified and sampled separately. Often carried out using a line or belt transect.

systole the stage of the cardiac cycle in which the atria contract, followed by the ventricles, forcing blood out of the right side of the heart to the lungs and the left side of the heart to the body.

T helper cells T lymphocytes with CD4 receptors on their cell-surface membranes, which bind to antigens on antigen-presenting cells and produce interleukins, a type of cytokine.

T killer cells T lymphocytes that destroy pathogens carrying a specific antigen with perforin.

T lymphocytes lymphocytes which mature in the thymus gland and that both stimulate the B lymphocytes and directly kill pathogens.

T memory cells T lymphocytes that live a long time and are part of the immunological memory.

T regulator cells T lymphocytes that suppress and control the immune system, stopping the response once a pathogen has been destroyed and preventing an autoimmune response.

tachycardia a fast heart rhythm of over 100 beats per minute at rest.

taxonomic group the hierarchical groups of classification – domain, kingdom, phylum, class, order, family, genus, species

telophase fourth stage of mitosis when chromosomes assemble at the poles and the nuclear envelope reforms.

temperature coefficient (Q_{10}) a measure of how much the rate of a reaction increases with a 10 °C temperature increase.

template strand the antisense strand of DNA that acts as template during transcription so that the complementary RNA strand formed carries the same code for a protein as the DNA sense strand.

tertiary structure further folding of the secondary structure of proteins involving interactions between R-groups.

tidal volume the volume of air which moves into and out of the lungs with each resting breath.

tissue a collection of differentiated cells that have a specialised function or functions in an organism.

tissue fluid the solution surrounding the cells of multicellular animals.

tonoplast membrane forming a vacuole in a plant cell.

total lung capacity the sum of the vital capacity and the residual volume.

totipotent a stem cell that can differentiate into any type of cell and form a whole organism.

trachea the main airway, supported by incomplete rings of cartilage, which carries warm moist air down from the nasal cavity into the chest.

tracheal fluid fluid found at the ends of the tracheoles in insects that helps control the surface area available for gas exchange and water loss.

transcription the process of copying sections of DNA base sequence to produce smaller molecules of mRNA, which can be transported out of the nucleus via the nuclear pores to the site of protein synthesis.

transfer (t)RNA form of RNA that carries an amino acid specific to its anticodon to the correct position along mRNA during translation.

translation the process by which the complementary code carried by mRNA is decoded by tRNA into a sequence of amino acids. This occurs at a ribosome.

translocation the movement of organic solutes around a plant in the phloem.

transmission electron microscopy (TEM) an electron microscope in which a beam of electrons is transmitted through a specimen and focused to produce an image.

transpiration the loss of water vapour from the stems and leaves of a plant as a result of evaporation from cell surfaces inside the leaf and diffusion down a concentration gradient out through the stomata.

transpiration stream the movement of water through a plant from the roots until it is lost by evaporation from the leaves.

transport system the system that transports required substances around the body of an organism.

triglyceride a lipid composed of one glycerol molecule and three fatty acids.

triplet code the genetic code is a sequence of three nucleic acid bases, called a codon. Each codon codes for one amino acid.

turgor the pressure exerted by the cell-surface membrane against the cell wall in a plant cell.

ultrastructure the ultrastructure of a cell is those features which can be seen by using an electron microscope.

undifferentiated an unspecialised cell originating from mitosis or meiosis.

vaccine a safe form of an antigen, which is injected into the bloodstream to provide artificial active immunity against a pathogen bearing the antigen.

vacuoles membranous sacs used to transport materials in the cell.

vascular bundle the vascular system of herbaceous dicots, made up of xylem and phloem tissue.

vascular system a system of transport vessels in animals or plants.

vector a living or non-living factor that transmits a pathogen from one organism to another, e.g. malaria mosquito.

Ventilation rate is the total volume of air inhaled in one minute. Ventilation rate = tidal volume × breathing rate (per minute).

vital capacity volume of air that can be breathed in when the strongest possible exhalation is followed by the deepest possible intake of breath.

V_{max} maximum initial velocity or rate of an enzyme-catalysed reaction.

water potential (Ψ) measure of the quantity of water compared to solutes, measured as the pressure created by the water molecules in kilopascals (kPa).

xerophytes plants with adaptations that enable them to survive in dry habitats or habitats where water is in short supply in the environment.

xylem plant transport tissue that carries water and minerals from the roots to the other parts of the plant as a result of physical forces.

zygote the initial diploid cell formed when two gametes are joined by means of sexual reproduction. Earliest stage of embryonic development.

Answers

2.1

History of the light microscope and development of cell theory

1 *idea that* cell is unit of life (1) / many organisms unicellular (1) / (most) cells are too small to see without microscope (1) / cell components / organelles, are even smaller (1) / *idea* that need to see organelles to determine function (1)

2 Prior to the mid-19th century microscopes were of too low a magnification (1) to see and identify cells (1) and cell components (1).

Sample preparation

1 a so light can shine through it (1) / details can be seen (1)

   b reduce / prevent, diffraction between liquid and glass (1) / prevent / reduce distortion of image (1)

   c reduce / prevent air bubbles being trapped (1)

Using staining

1 gram negative have thinner cell wall (1) / penicillin disrupts cell wall formation (1) / less cell wall formation (in gram negative) / membrane (around gram negative) prevents entry of penicillin (1)

2 avoid skin / eye contact (1) / wear gloves / goggles (1)

Less is more

1 a shading / label lines not touching relevant object (1) label lines not parallel with top of page (1) no magnification stated (1)

1 Both plant and animal tissue is composed of cells (1); cells are the basic unit of all life (1); cells only develop from existing cells (1)

2 Staining provides contrast (1) / different structures/organelles absorb stain differently allowing identification (1).

3 Objective lens and eyepiece lens (1); objective lens magnifies the specimen (1); eyepiece lens magnifies image (from objective lens) (1); higher magnification (produced than with just one lens) (1)

4 a i $0 \times 10 = 100$ (1)

   ii $10 \times 40 = 400$ (1)

   b diameter of field of view = 2000 μm (1) / 2000 / 60 (1) / number of whole cells = 33 (1)

2.2

Using a graticule to calibrate a light microscope

1 graticule / stage micrometer, eyepiece graticule (1)

2 20 divisions of eyepiece graticule = 9.5 micrometer divisions (1); 1 micrometer division = 10 μm (1); 20 graticule units = 95 μm (1); so 1 graticule unit = 95/20 = 4.75 μm (1); calibration factor of the ×10 lens = 4.75.

3 a 14 (1)   b 52.25 (1); 76.0 (1); 66.50 (1)

   c $52.25 + 76.0 + 66.5 = 194.75/3 = 64.9$ μm

   d Scale on eyepiece graticule always same (1); but magnification of other lenses changes (1); need to calibrate eyepiece graticule for each lens to know actual measurements represented by eyepiece graticule at different magnifications (1); necessary to calculate real size of objects seen.

1 simplifies calculation (1) / reduces errors (1)

2 $3846 \times 10 \times 1000 \times 1000$ (1) / $= 3.846 \times 10^{10}$ (1)

3 Contrast is difference in colour/shade between two objects (1) Resolution is the smallest distance between two objects that can still be seen as separate (1)

4 Approximately ×366, when the diameter is 22mm (2).

5 diffraction happens when light passes through structures (1) / light waves spread out (1) / (light waves) overlap (1) / individual objects do not appear separate (1) / causes blurring (1)

6 (eyepiece graticule is) arbitrary scale / calibrated for each lens (1) / using stage micrometer (1)

2.3

Sample preparation for electron microscopes

fixation stabilise sample / prevents decomposition (1) / *dehydration* prevent vaporisation of water in vacuum (1) vaporisation would damage sample (1) / *embedding* allows thin slices to be obtained (1) *staining with heavy metals* creates contrast (1) in electron beams (1)

Scientific drawings from electron micrographs

a is best (1) no shading (1) / label lines parallel with top of page (1)

Identifying artefacts

Evidence supports artefact theory (1) / not present normally (1) / *idea* that antibiotics responsible for appearance (1)

Fluorescent tags

a ability to see individual objects as separate (1)
b resolution the same (1) / resolution limited by wavelength of light (1) / fluorescence is light emitted (1) / super resolved fluorescent microscopy has higher resolution (1)

Atomic force microscopy

1 image not formed by light (1) / (image formed by) deflections of, tip / probe (1) / (as tip / probe) moves across surface of specimen (1)
2 higher resolution (than electron microscope) (1) / magnification depends on resolution (1)
3 (AFM) only scans surface (1) / *idea that* cannot see into cells (1) / *idea that* need to see how organelles are related to understand function (1)

Super resolved fluorescence microscopy

1 a specimen preparation kills cells (1); detail e.g. fixation (1)
   b single molecules can fluoresce (1); multiple images obtained (1); different molecules fluoresce in each image (1); images superimposed (1); idea of individual molecules seen in relation to each other interacting (1)

1 Electron microscopes use electrons instead of light and electrons have a shorter wavelength than light (1) which produces images with a higher resolution (1).

2 a An artefact is a visible object (1) or distorted cell structure (1) present in an electron micrograph (or other micrograph) due to the sample preparation process (1).
   b more sample preparation (in electron microscopy) (1) / (leads to) more damage to specimen (1) / damage results in artefacts (1)

3 a Left: transmission electron microscope Right: scanning electron microscope (1)

b organelles visible in *a* (1) / *a* has greater magnification (1) b shows surface detail (1)

c *TEM advantages* greater, magnification / resolution (1) / more detail (1) *TEM disadvantages* 2D image (1) / very thin specimens needed (1) more preparation so more artefacts (1) *SEM advantages* specimens do not need to be thin (1) 3D image (1) *SEM disadvantages* lower, magnification / resolution (1) (max 6)

4 a Answers a emission of light (1); (that has been) absorbed (1)

b increase intensity (of light) (1)

c scattered light / light from outside the focal plane (1); is eliminated (1); reduces blurring / increases resolution (1)

d idea of light penetration (of sample) is limited (1)

2.4

Cell movement

1 microtubules (and microfilaments) are involved (1) / undergo polymerisation and hydrolysis (1) / intermediate fibres are not involved (1) (have) role in cell stability (1)

1 Lysosomes are specialised vesicles (1) that contain hydrolytic enzymes (1) for breaking down waste material. The membrane that forms lysosomes has an important role in compartmentalising these enzymes away from cell structures that could be damaged by activity of the enzyme (1).

2 Incompatible reactions / catabolic and anabolic reactions require different conditions / damage due to hydrolytic enzymes (3) three named examples (e.g. nucleus, vesicle, lysosome, mitochondrion, Golgi body, endoplasmic reticulum, chloroplast) (1).

3 Rough ER has ribosomes attached **and** smooth ER does not have ribosomes attached (1); *rough ER* protein synthesis (and modification) (1); *smooth ER* lipid synthesis (1).

4 The cytoskeleton has three components: microfilaments (1) are contractile fibres made of actin that bring about cell contraction during cytokinesis (1); microtubules (1) are formed from the cylindrical protein tubulin and form scaffold-like structures used both in the movement of organelles and vesicles and as spindle fibres in the segregation of chromosomes/chromatids in cell division (1); intermediate fibres give mechanical strength to cells (1).

5 a $7 \times 10^7 \times 0.34 \times 10^{-9}$ (1) / = $2.38 \times 10^{-2} \times 46$ (1) 1.09 m (1)

b coiled / wrapped (1) / around histones (1) / further coiling (1) / formation of chromatin (1)

6 microfilaments composed of actin (1) / (actin is) contractile (1) / microtubules composed of tubulin (1) / (tubulin) polymerises (1) / (contraction and polymerisation lead to) change in length of filaments (1) / change in length (of filaments) results in movement of cell (1) / intermediate fibres have fixed length (1) / for stability (1)

2.5

1 cell wall (1) / chloroplast (1) / plant cell (1) / presence of, chloroplast / cell wall (1).

2 a plant cell walls contain cellulose (1)

b prevent cells bursting (1) / allows turgidity (1) / *idea that* keep plants upright (1).

3 *both have* three named organelles (e.g. nucleus, cell surface membrane, mitochondria, ribosomes, Golgi body, endoplasmic reticulum) (1) / *only plants have* two named organelles (e.g. chloroplasts, cell wall, large (central) vacuoles) (1) / centrioles present in animal cells but not flowering plants (1).

2.6

> ### Endosymbiosis
>
> **1** *mitochondria / chloroplasts, are* (about) the same size as bacteria (1) / have a double membrane (1) / (second membrane) acquired upon entry to cell (1) / contain DNA (1) / necessary for protein synthesis (1) / (and) replication (1)

> ### Prokaryotic cell study
>
> **1 a** correctly drawn scientific diagram (1) / *showing* cell wall (1) / DNA / chromosome (1) / cytoplasm (1)
>
> **b** *idea of* many more structures (1) / membrane bound organelles (1) / three named structures (e.g. nucleus, mitochondria, endoplasmic reticulum, Golgi body) (1)

1 prokaryotic cells: no nucleus / no membrane bound organelles, e.g. mitochondria / smaller / 70s ribosomes / plasmid / extra chromosomal DNA / peptidoglycan / murein cell wall. (Any 3). Accept reverse arguments for eukaryotic cells.

2 Prokaryotic cells have ribosomes (1), which are needed for protein synthesis (1). Ribosomes are not membrane bound (1).

3 Eukaryotic cells do not have peptidoglycan (1) cell walls (1) and these antibiotics do not damage any other cell components (1) named example (e.g. nucleus , ribosomes, mitochondria) (1)

3.1

1 Atoms form bonds with each other when pairs of electrons are shared (1) according to the bonding rules (1).

2 A cation is an ion with a net positive charge (1), i.e. it has lost one or more electrons (1). An anion is an ion with a net negative charge (1), i.e. it has gained one or more electrons (1).

3 *water* – one oxygen atom binds to two hydrogen atoms (1), oxygen can form two bonds, each hydrogen can only form one bond (1). *carbon dioxide* – one carbon atom and two oxygen atoms (1), carbon can form four bonds, each oxygen atom can form two bonds, therefore carbon forms a double bond with each oxygen atom (1).

4 a X ray diffraction does not involves lenses (1) / electron microscope uses electromagnetic lenses (1) / beams focused in electron microscopy to produce image (1).

b Cells are larger than ribosomes (1) / cells are larger than, half the wavelength / resolution limit, of light (1) / electron microscopes have greater resolution (than light microscopes) (1) / *idea that* molecules are smaller than resolution limit of light and larger than resolution limit of electron beam (1).

3.2

1 Oxygen and hydrogen share electrons unequally when they bond. Oxygen, has a greater share/ is more negative (1). Hydrogen, has a smaller share/is more positive (1). The more negative oxygen atom is attracted to the more positive hydrogen atom (1).

2 Water is composed of hydrogen and oxygen atoms and bonds between oxygen and hydrogen involve unequal sharing of electrons (1) in bonds resulting in the oxygen atom being more negative and the hydrogen atoms being more positive (1).

3 Liquid so transport medium (1); polar solvent (1); (many) biological molecules / examples (e.g. enzymes, glucose), are polar (1); ions are charged (1); coolant so (relatively) resistant to temperature change (1).

4 (water is) liquid (1) / allows movement of substrates **and** enzymes (1) / **idea that** this is necessary for reactions to occur (1) / (water is) a polar solvent (1) / substrates / enzymes / products, are, polar / ionic (1) / (water is) substrate for (some) reactions (1).

3.3

1 hydroxyl group on carbon 1 is in a different position (1); in alpha glucose it is below the ring in beta glucose it is above the ring (1).

2 Bond formed between two glucose molecules (1) – hydroxyl group of carbon 1 on one molecule (1) and carbon 4 (1) on the other interact in a condensation reaction/removal of water molecule (1) to form an 'oxygen bridge'.

3 Cellulose is straight chain molecule (1) with many hydrogen bonds between individual chains (1) and staggered ends (1). This confers strength to the fibres (1).

4 In beta glucose the hydroxyl group at carbon 1 is above the ring (1) so alternate glucose molecules must rotate 180 degrees (1) so the hydroxyl groups on carbon 1 and carbon 4 are close enough to react (1) condensation reaction (1) forming a glycosidic bond (1). The rotation of molecules produces a straight chain molecule (1) – cellulose.

3.4

Quantitative methods to determine concentration Colorimetry

1 100% – transmission % = absorbance % (1)
2 (to) maximise absorption (1); complementary colour / red for Benedict's solution (1)
3 use distilled water (1); set colorimeter to 100% (1)
4 unreacted Benedict's solution (1); supernatant (1)
5 correct axes (1); correct plots (1); line of best fit (1)
6 concentration of glucose at 44% absorbance (1); units (1)

Biosensors

1 biological detector (1); presence of (toxic) gas causes a change (1); distress of bird is display (1); canary in a cage is a biosensor (1); *disadvantage* not specific to one gas (1); ethical considerations of causing harm to an animal (1)

1 (enzymes have) active site (1); (active site) specific (1); to, substance / molecule, testing for (1)

2 Reducing sugars react with copper ions in Benedict's reagent resulting in the addition of electrons to blue Cu^{2+} ions (1), reducing them to Cu^+ ions which form a brick red precipitate (1).

3 In an iodine test a purple/black colour indicates the presence of starch (1). Starch is a product of photosynthesis (1). The test shows that starch is produce when light is available to the plant, but not when the plant is kept in the dark (1).

4 Reagent strips are quantitative (1); they can be used to estimate the concentration (1) of reducing sugar (glucose) (1) in the blood. They are simple to use and interpret (1).

3.5

Fats in our diet

1 hydrogenation / addition of hydrogen (1); removes double bonds in fatty acids (1); closer packing of molecules (1)
2 unsaturated **and** saturated fat have high energy content (1); excess energy intake leads to obesity (1)

1 Oils are (usually) unsaturated (1); unsaturated fatty acids contain double bond(s) (1); molecules cannot pack closely (1); fats are usually saturated so fatty acids have no double bonds (1).

2 Hydroxyl group from glycerol (1); hydroxyl group from fatty acid (1); condensation reaction (forms ester bond) (1); *idea that* hydrolysis is reverse of the process described (1).

3 a A

b both have phosphate group attached to glycerol (1); both have fatty acid (tail) (1); cross links between fatty acid tails in A (1) ; no oxygen attached by double bond (on A) / ester bond not present (on A) (1).

c cross links (1); stabilise membrane (1)

4 *procedure for emulsion test* statement 2, sample / lipid, dissolved in ethanol (1); water is mixed with ethanol (and lipid) solution (1); statement 3, *idea that* water displaces lipid from ethanol forming suspension (1); statement 1, (suspension forms because) lipids not soluble in water (1)

3.6

Separating amino acids using thin layer chromatography

1 a to prevent contaminating stationary phase (1); *idea of* biological material (on skin) (1)

b testing unknown compounds (1); not known whether, polar / non-polar (1); *idea that* the different solvents will dissolve both polar and non-polar compounds (1)

c so the concentrated spots were not covered (1)

d (so) air inside jar is saturated with (solvent) (1); prevents evaporation of solvents (1)

2 a from bottom to top glycine, proline, phenylalanine (1)

b from bottom to top alanine, methionine (4)

Identification of proteins

1 mauve / lilac / purple (1)

2 no peptide bonds present (as no protein) (1); test is negative (1); solution (remains) blue (1); as copper sulfate solution is blue (1)

3 Biuret test identifies peptide bonds (1); degree of colour change dependent on number of peptide bonds (1); different proteins have different numbers of peptide bonds (1); *idea* that different degrees of colour change could indicate different proteins not different quantities of protein (1)

1 Diagram showing the amine group (1), carboxylic acid group (1) and variable group (1) in correct positions. See Figure 1 in Topic 3.6, Structure of proteins.

2 Condensation reaction (1) between amine group of one amino acid (1) and carboxylic acid group of another (1), forming a water molecule.

3 a

b Oxygen is relatively negative **and** hydrogen (attached to nitrogen) is relatively positive (1); oxygen and hydrogen are attracted to each other (1).

c Secondary structures are (simple) repeating structures (1); globular protein / haemoglobin, has a tertiary structure (1); tertiary structure is formed from complex folding of secondary structure (1).

4 R groups on amino acids interact (1); *tertiary structure* – interactions <u>within</u> a protein molecule (1) determines shape of molecule (1); *quaternary structure* – interactions <u>between</u> protein molecules (1); holds molecules together (1). Both involve the same interactions (1), i.e. hydrogen bonds, ionic bonds, disulfide bonds and hydrophobic and hydrophilic interactions.

3.7

The structure of fibrous proteins – Elastin and collagen

1 strength, non-elastic

2 (collagen is a) large molecule so unlikely to enter skin (collagen), has a complex structure, *idea of* individual components arranged in hierarchical structure, *idea that* new molecules would not incorporate into existing collagen

3 B

1 Conjugated proteins contain a non-protein group (1) called a prosthetic group (1), simple proteins do not (1).

2 *insulin* globular protein (1); soluble (1); specific shape (1); binds to receptor (1); chemical messenger / described (1); *keratin* fibrous protein (1); strong (1); structural function / example (e.g. hair, nails)

3 *globular proteins* hydrophobic R groups, in the centre (of the molecule) not in contact with water (1); hydrophilic R groups, on the outside (of the molecule) / in contact with water (1); hydrophobic R groups are repelled by water / hydrophilic R groups are attracted to water (1); fibrous proteins have R groups on the outside of the molecule (1)

4 *similarities* globular (protein) (1); alpha helices (1); prosthetic group (1); hydrophobic R groups positioned towards the centre (of the molecule) (1); *differences* single polypeptide not four polypeptides / *myoglobin* tertiary not quaternary (1); no beta chains (1)

3.8

DNA extraction

1 reduce activity of enzymes (1); reduce breakdown of DNA (1)

2 disrupts membrane structure (1); phospholipids form suspension in aqueous solution (1)

1 *DNA nucleotide* – deoxyribose sugar, thymine base (1); *RNA nucleotide* – ribose sugar, uracil base (1)

2 A pyrimidine base always pairs with a purine base (1). Adenine and thymine/uracil always hydrogen bond together (1) and cytosine and guanine always hydrogen bond together (1).

3 Polymer so contains a lot of information (1); idea that base sequence is used as a code (1); double stranded so molecule is stable (1); double stranded so accurate replication (1).

4 Adenine always base pairs with thymine, so same amount of thymine (17%) (1). Adenine and thymine together is 34%, so cytosine and guanine must be 100 – 34 = 66% (1). Cytosine always base pairs with guanine so cytosine amount equals guanine, 66/2 = 33. Therefore cytosine 33% and guanine 33% (1).

3.9

> **Continuous and discontinuous replication**
>
> **1** continuous replication – DNA polymerase binds to the end of a strand, free DNA nucleotides added without any breaks; discontinuous replication – DNA polymerase cannot bind to the end of a strand, free DNA nucleotides added in sections, sections then joined .
> **2** enzymes are (substrate) specific (1); DNA polymerase catalyses the joining of nucleotides (1); nucleotides have a different shape to Okazaki fragments (1)

1 Semi-conservative means 'half the same' (1). When DNA replicates the double helix unwinds into two separate strands (1) Free nucleotides pair with their complementary bases (1) Two new molecules of DNA are produced (1), each with one old strand and one new strand (1).

2 The triplet code is a particular sequence of three bases (1) that codes for a specific amino acid (1).

3 A mutation in the DNA changes the triplet code (1), meaning different amino acids are incorporated into the protein/enzyme (1) that the DNA codes for. If such a change affects the precise structure of the active site (1) a substrate may not be able to bind (1), rendering the enzyme non-functional.

4 The triplet code of DNA is degenerative (1), there are 64 different triplets/codons but only 20 amino acids (1), therefore an amino acid can be coded for by more than one codon (1), so more opportunity for differences in DNA sequence than amino acid sequence (1).

3.10

1 From column left to right:
UAC CGG AGU GCA

2 mRNA – copies gene from DNA(1), takes copy to ribosome (1); tRNA – brings amino acid to ribosome (1); rRNA – formation of ribosome (1)

3 a catalyse the formation of bond between two amino acids (1); peptide bond (1)

   b bind to tRNA (1); complementary base pairing (1)

   c free floating ribosomes produce proteins for use in, cell / cytoplasm (1); bound ribosomes produce proteins for export from the cell (1)

4 a role of protein dependent on structure (1); shape / 3D structure, dependent on primary structure / sequence of amino acids (1); base triplets / codons, on mRNA, code for amino acids (1); introns would code for, unnecessary amino acids / stop signal (1); codons could cause frameshift (1)

   b different proteins produced from one gene (1)

   c idea that originally functional gene(s) (1); mutation/s (1); base sequence/s changed (1); no longer code for (useful) amino acids (1)

3.11

1 sugar/ribose sugar (1) joined to a base/adenine (1) and to three phosphates (1)

2 It is present in all cells (1), it is present in all organisms (1). It releases energy in, small/ manageable quantities (1).

3 (fat is) long term energy store (1); *idea that* fat is stable molecule **and** ATP is unstable molecule (1); fat has other uses (1); e.g. insulation (1)

4 a bond formation releases energy (1); bond uses energy (1)

   b ATP provides energy for, reactions / processes (1); ATP is present in all living organisms (1); *idea that* there is no other equivalent molecule (1); therefore statement is valid (1)

4.1

1 a protein (1)   **b** amino acids (1)

   c specific, 3D shape / tertiary structure (1); (formation of) active site (1); binds to substrate(s) (1); catalyses reaction (1)

2 Catabolism is breaking down of molecules (1); anabolism is building of molecules (1); reactions involve breaking down and building of molecules (1); *idea of* metabolism is sum of all reactions (1).

3 a simple / easy to understand (1); representation (1)

b *Both models* substrate interacts with R groups in active site (binds) (1); (leading to) bond strain in substrate molecule (1); *Lock and key* substrate is complementary to active site (of enzyme) (1); *Induced* fit active site is flexible (1); (active site) changes shape as substrate binds, closer fit between active site and substrate.

4 a energy required (1); to start reaction (1)

b bonds are broken (1); in substrate (1); energy required (1); energy (of system (increases) (1)

c *idea* of improved technology (1); *idea* of continually investigated (1); more evidence (1); more accurate representation (1).

4.2

Enzymes in action

extremities are at, same temperature as rest of the body / higher temperature than normal, in womb (1); enzyme / tyrosinase, not denatured (1); pigment is broken down (1)

Investigations into the effects of different factors on enzyme activity

1 easy to obtain (1); contains catalase (1)
$2H_2O_2 \rightarrow 2H_2O + O_2$

2 reactant (1); products (1)

3 volume of oxygen released increased (1); figures quoted (1); enzyme / catalase, catalysed reaction (1) **4** enzyme / catalase, is a protein (1); high temperature denatures protein (1); shape of active site changed (1); (active site) no longer complementary to substrate (1); reaction not catalysed so no oxygen produced (1)

4 enzymes are proteins (1); (boiling) denatures protein (1); tertiary structure (of protein) changed (1); active site no longer complementary to substrate (1); fewer enzyme-substrate complexes formed (1); decreased reaction rate (1)

5 same apparatus (1); range of hydrogen peroxide concentrations (1); temperature kept constant (1); ruler used to measure change in water height (1); readings taken at set time intervals (1); radius/diameter, of test-tube used to calculate volume (of gas collected) (1), repeats (1)

6 a *independent variable* – concentration (hydrogen peroxide) dependent variable – volume of gas collected (1)

b *controlled variables* – concentration of hydrogen peroxide (1); volume of hydrogen peroxide (1); mass of liver tissue (1); surface area of liver tissue (1)

7 only independent variable / temperature, is only factor that is changed (1)

8 a reading that, lies outside the normal range / does not follow trend (1)

b 65 test 2 100 at 20 seconds

9 reliability / identify anomalies (1)

10 correct graph – axes including units (1); plots (1); lines of best fit (1); use of graph paper (1)

11 (initial) rate of reaction higher with higher concentration (of substrate) (1); more substrate present (1); more enzyme-substrate complexes formed (per unit time) (1); shown by steeper line (1); rate of reaction, slows down / plateaus, as substrate used up (1); lower plateau at lower substrate concentration (1)

12 *results are valid because* – reliability is good as repeats are (relatively) consistent (1); expected trend observed (1); only one anomaly (1); many intermediates (1) *results may not be valid because* – only three different concentrations (of substrate) used (1); temperature, may have changed / not controlled (1); pH, may have changed / not controlled (1); may have been timing errors (1)

Serial dilutions

serial dilution (1); *described* e.g. 1ml of stock solution and 9ml of distilled water (1); 2 mmol dm^{-3} solution (1); repeat with diluted solution (s) (1); 0.2 mmol dm^{-3} and 0.02 mmol dm^{-3} and 0.002 mmol dm^{-3} (1)

1 R-group interactions are disrupted (1); change in tertiary structure (1); change in 3D shape of active site preventing binding with substrate (1).

2 Curve A (1): low/acidic pH is optimum pH and the stomach contains acid/has a low pH (1).

3 Bacterial enzymes have high optimum temperatures, human body temperature is lower (1). Enzymes will have low activity (1) and bacteria will not thrive (1).

4 (at low temperatures) kinetic energy is low (1); substrates / enzymes, move slowly (1); (so) fewer collisions (1); collisions have less energy (1); increased flexibility of active site (1); increases chances of successful collision (1)

4.3

1 A non-competitive inhibitor binds to an enzyme away from the active site (1) at an allosteric site (1), which has a different shape than the active site (1).

2 Inhibitor will always be present (1); some enzymes always inhibited (1).

3 End-product inhibition regulates rate of reaction (1); concentrations of substrate **and** product determine reaction rate (1); (so must be) competitive (1); substrate concentration has no effect in non-competitive inhibition (1); e.g. ATP and PFK in respiration (1).

4 Ethanol has similar shape to ethylene glycol (1); (ethanol) binds to active site of enzyme which breaks down ethylene glycol (1); competitive inhibition (1); less ethylene glycol broken down (1); more (ethylene glycol) leaves body unchanged (1); fewer toxic effects (1).

4.4

> ### Enzyme activation and the blood clotting-mechanism
>
> enzymes responsible for blood clotting are present as precursors / described [1]; e.g. Factor X, prothrombin [1]; prevents clotting unless required [1]

1 Transfer, atoms / groups, between reactions (1); form part of active site (1).

2 Coenzymes bind loosely to enzymes (1); e.g. NAD (1); prosthetic groups are a permanent feature of / bind tightly to, proteins / enzymes (1); e.g. iron ion in haemoglobin (1).

3 Presence of cofactor (1); e.g. vitamin K and Factor X (1); change in tertiary structure / described (1); e.g. (activated) factor X catalyses the breaking of bonds in prothrombin (1); forming thrombin (1); thrombin catalyses the conversion of fibrinogen to fibrin (1).

5.1

1 Membranes form cells and separate areas within cells (1), isolating each area from its external environment (1).

2 *intrinsic protein* – embedded in both sides of the bilayer (1). For example a channel protein or carrier protein (1). *extrinsic protein* – embedded in one side of the bilayer (1). For example a glycoprotein or enzyme (1).

3 Lipid soluble molecules can pass through membranes (1); (by) simple diffusion (1); (so) diffuse quickly through (whole) body (1).

4 Process occur within/across, membranes (1) process is enzyme controlled (1) Folding gives increased surface area (1) (so) more enzymes (1) increased rate of reaction(s) (1); and therefore an increased rate of ATP production (1).

5.2

> ### Investigating membrane permeability
>
> 1 a to remove all surface pigment released from damaged cells [1]
>   b to allow the mixture to equilibrate [1]
>   c repeats for reliability [1]
>   d because the pigment is red [1]
> 2 more pigment molecules absorb more light [1]; light transmitted decreases [1].
> 3 The membrane was disrupted between 40 and 50 °C.
> 4 same procedure except temperature constant [1]; different (organic) solvents used [1]; e.g. ethanol [1]

1 water is a polar solvent (1); phospholipids will not dissolve in water (1)

2 use of colorimeter (1); detail (e.g. use of filter) (1); range of readings taken (at different temperatures) (1); graph (1)

3 Alcohol is lipid soluble and dissolves in membrane bilayer (1). This disrupts the bilayer and stops/reduces transport of materials (1) preventing the normal functioning (1) and may cause cell death. The liver is particularly affected due to its role in filtering substances from the blood (1). This may ultimately be fatal if the liver function is destroyed (1) or if impulse transmission is depressed, prevent involuntary reflexes (1) such as breathing and the gag reflex (which prevents choking).

5.3

Rate of diffusion and surface area

1

Cube size (cm)	Surface area (cm²)	Volume (cm³)	Surface area / volume	Diffusion distance (cm)	Rate of diffusion using distance (cm / min)	Rate of diffusion using volume (cm³ / min)	Rate of diffusion using volume per 64 cm3 agar
4 × 4 × 4	96	64	1.5	0.3	0.03	28.8	28.8
2 × 2 × 2	24	8	3.0	0.3	0.03	7.2	57.6
1 × 1 × 1	6	1	6.0	0.3	0.03	1.8	111.2

2 sodium hydroxide solution has diffused into the agar (1); sodium hydroxide is an alkali (so phenolphthalein indicator turns pink) (1)

3 larger width of pink colour in smaller blocks (1); smaller blocks have larger surface area to volume ratio (1); (so) sodium hydroxide has diffused further into the blocks (1)

Investigations into the factors affecting diffusion rates in model cells

1 *qualitative* detects the presence of, reducing sugar / glucose (1); *quantitative* colour change is estimate of concentration (of reducing sugar / glucose) (1);

2 tied dialysis tubing is, simplified / practical, representation of real cell (1); (tied dialysis tubing) has the same properties as membrane (of cell) (1); demonstrates diffusion across (cell) membrane (1);

3 a cell membranes more complex (1); ORA (or reverse argument) (cell membranes) have, carrier proteins / channel proteins (1); ORA active transport **and** diffusion across (cell) membranes (1); (cell membranes) have hydrophobic core (1); ORA (cell membrane permeability) determined by size and, polarity / charge (1); ORA

b (dialysis) tubing permeability based on pore size (1); dialysis tubing is not a barrier to (small) ions (1); phospholipid bilayer (of cell membrane) is barrier to ions (1); ions diffuse through channel proteins (in membrane) (1)

1 Increased temperature increases the kinetic energy of particles (1), causing the particles to move at increased speed (1).

2 Increased surface area (1) and reduced thickness (1).

3 Diffusion is described as passive because it does not require an external (metabolic) energy source (1). Diffusion relies on the energy from the natural random movement of particles (1).

5.4

1 Diffusion is always a passive process, it does not require a metabolic energy source (1). In facilitated diffusion a channel/co-transport protein aids diffusion (1).

2 Active transport requires metabolic energy in the form of ATP (1) produced in the mitochondria (1), so these cells have more mitochondria.

3 Plants need mineral ions (1); concentration of mineral ions higher in root hair cells (than soil solution) (1); mineral ions will diffuse out of root hair cells (1); energy required to move mineral ions against concentration gradient (1).

5.5

Osmosis investigations

1

Sugar Conc. mol dm⁻³	Original mass (g)	Final mass (g)	Difference in mass (g)	% mass change	Mean % mass change
0.0	3.0	4.0	1.0	33	34
	3.0	4.1	1.1	37	
	3.3	4.2	1.1	33	
0.1	3.0	3.5	0.5	17	15
	3.2	3.6	0.4	13	
	2.9	3.3	0.4	14	
0.3	3.0	3.0	0	0	1
	2.9	3.0	0.1	3	
	3.2	3.2	0	0	
0.5	3.2	2.8	−0.4	−13	−13
	3.0	2.6	−0.4	−13	
	3.1	2.7	−0.4	−13	
0.7	3.1	2.2	−0.9	−29	−30
	3.3	2.4	−0.9	−27	
	3.0	2.0	−0.1	−33	

1 One arrow from pure water to dilute solution (1). One arrow from dilute solution to concentrated solution (1).

2 The water potential of pure water is zero (1). Addition of solute decreases water potential (1). Therefore all solutions have negative water potential.

3 Where the graph line crosses the *x*-axis/where mass change is 0% (1). This is the isotonic value.

4 Electrolytes/solutes/minerals are necessary for many body processes (1) and help prevent excess water loss by osmosis (1) to help maintain correct fluid balance for reactions (1).

5 Pine kernel tissue has the highest solute concentration/lowest water potential (1) as it does not reach isotonic state even at the highest sodium chloride concentration (1). Pine kernels (sometimes called pine nuts) are the seeds of pine trees. Seeds store nutrients (1) for the seedling that will grow and they have low water content (1) while dormant, requiring uptake of water to germinate.

6.1

1 Mitosis is the process of replicating and diving the genome (1). Cytokinesis is the physical division of the cell (1).

2 DNA has been checked for errors (1); change in sequence of bases is a mutation (1); (leads to) change in amino acid sequence (1); function of protein dependent on, 3D shape / tertiary structure (1); tertiary structure dependent on primary structure (1); primary structure is sequence of amino acids (1)

3 a mutations occur during DNA replication (1); indefinite replication, increases chances of mutation / accumulation of mutations (1); increased chance of harmful mutation (1)

   b indefinite replication (1); cancer / formation of tumour (1);

4 a DNA is double stranded (1)

   b 3×10^9 / 50 (1); 6×10^7 (1)

   c many origins of replication (1); *idea of* simultaneous replication of different lengths (1)

   d (prokaryotic) genome, is shorter / has fewer genes / has no introns (1)

6.2

1 Chromosomes only become visible under the microscope during mitosis/meiosis (1). DNA needs to replicate for cell division. Chromosomes consist of two sister chromatids, which are identical copies of DNA (1).

2 So that each daughter cell has identical DNA after mitosis/cell division (1) and correct number of chromosomes (1) i.e. diploid after mitosis and haploid after meiosis.

3 *animal cells* cleavage furrow forms around middle of cell (1); furrow pulls inwards **and** fuses (1); *plants cells* furrow cannot form due to cell wall (1); vesicle assemble across centre of cell **and** fuse (1)

4 *prophase – 92 chromosomes have replicated G$_1$ – zero replication has not occurred yet*

5 Plant root tips continually grow at regions called meristems (1). Meristems are a good source of cells for studying mitosis as they are constantly diving (1). Plant cells are easy to obtain (1) and prepare for microscopy (1).

6.3

1 a Meiosis I/the first division is a reduction division as each daughter cell is haploid (1).

   b Gametes are the sex cells and two sex cells (one from each parent) must combine to produce a diploid offspring (1). Therefore gametes must contain only half the number of chromosomes/DNA, i.e. be haploid, otherwise with each new generation the number of chromosomes would increase (1).

2 A pair of same chromosomes, one from each parent (1), which have the same genes but can have different alleles of each gene (1).

3 Anther(s) from a flower should be used (1); prepare a squash slide (1); use stain (1); observe using microscope (1)

4 *crossing over* – homologous chromosomes pair up (in prophase), non-sister chromatids entangle (chiasmata) (1) and exchange genes/ alleles when they pull apart (at anaphase) (1). This produces new combinations of alleles. *independent assortment* – pairs of homologous chromosomes (meiosis I)/chromosomes (meiosis II) line up on the equator (at metaphase) and each (pair/chromosome) orientates independently (1) before being separated to opposite poles of the cell (at anaphase) (1). This produces new combinations of alleles.

5 a

(1 mark for each allele pair)

   b Creating different allele pairs during meiosis is an important source of genetic variation (1) in a population. Genetic variation is important for the process of natural selection (1), giving individuals in a population characteristics/traits that might confer an advantage (1) in changing environment (1), for example pathogen resistance. If there was no genetic variation in a population, the entire population would be vulnerable to such an external factor and there would be no opportunity for adaptation (1).

6.4

1 Squamous (1); flattened cells provide thin surface (1) e.g. (alveoli) in lungs (1); diffusion of gases (1); ciliated (1); (have cilia), for movement, of cell / liquid outside cell (1); e.g. trachea (1); movement of mucus (1).

2 *Any two appropriate examples with detail of structure related to function.* (2 marks each).

3 A tissue is a collection of cells (1) that work together (1), an organ is a collection of tissues (1) that work together (1).

4 The digestive system is a group of organs working together carry out a function (1). The pancreas produces digestive enzymes (1), the stomach contains acid for digesting food (1), the liver produces bile to aid the digestion of fats (1), the small intestine digests and absorbs soluble food (1), the large intestine absorbs water from undigested food, producing faeces (1).

6.5

Gene therapy using stem cells

1 Bone marrow contains stem cells (1); stem cells can differentiate (1); into T cells (1); immune system is functional (1).

2 Tissue (from) donor is not a good match (1); rejection (1); transplanted cells destroyed (1); patients own cells used in gene therapy (1); no chance of rejection (1).

3 Another gene damaged during process (1); mutation of (another gene) (1); (lead to) uncontrolled cell division (1).

Plant stem cells and medicines

Medicines are (often) derived from plants (1); many plants destroyed (in production of medicine) (1); using stem cells reduces number of plants destroyed (1).

1 *pluripotent* – stem cells that can form all tissue types but not whole organisms (1). Only present in embryos (1). *multipotent* – stem cells that can only form a range of cells within a certain type of tissue (1). For example, bone marrow is multipotent *(or any appropriate answer)* (1).

2 shoot tips / root tips (1); (meristematic tissue contains) dividing cells (1); (leading to) growth (1); new cells / stem cells can differentiate (1); (leading to) specialisation (1)

3 embryos left over from fertility treatment (1); discarded anyway (1); embryos now created (to supply stem cells) (1); embryos then destroyed (1); religious objections (1); life begins at conception (1); embryo has rights (1); ownership of genetic material (1); (incurable) diseases cured (1); improved quality of life (1)

4 a stem cells, divide **and** specialise (1); damaged tissue replaced (1)

   b progress of Parkinson's disease delayed by drugs (1); symptoms of Alzheimer's disease reduced using drugs (1); drugs are only short term measure (1); possible side effects of drugs (1); *idea of* stem cell therapy will lead to repair of tissue so, long term / permanent treatment (of both) (1); no / few, side effects (1)

7.1

1 Metabolic activity relatively low (1); so relatively little oxygen needed or carbon dioxide produced (1). SA:V is large (1); so diffusion distances small (1).

2 Large SA for exchange to overcome limitations of SA:V ratio of larger organisms (1); thin layers so distances substances have to diffuse short, making the process fast and efficient (1); Good blood supply so substances constantly delivered to and removed from exchange surface which maintains steep concentration gradient for diffusion (1); ventilation (for gaseous systems) maintains concentration gradients and makes process more efficient (1).

3 Radius 2 au = 3:2

   Radius 6 au = 1:2

   (1 for correct ratio, 1 for correct workings in each case)

   The SA:V ratio of smaller animal is three times bigger than that of larger animal; this illustrates how the SA:V ratios of larger animals are much smaller than those of smaller animals (1). As a result they need specialised exchange systems to get enough oxygen in, or carbon dioxide out of the system. (1)

7.2

Attacking asthma

1 Give immediate relief from symptoms (1); attach to receptors on membranes of smooth muscle cells (1); relax smooth muscles (1); dilate airways (1).

2 Steroids taken every day (1); reduce sensitivity of lining of airways to asthma triggers (1); reduce likelihood of an attack (1).

The first breath

1 Before first breath tissue has never been extended (1); drawing air in baby has to overcome the elastic recoil of the lungs (1); and the adhesion of the surfaces (1); 15–20 times more effort than next breath (1); lung surfactant stops alveoli collapsing and surfaces sticking together (1); means subsequent breaths easier (1).

1 a *Nose*: large SA with good blood supply warms the air to body temperature (1); hairy lining secretes mucus which traps dust and bacteria, protecting delicate lung tissue from irritation and infection (1); moist surfaces increase humidity of incoming air, reducing evaporation from exchange surfaces (1); produces air at similar temperature and humidity to air already in lungs (1). (max 3)

  b *Trachea*: wide tube, supported by incomplete rings of strong, flexible cartilage that stop tube collapsing (1); rings incomplete so food moves easily down oesophagus behind trachea (1); lined with ciliated epithelium with goblet cells between epithelial cells (1); goblet cells secrete mucus to trap dust and bacteria (1); cilia beat and move mucus and trapped particles away from lungs to throat to be swallowed and digested. (max 3)

  c *Bronchioles*: Small tubes spreading into both lungs (1); the smaller bronchioles (diameter 1mm or less) have no cartilage rings. The walls contain smooth muscle which contracts to close up bronchioles and relaxes to dilate them, changing the amount of air entering the lungs (1); lined with thin layer of flattened epithelium, making some gaseous exchange possible (1). (max 3)

2 a large SA of ~50–75 m² for gaseous exchange (1); thin layers so short diffusion distances (1); good blood supply with large capillary network supplying alveoli bringing carbon dioxide and picking up oxygen, maintains steep concentration gradient for carbon dioxide and oxygen between air in alveoli and blood in capillaries (1); good ventilation as breathing moves air in and out of alveoli, helping maintain steep diffusion gradients for oxygen and carbon dioxide between blood and air in the lungs (1).

  b Alveolar structure breaks down giving air sacs with much bigger radii (1); this reduces surface to volume ratio which makes them much less effective for gaseous exchange (1), e.g., *(Students can choose any radius they like – it will have a noticeable effect)*

3 Trachea lined with ciliated epithelium with goblet cells that secrete mucus (1); mucus traps dust and bacteria (1); cilia beat to move mucus and trapped particles away from lungs to throat to be swallowed and digested (1); in smokers cilia anaesthetised so do not beat (1); mucus with its load of bacteria and dust moves down into the lungs (1); more pathogens reach lungs so smokers more likely to get infections of breathing system than non-smokers with active cilia (1).

7.3

1 Record number of breaths for a timed period and repeat (1); calculate means of results under different conditions (1); use spirometer to observe breathing rate (1); *any other sensible suggestion.*

2 Ventilation rate is tidal volume of air breathed in at each breath, multiplied by number of breaths per minute (breathing rate) (1); units are cm^3 or litres per minute.

$VR = TV \times bpm$ (1)

Oxygen uptake closely related to ventilation rate, the more air is moved into the lungs, the more oxygen can be taken up by haemoglobin in blood (1); so as ventilation rate increases oxygen uptake also increases (1).

3 $VR = TV \times BR$ dog under stress so pants, breathes rapidly, but breathes shallowly so although breathing rate increases tidal volume falls (1); so the ventilation rate stays the same (1).

4 a $VR = TV \times BR$ (1) so $\frac{VR}{BR} = TV$ (1)

$TV = \frac{45000}{30} = 1,500 \ cm^3.$ (2)

Normal $TV = 500 \ cm^3.$ (1) So during strenuous exercise it is $3 \times$ higher (1)

b $VR = TV \times BR$ (1) so with infection
$VR = 300 \times 25 = 7,500$ (2)

Normal $500 \times 18 = 9000$ (1)

$9000 - 7500 = 1500$ (1)

$\frac{1500}{9000} \times 100 = 16.7\%$ fall in ventilation rate (1)

7.4

Discontinuous gas exchange cycles in insects

Look for thought and ingenuity on the part of the student along with recognition of the difficulties in these types of investigation and the need for safe and ethical handling of insects (up to 5 marks for each experiment suggested).

Dissecting, examining, and drawing gaseous exchange systems

1 Easy to make alterations and corrections (1); won't run if it gets damp/splashed (1)

2 For each diagram give marks for use of pencil; size; accuracy; and quality of drawing; accuracy of labelling.

The histology of exchange surfaces

1 For each diagram give marks for use of pencil, size, accuracy and quality of drawing, accuracy of labelling.

2 Show up more detail than can be seen with the naked eye (1); can use stains to show up specific aspects of tissues or organs (1)

1 In air gill filaments all stick together (1); SA for gas exchange is greatly reduced and so fish dies from lack of oxygen (1).

2 Table should compare key differences between humans, insects, and bony fish. For example the main organ of gaseous exchange, entry into exchange system, main site of gaseous exchange, and how ventilation occurs   (5 marks).

3 Fluid towards end of tracheole limits penetration of air for diffusion (1); when energy demands high lactic acid build up in tissues, water moves out of tracheoles by osmosis, exposing more surface area for gaseous exchange (1); tracheal system can be mechanically ventilated with air actively pumped into system by muscular pumping movements of thorax and/or the abdomen (1);movements change volume of body, changing pressure in tracheae and tracheoles so air drawn into trachea and tracheoles, or forced out, as pressure changes, making gaseous exchange more efficient (1). Some very active insects have collapsible enlarged tracheae or air sacs which act as air reservoirs, used to increase amount of air moved through gas exchange system (1); they are usually inflated and deflated by ventilating movements of thorax and abdomen (1).   (max 6)

4 Gills have large stacks of gill filaments carrying gill lamellae that have large surface area (1); good blood supply (1); and thin layers (1); needed for successful gaseous exchange. Constant flow of water maintained over gills so best possible diffusion gradient for the respiratory gases (1); tips of gill filaments overlap – increasing resistance to flow of water, slowing it down for more effective gaseous exchange (1); water and blood flow in opposite directions. Countercurrent exchange system maximises the potential exchange of gases (1).

An annotated diagram/diagrams to make any or all of these points would be acceptable.

8.1

1 Transports requirements for metabolism, e.g., oxygen, food molecules, to cells (1); removes waste products of metabolism from cells and carries them to excretory organs (1); transports materials made in one place to another place where they are needed (1).

2 Unicellular organisms have large SA : V ratio so diffusion distances small and metabolic demands low so diffusion can supply and remove substances quickly and efficiently enough (2). Multicellular organisms have small SA : V ratio, so long diffusion distances. Metabolic demands are high – diffusion alone can no longer supply all needs quickly and efficiently enough (2).

3 *similarities:* liquid transport medium (1), vessels to transport the medium (1), pumping mechanism to move transport fluid around system (1). *differences:* open has few vessels; closed has transport medium (blood) enclosed in vessels (1). In open transport medium is pumped into body cavity (haemocoel) under low pressure; in closed heart pumps blood around body under pressure (1). In open, transport medium is in direct contact with body cells; in closed transport medium has no direct contact with body cells (1). In open transport medium returns to heart through open ended vessel; in closed blood flows relatively fast and returns to heart all within vessels (1) (maximum of 6 marks)

4 *Land predators* top land predators hunt so need ability to move in fast bursts(1); they grow large and maintain own body temperature (1); need to support body against gravity (1); they may be pregnant and so have to support needs of growing fetus as well as own body needs (1) (max 3 marks)

high metabolic rate (1); they need a very efficient circulatory system supply. Double circulatory system supplies blood to lungs to be oxygenated and then returns it to heart to be pumped around body (1) so tissues receive a high level of oxygen and high levels of carbon dioxide can be removed. (1).

Aquatic predators such as pike need to hunt so also need efficient circulatory system (1); their single system less efficient than a double system (1); but bony fish have operculum so continuous flow of water over gills to oxygenate blood (1), countercurrent flow allows efficient oxygen uptake (1); they do not maintain their own body temperature and are supported by water (1); so demands of tissues much lower than those of an animal like a fox (1); so single circulation is adequate to supply their needs.

8.2

> ### Collagen, elastin and aortic aneurysms
>
> 1 A higher proportion of collagen:elastin increases likelihood that blood vessels will develop an aneurysm. (2); blood vessel less elastic (1); so less able to withstand surges of blood in aorta (1); more likely to stretch and bulge permanently (1).
> 2 Reduce high blood pressure (1) Regular screening of the aorta for signs of aneurysm developing (2).

8.2

1 Arterial blood under pressure from pumping of blood and elastic recoil of artery walls, so no tendency for it to flow backwards (1). After passing through capillary beds blood in veins under much lower pressure, there is no pumping from heart and little elastic recoil in veins so blood might flow backwards (1); as it moves back towards heart against gravity. Valves prevent this happening – they open as blood flows towards heart and close if it flows in opposite direction (1).

2 Arterioles have more smooth muscle and less elastin in walls than arteries, as they have little pulse surge (1); smooth muscle means they can constrict or dilate to control flow of blood into individual organs by preventing blood flowing into a capillary bed (vasoconstriction) or allowing it to flow (vasodilation) (1).

3 **a** *Diagrams are helpful in describing and comparing structures*

Large veins have thin walls as don't have to withstand high pressures of arterial system; large lumen as they contain large volume of blood; smooth muscle in veins contracts/ relaxes allowing constriction/dilation to change amount and pressure of blood; walls contain collagen and relatively little elastic fibre, so there is a limit to amount of blood that can flow through them; wide lumen and smooth lining mean blood flows easily. *(3 – for any three relevant points).*

Medium sized veins have similar structures and function to large veins but also have valves, which prevent backflow and move it through the venous system to largest veins and so back to heart (1).

Venules less structure in walls than veins; very thin walls with a little smooth muscle to allow blood to flow onto into veins; venules do not have valves so cannot control blood flow. (max 2)

b In all other areas of adult body veins and venules carry deoxygenated blood back from body to heart (1). In lungs they carry oxygenated blood from lungs back to heart (1).

8.3

1 Transport of oxygen (1) and carbon dioxide (1) (to and from respiring cells, respectively); transport of digested food from intestine to cells (1); transport of nitrogenous wastes from tissues to excretory organs (1); *transport of*: hormones (1); platelets (for clotting) (1); and antibodies (1); immune response (1); maintaining constant body temperature (1) and pH (1). (max 4)

2 **a** Platelets fragments of large cells called megakaryocytes (1); found in red bone marrow (1).

   b Carried around in body in circulatory system (10; involved in blood clotting mechanism (1); which prevents blood loss after injury.

   c 250 000 (1)/5257000 (1) ×100= 4.76% (1).

3 *Plasma*: straw coloured liquid which contains water, dissolved glucose and amino acids, mineral ions, hormones and large plasma proteins including albumin (important for maintaining osmotic potential of blood), fibrinogen (important in blood clotting) and globulins (involved in transport and immune system). (3 for any three points)

   Tissue fluid: Liquid contains same constituents as plasma except the plasma proteins – so no albumin, fibrinogen or globulins (1).

   Lymph: Liquid similar to tissue fluid but with less oxygen and digested food and more carbon dioxide and waste (it has been past the cells) and more fatty acids from small intestine. (2)

4 *Hydrostatic pressure* is the pressure from heart beat forcing liquid out through junctions of capillary, which at arterial end of capillaries is 3.2kPa (1); *oncotic pressure* is result of water potential in capillary from plasma proteins moving water into capillary which is –2kPa (1).

   At arterial end of capillary hydrostatic pressure higher than oncotic pressure, water is forced out of capillary and forms tissue fluid (1); as blood moves along capillary more fluid moves out and residual force from heart beat is lost. By venous end of capillaries hydrostatic pressure has fallen to 0.5kPa (1); plasma proteins don't leave capillary as they can't pass through loose junctions so oncotic pressure is still –2kPa (1);

as a result water now moves back into capillary by osmosis and by end of the capillary network around 90% of tissue fluid is back in capillaries again (1).

8.4

1 Biconcave shape gives large surface area for gaseous exchange (1); and makes it possible to move through capillaries (1); erythrocytes contain oxygen carrying pigment haemoglobin (1); mature erythrocytes have no nucleus so more room for maximum amount of haemoglobin (1); contains enzyme carbonic anhydrase involved in carriage of carbon dioxide in blood (1).   (max 3 marks)

2 **a**

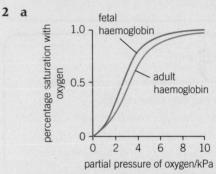

   (1 mark correct axes; 1 mark correct graph lines)

   b Foetal haemoglobin has higher affinity for oxygen than maternal haemoglobin (1); so foetal blood takes oxygen from maternal blood (1); enables foetus to survive and grow (1).

3 **a**

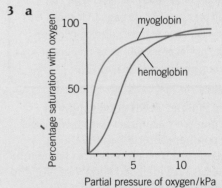

   (1 mark correct axes; 1 mark correct graph lines)

   b Myoglobin in muscles has higher oxygen affinity than haemoglobin in blood (1); so muscles can take oxygen from haemoglobin in blood, enabling muscles to get extra oxygen when they are contracting during exercise (1).

4 Flow chart should clearly shows main stages of carbon dioxide transport. It should not include plasma transport as specifically states red blood cells (6).

8.5

Dissecting a heart

Atria not very clearly displayed – either lost at butchers OR very small compared with ventricles so not easy to see; lack of major blood vessels e.g.. pulmonary vessels, aorta – lost with removal of atria; blood vessels not as clear as on Figure 1 and 2, valves not as clearly defined as in Figure 2; difficult to pick up heart wall from ventricular lining; Figure 1 and 2 schematic designed to show the principles of the structures and how they are related to each other and to their function – NOT an accurate anatomical representation; any other sensible point.

A hole in the heart

Small hole means blood flows past, and two sides of heart remain effectively separate (2)
Large hole means deoxygenated blood from right side of heart mixes with oxygenated blood from left side of heart so blood does not carry enough oxygen to tissues (3).

Blood pressure

Weak heart does not beat strongly so blood leaves heart at a lower pressure than in a healthy person – giving low blood pressure (3); damaged, closed or less elastic vessels – lumen is narrower so blood flowing through is under higher pressure (same amount of blood flowing through a smaller space) (3).

1 Heart cardiac muscles needs good supply of oxygen and glucose to contract with a regular rhythm (1); coronary arteries supply blood carrying glucose and oxygen to heart (1); healthy coronary arteries provide good supply of blood to heart muscle so it can continue to beat (1).

2 **a** *First heart sound* – blood hitting against atrio-ventricular valves (1). *Second heart sound*: sound of semilunar valves as they close to prevent a backflow of blood (1).

   b Pressure difference between atria and ventricles as atria empty and ventricles s start to contract means blood is forced against the atrio-ventricular valves which close to prevent backflow of blood into the atria (3); pressure difference between blood in artery and ventricles as they empty means blood hits semilunar valves which are closed to prevent backflow of blood into heart (3).

3 **a** *Bradycardia* is the slowing of the heart (1); when animals dive they need to conserve their oxygen and food to last for whole dive (1); they undergo bradycardia as part of slowing down metabolism to enable them to stay under water as long as possible (1).

   b *Tachycardia* is speeding up of the heart (1); at altitude there is less oxygen available in air – this means there is less oxygen available in blood (1); heart speeds up to compensate and carry more oxygen to tissues, even if it isn't effective because of low oxygen atmosphere (1).

4 **a** i)   Approx 63 bpm (2)

       ii)   Approx 48 bpm (2)

       iii) Approx 107 bpm (2)

   b Normal 0 (1); 23% decrease (1); 69% increase (1).

9.1

Observing xylem vessels in living plant stems

1 Less detailed, only gives position of xylem, not other tissues, but shows the movement of water happening in living tissue.

2 Can only see xylem – cannot be modified to show phloem. To get good sections where xylem vessels can be clearly seen is largely dependent on how sharp blade is and having a steady hand. If longitudinal sections are cut in wrong place in stems then you won't see any xylem.

9.1

1 Too big for diffusion alone to supply needs as SA : V ratio to small for diffusion to be effective means of transport (1); transport system required for transporting oxygen and glucose for respiration (1); waste product removal (1); water and mineral ions from roots to all the cells (1); hormones made in one part of a plant to the areas where they have an effect (1).

2 In plants no heart to act as central pump, whereas many multicellular animals have heart (1); in plants one type of vessel is made of dead cells – all animal vessels made of living tissue (1); in plants there are two different transport systems carrying different materials – animals have different types of vessels but the same transport medium in both (1).

3 *Stem* – vascular bundles around outside (1); helps give strength and support to structure (1); *roots* – vascular bundles in centre (1); to help give strength against tugging forces when plant blown by wind (1); *leaves* – large central vein containing vascular tissue (1) gives supports to broad structure of leaf (1).

4 *Similarities* both transport materials around plant; both made up of cells joined end to end forming long, hollow structures, *any other sensible points*.

Differences Xylem largely non-living tissue, phloem living; xylem transports water, mineral ions, and supports plant, phloem transports organic solutes around plant from leaves; in xylem flow of material from roots to shoots and leaves, in phloem flow of material up and down; xylem cell walls lignified, phloem not; xylem have wide lumen; mature phloem cells have no nucleus; other cells associated with xylem in herbaceous dicots include xylem parenchmya and xylem fibres, equivalent in phloem include fibres and scleroids. Any other sensible points. (max 6)

9.2

1 Very small to penetrate soil particles (1); large SA:V ratio for water absorption (1); thin layers for ease of diffusion (1); high solute concentration in cytoplasm gives low water potential so water moves in from soil by osmosis (1).

2 *Symplast pathway*: relies on osmosis as water moves through cell membranes and cytoplasm (1). Water moving in from soil by osmosis into root hair cell raises water potential compared to next cell so water moves again by osmosis (1). Active transport of ions needed to move water from endodermis to xylem by osmosis (1).

Apoplast pathway: water moves through cellulose cell walls by cohesive forces between water molecules and as result of transpiration pull up xylem (1); moves into symplast pathway in endodermis as a result of Casparian strip (1); and needs active pumping of ions into xylems followed by osmosis (1) before water moves back to apoplast pathway in xylem.

3 a Temperature increase increases root pressure BUT although increase in temperature increases rate of chemical reactions (1) it also increases rate of passive processes such as diffusion and osmosis (1).

b Active transport needs energy in form of ATP (1); cyanide poisons mitochondria where cellular respiration takes place so cyanide prevents ATP formation (1); Oxygen and respiratory substrates needed for cellular respiration and ATP production (1).

Factors which interfere with ATP production also interfere with development of root pressure (1); suggesting that development of root pressure is an active process requiring ATP (1).

9.3

Measuring transpiration

1 5cm (1)

2 Water vapour lost from plants through open stomata in transpiration (1); most water taken up by plant is used for transpiration (1); as the rate of water uptake slows down significantly with vaseline on lower surface this suggests that most stomata are on under surface of leaf (1); and once covered very little water vapour could be lost (1); so very little water taken up by shoot (1).

1 *Transpiration*: evaporation of water from surface of a leaf (1). *Transpiration stream*: flow of water moved up from soil into root hair and through root cortex by osmosis, into xylem and up through stem by cohesion of water molecules, cross leaf cells by osmosis and out of leaf by evaporation and diffusion (1).

2 *Root pressure*: active movement of solutes followed by passive movement of water by osmosis, which gives a positive pressure forcing water up the xylem (1).

Transpiration pull: pulling of a constant stream of water molecules up xylem held together by cohesive forces as a result of evaporation of water from surface of spongy mesophyll cells in leaf (1); a passive process and a negative pressure (1).

3 All conditions except air movements kept the same (1); sensible suggestions as to how to investigate effect of air movements, e.g., use of fan for set time (1); fan placed at different distances from potometer (1); control readings in still air allow plant recovery time between different distances (1), more than one reading for each distance (1) etc. Total of 4 marks from these or any other sensible suggestions.

4 a 0.01 cm/s (1); dark so no photosynthesis (1); and most stomata closed (1); little gas exchange and so little water lost by transpiration (1); so little water movement up stem to replace it (1).

b 0.14 cm/s (1); photosynthesis taking place (1); quite a few stomata open for gas exchange (1); so water lost in transpiration (1); moves up stem to replace it (1).

c 0.2 cm/s (1); bright light – maximum photosynthesis (1); so many or all stomata open for maximum gas exchange (1); lots of water lost by transpiration (1); and lots taken up to replace it (1).

d 0.3 cm/s (1); bright light – maximum photosynthesis so many or all stomata open for maximum gas exchange so more water lost in transpiration (1); when windy water vapour moves away from leaf as soon as it appears reducing humidity(1); and increasing diffusion gradient for water out of leaf (1); even more water lost in transpiration (1); so more needed to replace it giving rapid uptake of water (1).

5 a Water molecules evaporate from surface of mesophyll cells into air spaces in leaf and move out of stomata into surrounding air by diffusion down concentration gradient (1); loss of water by evaporation from mesophyll cell lowers water potential of cell (1); water moves into cell from adjacent cell by osmosis, along apoplast and symplast pathways (1); repeated across leaf to xylem. Water moves out of xylem by osmosis into cells of leaf (1); water molecules form adhesive bonds with walls of narrow xylem vessels and also form hydrogen bonds and so tend to stick together – cohesion (1); as a result of these cohesive forces, along with adhesion to walls of xylem, water is pulled up xylem in continuous stream to replace water lost by evaporation, this is the transpiration pull (1); transpiration pull results in tension up xylem which helps to move water across roots from soil, movement is partly by osmosis (change in water potential in cells across root) and partly by cohesion through apoplast pathway (1).

b (i) Cohesive forces in xylem cause negative pressure (1); that draws tissues in and reduces diameter of tree (1); in daytime when maximum photosynthesis and so maximum transpiration is taking place (1); minimum transpiration takes place during night so less tension in the xylem tissue and it expands (1).

   (ii) Tension and negative pressure in xylem means air may be pulled into xylem when stem is cut (1); this breaks cohesive stream of water molecules (1); and prevents more water moving up stem (1); as a result flowers droop and die very quickly, as air bubble is pulled up the stem (1).

9.4

1 *source*: supplier of carbohydrates needed by cells of plant (1); e.g., leaves, stems, storage organs, seeds (1); *sink:* area of plant that needs assimilates in phloem sap (1), e.g., roots, meristems, developing fruit, seeds and storage organs (1).

2 Aphids pierce plant tissues and push their stylet directly into phloem vessels (1); if aphid removed, leaving stylet in place, pressure in phloem vessel continues to force sap out of stylet (1); this can be to show presence and concentration of assimilates in phloem (1); pressure in phloem (1) and how these things change with manipulation of other factors such as light intensity (1).

3 a Sucrose moves from cells to companion cells and sieve tube elements by diffusion along a concentration gradient, is moved into companion cells and sieve tube elements by an active process (1); hydrogen ions (H^+) actively pumped out of companion cell into surrounding tissue using ATP (1); they return into companion cell down a concentration gradient via a co-transport protein carrying sucrose as well (1); build-up of sucrose in companion cell and sieve tube element means water moves in by osmosis causing a build-up of hydrostatic pressure (1). Water carrying assimilates moves into tubes of sieve elements, reducing pressure in companion cells and moving up or down the plant by mass flow (1). Sucrose moves out of phloem into cells which need it by diffusion, followed by water by osmosis.

b Companion cells have membrane folding to give large surface area for transport of sucrose (1); also have many mitochondria to produce ATP needed for active transport (1); if mitochondria in companion cells are poisoned, translocation in phloem stops (1); the pH of the companion cells is higher (more alkaline) than the surrounding cells supporting idea of hydrogen ion pump (1); the flow of sugars in the phloem is about 10 000 times faster than it would be by diffusion alone, suggesting an active process is driving mass flow (1).

9.5

Investigating stomatal numbers

1 Expect few stomata in xerophytes (1); and expect them to be hidden in pits, or grooves on leaf (1); covered in hairs (1); inside rolled leaves (1); etc to maintain a still, moist microclimate (1); and reduce loss of water by transpiration (1).

2 The leaves are reduced to spines, so few stomata and difficult to find (1); the stomata on the stems are protected by spines (1).

1 *Students can choose any three from:* thick waxy cuticle, sunken stomata, reduced leaves, hairy leaves, curled leaves, succulent leaves. NOT leaf loss as that is not a structural adaptation of the leaf.

Students should give a clear and full explanation of the relationship between the structural adaptation and the function in conserving water. (1 mark for each)

2 *Students can chose any three adaptations and give a clear and full explanation of the relationship between the structural adaptation and the function in conserving water.* (1 mark for each)

3 *Dry conditions*: conflict between need to open stomata for gaseous exchange and loss of water by evaporation from open stomata. Hot dry air so evaporation will take place fast. Ground dry or frozen so little water available. Tissues need water to maintain turgor and carry out photosynthesis. *Any other sensible points.* (up to 2 marks)

In water: waterlogging of tissues – no access to oxygen. Sinking – need to be near the surface to get light for photosynthesis. Slow diffusion of oxygen in from water. *Any other sensible points.* (up to 2 marks)

4 *Characteristics should should be well argued and make sense in terms of either withstanding drought or withstanding flooding. They should be able to explain the advantages of the characteristics they have chosen and why they think they would be particularly useful to the crop.*

Example for drought: Hairy leaves (1): these trap a microclimate around the stomata and reduce water loss by transpiration. Only suitable for crops where the leaves aren't eaten.

Example for flooding: Resistance to rotting: (1) Plants in flood conditions often rot so resistance to fungi or bacteria which cause rot would be very useful.

10.1

The development of classification systems

The earliest classification systems were based on visual similarities between organisms. Advances in our understanding of the biological or genetic make-up of organisms has provided evidence for how organisms are linked. This information informs revisions to the system of classification which is used.

1 *Any 2 appropriate reasons. For example:*

Enables scientists to share information / makes communication easy (1); provides information about an organism, based on members of the same group (1); allows accurate identification of an organism (1)

2 Ligers cannot reproduce to produce more ligers therefore they are not a species (1). Both lions and tigers reproduce to produce fertile offspring, therefore they are species (1).

3 **a** *Phylum* (1); **b** *Erithacus* (1); **c** *rubecula* (1)

4 Both parents are members of the same genus (*Rubus*) (1) but different species (*ursinus* and *idaeus*) (1). Two different species cannot produce fertile offspring / according to the taxonomic classification system the loganberry should not be fertile (1).

10.2

1 *Any two from*:

Plants have chloroplasts / chlorophyll, whereas fungi do not (1); plants are autotrophs, whereas fungi are heterotrophs (1); fungi may be unicellular, plants are always multicellular (1); fungi store food as glycogen, whereas plants store food as starch (1); plant cell walls are composed of cellulose, whereas fungi cell walls are composed of chitin (1) *Or other suitable example.*

2 **a** Prokaryote (1), **b** Fungi (1), **c** Protoctista (1)

3 *Any three from*:

Advances in biological techniques have identifies large differences in composition (1); ribosomes/rRNA differ (1); cell walls differ – peptidoglycan not found in archae (1); old classification does not show correct phylogeny (1)

4 *Any six from*:

Living organisms classified into two kingdoms based on major differences in characteristics (1) For example, those that moved and ate (animals) and those that didn't (plants) (1); scientific advances/use of microscope allowed smaller details to be observed (1); organisms divided into five kingdoms (1); Plants, animals, fungi, protoctista, prokaryotes (1); Advances in science allowed DNA and proteins to be studied (1); Provided evidence for evolutionary relationships (1); Three domain system proposed (1); Relevant scientists mentioned (Linnaeus, Whittaker, Woese) (1).

10.3

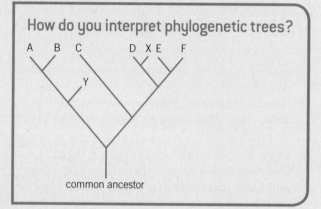

How do you interpret phylogenetic trees?

common ancestor

1 Historical classification systems based on physical characteristics / niche occupancy, whereas phylogeny based on evolutionary relationships (1).

2 It takes into account evolutionary relationships that might not be obvious by just looking at characteristics (1). It forms a continuous tree so organisms do not have to be forced into groups (1). It is not hierarchical therefore different groups on the tree are represented according to their evolutionary position – and can thus be compared (1).

3 a snakes (1)

b They have become extinct / not present in the world today. (1) They are placed along the timeline at the point they existed in time (1).

c The bird and crocodile branches are closer together than the bird and turtle (1). The birds shared a common ancestor with the dinosaurs. This organism shared a common ancestor with the crocodiles (the common ancestor with turtles is much further back in history) (1).

10.4

Evolutionary embryology

Very unlikely that an embryo could form a fossil, as it is made up only of soft tissue (only in rare cases, such as encasement in tar or resin, are bodily tissues preserved).

1 A diagram used to show evolutionary relationships between organisms (1); the closer the branches of the tree the closer the evolutionary relationships (1).

2 *advantages* (2): for example – radioisotopes can be used to date fossils / changes can be tracked over time / chronological order apparent in rock strata.

disadvantages (2): for example – many organisms decompose quickly before they have a chance to fossilise/destroyed by volcanoes/ destroyed by earthquakes.

3 *1 mark for scientist, 1 mark for contribution* (max 6)

Lyell – suggested that fossils were actually evidence of animals that had lived millions of years ago. *Hutton* – proposed theory of uniformitarianism.

Darwin – came up with theory of evolution by natural selection through observations in Galapagos islands / jointly published theory.

Wallace – came up with theory of evolution by natural selection in Borneo/jointly published theory.

4 *Any three from*:

Study of similarities and differences in proteins and nucleic acid/DNA of an organism (1); changes in highly conserved molecules can help identify evolutionary links (1); such as cytochrome C / ribosomal RNA (1); species that are closely related have the most similar DNA and proteins /distantly related have far fewer similarities. (1)

10.5

Studying variation in identical twins

1 They have no genetic variation therefore all variation is the result of the environment.

2 They have been exposed to greater amount on environmental influences.

3 Eye colour is determined solely by genes as both identical, ear piercing determined solely by environment as they have opposite results, mass and height controlled by a combination of both genetic and environmental factors as results vary, mass controlled more by the environment than height as greater variation shown.

1 Interspecific variation is differences between individuals of different species whereas intraspecific is differences between individuals of the same species (1).

2 a *Any two suitable examples*, e.g. scar, tattoo, dyed hair (1)

b *Any two suitable examples*, e.g. eye colour, blood group, lobed or lobeless ears (1)

3 Caused by a combination of genetics and the environment (1). Genes determine the natural colour of hair and texture, e.g. curly/straight (1). Environment affects final appearance, e.g. if hair is cut, dyed, or lightened by sunlight (1).

4 *Any 3 from*: Individuals produced by asexual reproduction are clones/genetically identical to parents (1). No fertilisation so no mixture of genetic material (1). Meiosis does not take place/no production of gametes (1). DNA can only be altered as a result of mutation (1).

10.6

Flagella length variation in *Salmonella*

1 2.4μm

2 0.49

3 1.91—2.89μm (68% of population will fall within standard deviation of the mean).

4 Continuous variation, the length of flagella can take any value within a range.

1 *Continuous* – **b**, **c** (1). *Discontinuous* – **a**, **d** (1)

2 Characteristics which show discontinuous variation are purely controlled by genetics/no environmental influence (except scars/tattoos just environment) (1). Normally controlled by a single gene (1). Characteristics which show continuous variation are controlled by a combination of genetic and environmental causes (1). Controlled by a number of genes/polygenes (1).

3 *Any two from*: The values of a characteristic which shows discontinuous variation fall into discrete categories (1) if a mean is calculated it may produce a value which does not fit into a category (1) many of the characteristics do not have a numerical value (1).

4 *Any four from*: Continuous variation (1) as controlled by both genetic and environmental causes (1). Normal distribution (1) very few rabbits would be extremely large or extremely small (1) most would be within one standard deviation of the mean (1).

5 a

Diameter of stem / mm	Rank	Number of thorns per unit length	Rank
1	1	8	1
2	2	11	3
3	3	9	2
5	4	12	4.5
8	5	12	4.5
10	6	27	7
11	7	23	6
14	8	30	8

(2) – 1 mark for each ranked column correct

$\Sigma d^2 = 4.5$ (1)

$r = 1 - \dfrac{6\Sigma d^2}{n(n^2-1)}$ (1)

$r = 1 - (6 \times 4.5) / (8 \times 63)$ (1)

$r = 0.946$ (1)

b df =6 (1); ρ = 0.946 > 0.881 therefore >99% confidence. Therefore there is very little (<0.1%) likelihood that the null hypothesis is true (1).

10.7

Classification of giant pandas

1 Giant pandas have: a large body mass, shaggy fur, a pseudo-thumb, the ability to climb

2 Student's own answer. Any conclusion drawn should be consistent with the evidence available: for example, a student may conclude that the giant panda is a panda, due to the similarities in its teeth, snout and paws to the red panda.

3 The molecular sequence of a particular molecule is compared, by looking at the order of DNA bases or at the order of amino acids in a protein.

Species that are closely related have similar DNA and proteins, whereas those that are distantly related have far fewer similarities.

1 *Anatomical* – camouflage, sharp canine teeth (1); *Physiological* – melanin production, production of toxins (1); *Behavioural* – migration, courtship dance (1).

2 Analogous structures are structures that have adapted to perform the same function but have a different origin whereas homologous structures appear superficially different but have the same underlying structure (1).

3 a Insect and bird wing – both have evolved to fly to escape predators/hunt for food (1).

4 *2 marks for named adaptation and suitable explanation. 2 marks for correctly naming the adaptation as behavioural/anatomical/physiological.*

5 They have analogous structures – anatomical features that perform the same function in different organisms, but have a different origin (1).

Any two from: Both burrow through soft soil to find insects (1). Both have a streamlined body shape, and modified forelimbs for digging (1). Both have velvety fur which allows smooth movement through the soil (1).

10.8

Anolis lizards

1 When few individuals of a species colonise a new area their offspring initially experience loss in genetic variability, resulting in individuals that are physically and genetically different from their source population (1).

2 Have no lizard populations.

3 Area covered in scrub/short vegetation, short hind limbs provide stability to walk along narrow perches.

4 *Any four from:* released pairs of lizards on islands with no lizards but the same vegetation; measured genetic variation; measured hind leg length; hind leg length shortened over time providing evidence for natural selection; founder effect would produce random leg length.

5 Decreases fitness of population and their ability to survive and reproduce, as a result of less variation, increased chance of recessive disorders.

1 *Any three from*: availability of light / water / nutrients / carbon dioxide / space, risk of being eaten, disease, ability to cross-pollinate, or other suitable example (1).

2 Variations exist within a population (1); those with the best characteristics survive AND reproduce (1); characteristics are passed onto their offspring through genes (1).

3 A mutation occurred / existed in the mosquitos DNA which made them DDT resistant (1) these organisms survived exposure to DDT and reproduced (1); mutation which caused resistance is passed onto their offspring (1); frequency of the DDT-resistant allele increases in the population (1).

4 *Any six from (or other appropriate examples):* Flavobacterium digests nylon waste (1); positive – used to clean up factory waste (1); bacteria e.g. MRSA – antibiotic resistance (1); negative – no longer killed using current medical treatment

(1); sheep blowfly – insecticide resistant (1); negative – no longer killed by insecticide so increased sheep death (1).

11.1

1 *Species richness* – the number of different species living in a specific area (1).
Species evenness – the number of individuals within the species living in a community (1).

2 Habitat biodiversity: desert – low; coastline – high (1). Suitable habitat examples given (1). Species biodiversity: desert – low; coastline – high (1).

3 *Answer must include:* (max 4 marks) Greater genetic variation / wider range of alleles (1); therefore increased likelihood some organisms are suited to a habitat change (1). *Plus up to two from:* some organisms may be suited to different habitats (1); therefore more areas may be colonised by the species (1); some individuals will be resistant to a new disease (1); therefore lower probability of all organisms being killed by the disease (1); some organisms will be better adapted to avoid new or adapted predators / catch prey (1); therefore less chance of being eaten / starvation (1); other suitable example with consequence (2).

11.2

1 Random – all organisms have an equal chance of selection; non-random – different organisms have higher / lower probabilities of being selected (1).

2 Use random sampling (1); removes sampling bias (1); use as large a sample size as possible (1); removes the effects of chance (1).

3 **a** Systematic sampling (1); abiotic conditions vary as you travel downstream affecting the type and abundance of organisms present (1).

   b Random sampling (1); any one from: environment easy to study/fairly uniform (1); reduces sample bias/increases reliability (1).

11.3

Belt transect in a National Park

1 Mark out line across sampling region (1); quadrats are placed at fixed intervals (1m apart), and used to measure percentage cover in that area (1).

2 Students were trying to investigate specifically how the distribution of plant species varied according to land use. Random sampling would not sample required region in a systematic manner (1).

3

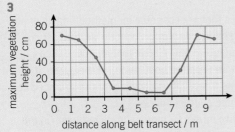

Correct axes, labels, and suitable scale (1)
Correct plot and line (1)

4 Any 4 from: Bare ground is the path (1); path is 3 / 4m wide / is found between positions 3 / 4 and 6 (1); where the ground is trampled, grass is the dominant species (1); there appears to be a direct correlation between the maximum height of vegetation and the percentage cover of grass (1); untrampled regions have greater species biodiversity (1); larger species are found away from the path region (1); other relevant points from the data (1)

5 Suggested abiotic factor e.g. soil pH (1); explanation of how the factor could be measured at each station along the transect in a systematic manner e.g. measured using a pH meter positioned at the centre of each quadrat position (1)

6 Only one set of data was collected along the line (1) which may not be representative of the whole region (1); not all samples sum to 100% percentage cover (1) therefore errors were made /some unidentified species were not recorded (1); therefore the conclusions drawn are tentative (1)

1 a pooter (1).   **b** sweep net (1).
 c pH probe (1).   **d** quadrat (1).

2 *Any two from:* likely to have higher resolution; rapid changes can be monitored; less possibility for human error; data can be stored on a computer (2).

3 *At least one advantage and disadvantage should be included for each sampling measure. Population density* – advantages: accurate; disadvantages: time consuming, can only be used when individual members of a species can be identified (1) *Frequency* – advantages: rapid, can be used when individual members of a species cannot be identified; disadvantages: only gives an approximate result (1) *Percentage cover* – advantages: allows a lot of data to be

collected quickly; disadvantages: only gives an approximate result; least precise sampling technique (1).

11.4

1 *Any three from:* large number of successful species; ecosystem fairly stable; many ecological niches available; environment unlikely to be hostile; complex food webs exist (3 max).

2 a Pond B because it has the higher value of Simpson's Index of Biodiversity (1).

 b Pond A was more polluted. It has the lower value of Simpson's Index of Biodiversity (1).

Fewer species can survive (1) the harsher environmental conditions (1).

3 N = 28 (1)

Organism	$\dfrac{n}{N}^{2}$
Bird's-foot trefoil	0.01
Crested Dog's-tail	0.04
Meadow buttercup	0.14
Oxeye daisy	0.09
Rough hawkbit	0.01
Smaller cat's-tail	0.02

(1)

$$\sum \left(\frac{n}{N}\right)^{2} = 0.31 \ (1)$$

Simpson's Index of Diversity = 0.70 (1)

11.5

1 *Any two from:* number of alleles in a population must increase (1); mutations can create new alleles (1); gene flow can introduce new alleles into population (1).

2 More genetically biodiverse a species is the greater variation in DNA/number of alleles present (1); species more likely to survive a change to the environment (1) as there is a higher probability that some members of the species will have the allele to survive the change and reproduce (1).

3 Species A 0.48/48% of genes were polymorphic (1). Species B 0.6/60% of genes were polymorphic (1). Species B is more genetically diverse as it has a higher percentage pf polymorphic genes (1) therefore more alleles are present (1).

11.6

> ### Loss of biodiversity in the UK
>
> 1 Economic gain (1)
> 2 Twice the proportion of chalk grassland (compared to lowland mixed woodland) (1).
> 3 300 000 hectares (1)
> 4 *Any two from:*
> Replanting hedgerows (1) – provides an additional habitat allowing more species to be supported (1).
> Mixed planting/crop rotation (1) – greater range of plant species enabling more animal species to be supported (1).
> Reduced/no chemical use (1) – pests/weeds not destroyed which provide food source for a range of species (1).
> *Any other suitable reason and explanation.*

1 a Only one type of plant is grown (1) this will be the food source for only one / a few species of animal (1) or other relevant reason, correctly justified.
  b Deforestation / removal of habitat / reduction in plant diversity (1) leading to reduction in animal diversity because more limited range of foods / shelter available (1) or other relevant reason, correctly justified.
  c Pesticides remove pest species (1) which reduces the food source for the animals that live off these species (1) or other relevant reason, correctly justified.

2 *Any two from*: Deforestation removes the food source for a / many species, which causes starvation for the predator species which feed on them (1); loss of habitat removes the shelter used by organisms, causing local extinction / migration (1); migration of species can lead to reduction in species diversity in adjacent areas (1).

3 *Any three from (effect and explanation required for each mark)*: Polar ice caps melting / reducing in size leads to reduction in size of ice sheets, removing the habitat of the native species (1); rising sea levels leads to loss of habitats in low lying land / salination of rivers further upstream, affecting the species which inhabit these areas (1); higher temperatures can lead to drought in a region, affecting non-drought resistant species (1); insect life cycles and populations may change, spreading diseases to new regions, affecting the animal organisms which live in these areas (1). *Other suitable suggestion with explanation of its effect on a population.*

11.7

> ### Keystone species
>
> 1 Species that are essential for maintaining diversity – they have a disproportionately large effect on their environment relative to their abundance (1).
> 2 Predators keep the populations of their prey at a consistent level, thus allowing for the balanced co-existence of other species (1).
> 3 *Up to two marks for stating how a reduction in the population of prairie dogs would lead to a reduction in the population of purple coneflowers; up to two marks for a linking explanation of the effect.*
> Less aeration of the soil (1) leading to fewer decomposers redistributing nutrients (1).
> Less animal droppings / movement of soil (1) less redistribution of nutrients (1).
> Water being lost (by evaporation) / not channelled into the water table (1) lack of sufficient water for growth / photosynthesis (1).

1 *Aesthetic* – how the region appears; *economic* – how the region can provide an income; *ecological* – the species that can be supported in the region (1).

2 Any two sensible suggestions, for example: All living organisms have right to survive and live in the way they have become adapted (1); habitat and biodiversity loss prevents many organisms living where they should (1); moral responsibility to conserve for future generations (1).

3 a Only two potato varieties plants in Ireland so little genetic diversity (1); when new disease introduced (Phytophthora infestans) no potato had resistance to disease (1).
  b Plant wider range of crops to increase genetic biodiversity within populations (1); if new disease/climate change/pests introduced some will have resistance/ ability to tolerate changing conditions, so some crops will survive (1).

11.8

1 Captive breeding – animals reared and bred in human-controlled environments (1). Seed banks – collections of seeds are stored (at low temperatures) for future use (1). Botanic gardens – collections of plants are grown in controlled environments (1).

2 *Up to two advantages, and two disadvantages:*
Advantages – allows for an individual of a species to survive; maintains or increases endangered populations (1); allows for reintegration of a species into its natural habitat (1) *Disadvantages* – leads to loss of genetic diversity (1); leads to organisms not learning behaviours of their

wild counterparts (1); can decrease the disease resistance of a population (1); genetic problems can occur in offspring due to in-breeding (1).

3 *Any four from:* controlled grazing allows species time to recover, rather removing them entirely from habitat (1); restricting human access to prevent poaching / avoid plant species being trampled (1); feeding animals to ensure native populations able to survive to reproductive age (1); culling/removal of invasive species to ensure native species able to access resources / to remove competition for resources (1); halting succession ensuring the habitat remains in its current state, so appropriate resources are available for current species (1).

4 *Any four from*: Landowners/countries have economic/cultural reasons for exploiting natural resources (1); exploitation leads to loss of biodiversity (1); (financial) incentives often needed to replace income exploiting a resource would provide; animals do not respect nation's boundaries (1); agreements between nations needed to manage endangered populations and help limit trade of controlled/protected species (1); to preserve number of species in natural habitat (1).

12.1

1 a Disease that can be passed from one organism to another. (1)

b Pie chart should show: 9% injuries (1); 23% communicable diseases (1); and 9% injuries (1).

2 Table should summarise key comparisons between the four. For example, whether they are prokaryote/eukaryote, whether they have plasmids or not, whether they are heterotroph/ autotroph, whether they are beneficial/ pathogenic/neutral, comparison of cellular structure, whether they have a nucleus or not. *Extra credit for students who remember 70s and 80s ribosomes for prokaryotes/eukaryotes.*

3 13000000 is 23% of all deaths

68% of all deaths are caused by non-communicable diseases (1)

68% is $\frac{13000000}{23} \times 68$ (1) = 36-37 million (1) then approximate because working with approximate numbers (1)

4 Take over cell metabolism, viral genetic material inserted into host DNA (2); take over cell and digest contents (e.g., some Protista) (1); completely digest living cells and destroy them (e.g., fungi) (1); produce toxins which poison or damage host cells, some toxins break down cell membranes or inactivate enzymes or prevent cell division (e.g., most bacteria) (2).

5 a Viruses insert genetic material into host DNA (1); and take over cell metabolism to make new viruses before breaking out of cell(1); protists take over cells and feed on cell contents (1) and divide before breaking out of the cell(1)

b viruses only active when inside a host cell (1); they have little structure and take over the whole host cell (1).

12.2

The threat to English Oak trees

Acute oak decline caused by bacterium found on or in oak jewel beetles showing they may transfer disease (2); trees become infected in regions where no oak jewel beetles so they are not vectors/cause of disease or not the only vector/cause of disease (2). Any other sensible suggestions.

Banana diseases and food security

1 30,000,000,000 kg bananas (1)

2 Increased malnutrition as people deprived of their main staple food (1); increased disease and death as people less able to resist disease due to malnutrition (1); shortages of other foods as people try to buy other staples (1); increased food prices as a result of short supplies driving up demand (1) *any other sensible point*

3 a lack of awareness of causes of disease (1); lack of biocontrol on farms (1); contamination between farms (1) (max 2)

b cloned Cavendish plants so if one plant susceptible to disease, they all will be and therefore whole plantations can be wiped out (1)

1 *Bacterial:* Ring rot (caused by *Clavibacter michiganense*) affects potatoes, aubergine, and tomatoes. TB (*Mycobacterium tuberculosis* and *M. bovis*) affects humans, cows, badgers, deer. Bacterial meningitis affects humans. *Viral:* Tobacco mosaic virus affects tobacco plants and 150 other species. HIV/aids affects humans and some apes. Influenza (*Orthomyxoviridae* spp affects mammals including humans, pigs, and birds. Protist: Potato blight (*Phytophthera infestans*) affects potatoes and tomatoes. Fungal: Black sigatoka (*Mycosphaerella fijensis*) affects bananas and plantains. Ring worm (*Trichopyton verrucosum*) affects cattle (other spp. Affects most animals including people). Athlete's foot (*Tinia pedia*) affects human feet (max 6).

TB, cows and badgers

1. Wild animals such as badgers and possums which are infected with TB regularly use the same pastures and so the bacterium is around in the grass (2).
2. It isn't possible to tell if animal is infected with TB or has been vaccinated so would be impossible to protect consumers against infected milk products etc (2).
3. Any sensible suggestions for example: People's perception of badgers are different to that of cattle; some people put a higher value on wildlife than the health of farm animals and people; some people feel that there are more humane alternatives that would protect both badgers and cattle (max 2).
4. Catching the animals to deliver vaccine would be difficult and stressful for animals and people; would never know what proportion of the population was vaccinated; difficult to tell if an animal is vaccinated or not; if vaccine in food difficult to quantify how many badgers eat food and impossible to control dose. Any other sensible point (4).
5. Look for clear evidence of good research skills, balanced approach, citing sources etc.

Zoonotic influenza

1. Viruses not affected by antibiotics so cannot be cured (1).
2. Enables scientists to identify cause of outbreak (1); track its spread (1); and helps in the development of vaccines (1) and (if bacterial) medicines to treat the disease (1).
   a. they were much younger than usual flu victims – up to 80% were under 65, whereas in normal flu outbreaks, around 90% of deaths are in people over 65
   b. H1N1 was a new strain of flu which crossed species barrier from animals to people. It has happened before – older people may have met a similar virus earlier in their lives whereas people under 65 had not encountered a similar virus before and so it was extremely damaging to them (3).

Identifying pathogens

1. Appropriate treatment can be used (1); and appropriate steps taken to prevent the spread of the pathogen (1).
2. *Benefits:* Relatively cheap (1); available in all

hospitals (1); special stains show up classes of organisms easily (1). *Any other sensible point.* *Limitations:* can take some time for culture to grow (1); some organisms e.g. viruses can be very difficult to culture (1); light microscope limited e.g. viruses not visible (1).

3. Look for evidence of good research skills, clear explanations, citing of sources etc

2. *Show evidence of understanding of the ways in which bacteria attack animal and plant populations and the different effects they have.* For example: ring rot (plants) and TB (animal), any diseases may be chosen as long as bacterial (1); both caused by bacteria, both cause tissue destruction, both remain infective in the environment (3); animal disease can be cured by antibiotics/prevented by vaccination, no treatment or vaccine for plant disease (2).

3. *Students should show awareness of the variety of ways in which the diseases they have chosen can be spread for example:* For animals direct transmission from one animal to another via direct contact, inoculation, and ingestion and indirect transmission e.g., droplet infection, fomites vectors etc. For plants direct transmission plant to plant and indirect transmission including soil contamination and different types of vectors e.g., wind, water, animals, humans. 3 for any three correct animal methods, 3 for any three correct plant methods including up to 2 for specific types of vectors.

12.3

Preventing the spread of communicable diseases in humans

1. Removes and destroys pathogens so not transmitted through direct contact, ingestion or leaving on fomites
2. Living close together increases droplet infection risk and contagion (2); poor sanitation increases risk of contagion, contaminated water and food, and vectors (2) and poor nutrition means people are more vulnerable to infection (1).
   All increase the risk of the spread of communicable diseases so improving them lowers the risk (1).
3. Mosquitos breed in water (1); mosquitoes carry malaria (1); any waste container which holds water provides a breeding space for mosquitos (1); and so increases malaria (1); removing the waste reduces breeding opportunities for mosquitos (1); and so reduces the incidence of malaria (1).
4. Any sensible well-structured answer using variety of sources.

Preventing the spread of communicable diseases in plants

Points could include: If pathogen getting to plant prevented there will be no disease so methods of reducing spread of pathogens would be key (e.g., human hygiene not spreading spores, disposing diseased plant tissue carefully, removing all traces of damaged plant tissue, insect vector control, any other sensible points); breed plants that are not susceptible to infection or disease resistant; if conditions favourable for healthy plant growth plants will increase disease resistance (so good management e.g., soil fertilising, pest control); need to balance all three elements to avoid disease. *Any other sensible points*.

1 *direct* pathogen is spread directly from one organism to another (1); *indirect* – the pathogen is spread from one organism to another through another medium, e.g. the air, a vector (1). *Show awareness of the differences between organisms that can move around and organisms that cannot.*

2 *Similarities*: being crowded close together increases risk of direct and indirect transmission. Weakened individuals more at risk of infection. Damage to protective outer layers can allow pathogens in (2). *Differences*: Animals actively exchange body fluids (sex, kissing, bites) plants don't. Animals transfer food and drink into body through mouth which plants don't (2). *Any other sensible points*.

3 *Show awareness of the differences between organisms that can move around and organisms that cannot.* *Similarities*: animals, wind and water can act as vectors, fomite, e.g. bedding, sacks, machinery can carry disease from one individual to another, soil contamination is common indirect method of disease spread in plants and can affect animals too. *Differences*: droplet infection from coughs and sneezes doesn't affect plants BUT droplets and splashes from one leaf to another can do. *Any other sensible points.*

4 Treating people to reduce pool of infection using medicines against disease (1); or using vaccines (no really effective ones developed yet); destroying mosquitoes that spread the disease (insecticide sprays on water and homes) (1); preventing mosquitoes breeding (1), draining swamps (1), removing waste filled with water (1); preventing mosquitoes reaching people (mosquito nets over beds, screens at doors and windows) (1) (max 5).

12.4

1 Receptors respond to molecules from pathogens (1), or to chemicals produced by the plant cell wall when it is attacked (1). These attach to receptors, stimulating the release of signalling molecules to switch on genes in the nucleus, triggering cellular responses (1).

2 *Diagram of table to include:* Production of defensive chemicals, e.g. insect repellants, insecticides, antibacterial compounds including antibiotics, antifungal compounds, anti-oomycetes, general toxins (2). Physical defences, e.g. callose barriers immediately, callose and lignin deposition in cell walls longer term, callose blocking sieve plates to prevent spreading through the phloem, callose deposited in plasmodesmata to prevent spread of pathogens from one cell to another (2). Sending alarm signals to uninfected cells so they can put defences in place (2).

3 *Show evidence of careful research from reputable sources and reference their sources.*

12.5

1

adaptation	How it prevents entry of pathogens
skin	Impermeable barrier between air/water and inside the body
sebum	Chemical produced by skin which inhibits growth of pathogens
mucus	Traps pathogens in nose/trachea etc uses lysosomes to destroy bacteria/fungal spores, phagocytes to engulf and digest pathogens
tears	Lysozymes break down pathogens
urine	Lysozymes break down pathogens
Stomach acid	Destroys most bacteria and fungal spores entering the gut
Expulsive reflexes	Coughing/sneezing expel pathogens from gas exchange system Vomiting and diarrhoea expel pathogens from digestive system
Blood clotting	Protects again the entry of pathogens through broken skin

2 Localised inflammatory response to pathogens at site of wound, mast cells activated and release histamines and cytokines (1); histamines cause vasodilation causing localised heat and redness (1); raised temperature helps prevent pathogens reproducing; histamines make blood vessels

leaky forcing tissue fluid out, causing oedema and pain (1); cytokines attract phagocytes to site which phagocytose pathogens; accumulation of dead phagocytes and pathogens forms visible pus layer (1).

3 a Pathogens produce chemicals which attract phagocytes (1); phagocytes recognise non-human proteins in pathogen (1); this is not a response to a specific type of pathogen, simply to a cell or organism which is 'not self'(1); phagocyte engulfs the pathogen and encloses it in a vacuole called a phagosome (1); phagosome combines with a lysosome to form a phagolysosome (1); enzymes from lysosome digest and destroy pathogen (1).

b *Cytokines* act as cell signalling molecules (1); that stimulate phagocytes to move to a site of infection or inflammation (1). *Opsonins* bind to pathogens (1); and tag them so they are more easily recognised by phagocytes (1); because phagocytes have receptors on their membranes which bind to common opsonins, e.g., antibodies (1).

12.6

1 immunoglobulins, Y shaped glycoproteins that bind to specific antigens on pathogens/toxins/foreign cells . Specific antibody for every antigen. Made up of two heavy and two light polypeptide chains with an active site made up of 110 amino acids which fits the antigen (1). Work by binding to antigen forming antigen-antibody complex which is then either engulfed by phagocytes or simply cannot function as a pathogen anymore (1).

2 *Similarities:* Both T and B cells form clones of active cells (1); both form memory cells which mean that when they meet a pathogen a second time there is a rapid response, destroying the pathogen before it can cause disease (1); *Differences:* T cells stimulate B cells (1); T cells destroy pathogens directly (1); B cells produce antibodies which act as opsonins stimulating phagocytes to engulf pathogens (1); T cells also regulate immune response so it stops once a pathogen is removed and doesn't turn against body cells (1).

3 In autoimmune disease immune system stops recognising 'self' and starts attacking healthy cells. Immunosuppressant drugs reduce activity of immune system (1); preventing/reducing destruction of healthy tissue BUT susceptibility to infection increases (1); as immune system less effective at recognising pathogens (1).

4 Humoral immune system responds to antigens outside of cells (1); bacterial and fungal cells present in body have antigens to which humoral system can respond (1); system makes antibodies to bacterial and fungal surface antigens, forms antigen-antibody complexes so macrophages readily engulf pathogen (1). Cell-mediated system

responds to changes in cells (1). Viruses get into body cells and take over cell metabolism (1) – not so obvious in blood presenting antigens e.g., bacteria. However, cell-mediated response detects changes in-virus infected cells and killer T cells attack and destroy them.

12.7

> ### Case study: Influenza
>
> 1 Vaccination (if available); isolating infected people as soon as any symptoms appear; preventing travel into and out of infected countries (1) any other sensible point.

1 *Flow diagram that covers every step of how artificial active immunity is induced and is clear and easy to follow as well as accurate and informative.* (4 marks)

2 *Any four sensible reasons including:* Some diseases so severe that patient killed before body can develop antibodies; some people do not have children vaccinated against diseases; if child immunocompromised, has a comorbidity, or neglected/malnourished will be more vulnerable to infections; bacterial disease may be resistant to current antibiotics (max 4).

3 a day 56 (1)   b Approximately 8600(1)
  c Approximately 800 (1)   d 86% (1)

4 a Most people vaccinated, so if bitten they are given course of injections to deliver antibodies (produced in another animal) directly into blood stream (1); antibodies form antigen-antibody complex with rabies virus, allowing phagocytes to destroy pathogen (an example of artificial passive immunity).

  b Epidemic occurs when communicable disease spreads rapidly to a lot of people at either local/national level (1); in vaccination, immune system stimulated to make antibodies to a pathogen by exposure to safe form of an antigen injected into blood stream (1); if an epidemic begins to build, mass vaccination (1); can protect people in the community by building immunity to infecting pathogen and prevent pathogen spreading disease into the wider population (1).

5 Health departments increase public health awareness decreasing levels of communicable disease (1); chlorinated water reduces water-borne infections and reduces number of deaths (1); public hygiene will mean fewer rats etc to act as vectors of disease (e.g., plague) (1); penicillin reduces deaths from bacterial diseases but does not affect viral diseases such as flu and polio (2); vaccines such as polio reduced deaths from infectious diseases where a vaccine has been developed (1); Any other sensible pint could be substituted for one of these.

Index

abiotic factors 269
activation energy 85
active transport 112, 208
    bulk transport 112–13
    evidence for the role of active transport in root pressure 200
Acute Oak Decline 297
adaptations 251–4, 290
    plants and water availability 210–14
adenine (A) 69
adhesion 45, 202
ADP (adenosine diphosphate) 81
aerenchyma 214
aesthetic arguments for biodiversity 283
agglutinins 312
agriculture 278, 279
    herbicides 279–80
    removal of hedgerows 279
    use of chemicals 279
AIDS (acquired immunodeficiency syndrome) 300
alleles 240
alpha helix 61
alveoli 155–6
amino acids 59
    separating amino acids using thin layer chromatography 60–1
    synthesis of peptides 59–60
amylopectin 48
amylose 47
anatomical adaptations 251–2
    analogous structures 253
    convergent evolution 253–4
animal defences 308
    non-specific defences 308–11
animal diseases 299–302
anions 41, 42
anti-oomycetes 307
anti-toxins 312
antibacterial compounds 307
antibiotics 295, 307, 322
    antibiotic-resistant bacteria 257, 322–4
antibodies 312

anticodons 77
antifungal compounds 307
antigens 312
apoenzymes 98
apoplastic pathways 199, 208–9
archaebacteria 230
arrhythmia 190
arteries 174
    aorta 187
    coronary arteries 185
    pulmonary artery 186
arterioles 174–5
artificial cloning 275
asexual reproduction 124, 275
assimilates 207
asthma 157
athlete's foot 302
atomic force microscopy 23–4
ATP (adenosine triphosphate) 80
atria 186
atrial fibrillation 190
atrio-ventricular node (AVN) 189
autoimmune diseases 315–16
autoimmune response 313
autotrophic feeders 228

B lymphocytes 313
bacteria 257, 294
    antibiotic-resistant bacteria 257, 322–4
    cell walls 295
bacteriophages 295
base pairing rules 70
behavioural adaptations 252, 290
belt transects 265, 269–70
beta pleated sheet 61
bicuspid valve 186–7
binomial nomenclature 225
biodiversity 220, 262
    biodiversity values 272
    genetic biodiversity 263, 274–6
    habitat biodiversity 263
    how to calculate biodiversity 271
    human activity versus biodiversity 285

human influence on biodiversity 278–81
    importance of biodiversity 262
    keystone species 285–6
    loss of biodiversity in the UK 282
    maintaining biodiversity 287–91
    measuring biodiversity 262
    reasons for maintaining biodiversity 283–6
    Simpson's Index of Diversity 271, 272–3
    species biodiversity 263
black sigatoka 298
blood 178
    deoxygenated blood 174
    functions of the blood 178
    oxygenated blood 174
blood cells 139
    counting blood cells 311
blood clotting 308
    enzyme activation 98
    wound repair 308–9
blood pressure 190–1
blood vessels 174
body coverings 251
Bohr effect 183
bonds 40, 44
botanic gardens 289
bradycardia 190
breathing 156–7
bronchioles 155
bronchus (bronchi) 155
bundle of His 189

callose papillae 306–7
capillaries 175–6
capillary action 45, 202–3
captive breeding 275, 289–90
carbaminohaemoglobin 183
carbohydrates 42, 46
    cellulose 49–50
    glucose 46–7
    starch and glycogen 47–9
    testing for carbohydrates 51–3
carbon 40
carbon dioxide 183–4

carbonic anhydrase 183
cardiac cycle 187–8
cardiac muscle 185
    myogenic muscle 189
cartilage 135, 154
Casparian strip 199
catalase 65
cations 41, 42
cell communication 104
cell cycle 120
    cell cycle regulation and cancer 123
    control of the cell cycle 121–2
    G1 checkpoint 121
    G2 checkpoint 121–2
    Go 121
    interphase 120
    mitotic phase 120–1
    spindle assembly checkpoint 122
cell-mediated immunity 313–14
cell membrane theory 102–3
cell signalling 104
cell specialisation 133
    organ systems 137
    organs 136
    specialised cells 133–4
    stem cells 138–42
    tissues 134–6
cell-surface membrane 27, 102–5
cell theory 8–9
cells 26–7
    cell movement 29
    cisternae 30
    compartments for life 27–8
    cytoskeleton 29–30
    endoplasmic reticulum (ER) 30
    Golgi apparatus 31
    lysosomes 29
    mitochondria 28
    nucleus and nucleolus 28–9
    ribosomes 30–1
    vesicles 28
cellular respiration 81
cellulose 33 49–50
centrioles 29–30
centromeres 124
centrosome 29
chloride shift 184
chloroplasts 34

cholesterol 105
chromatids 124
chromatin 28, 36
chromosomes 28, 36, 124
  diploid 128
  haploid 128
  homologous
    chromosomes 128
cilia 30
ciliated epithelium
  135, 154
circulatory systems 170–1
  closed circulatory
    systems 171–2
  double closed circulatory
    systems 172–3
  open circulatory
    systems 171
  single closed circulatory
    systems 172
cisternae 30
classification 220
  classes 222
  classification of
    humans 225
  classification systems
    222–3
  five kingdoms 227–31
  how are organisms
    classified? 224
  naming organisms 225–
    6
  Panda, Giant (*Ailuropoda
    melanoleuca*) 254–5
  phylogeny 232–3
  recent changes to
    classification
    systems 229
  three commonly used
    classification
    systems 231
  why do scientists classify
    organisms?
    223
climate change 278, 280–1
clonal expansion 314
clonal selection 314
Clostridium difficile 323
codons 74
coenzymes 97–8
cofactors 97–8
cohesion 45, 202
  cohesion-tension
    theory 203–4
collagen 65, 66, 155, 175
communicable diseases
  220, 294
  bacteria 294–5
  fungi 295–6
  preventing the spread
    of communicable

diseases in humans
  304
  protocista 295
  transmission 303–5
  viruses 295
comparative anatomy 237
  homologous structures
    237
comparative biochemistry
  237–8
compartmentalisation
  27, 102
complementary base
  pairing 70
condensation reactions
  46–7, 54, 59, 68, 71, 81
conjugated proteins 64–5
connective tissue 135
conservation 287
  conservation
    agreements 290–1
  ex situ conservation
    289–90
  in situ conservation
    287–9
contagions 303
continuous variation 243
  normal distribution
    curves 243–4
Convention on
  International Trade in
  Endangered Species
  (CITES) 291
convergent evolution
  253–4
coolants 45
correlation coefficients
  248–50
Countryside Stewardship
  Scheme 291
courtship 252
crenation 115
culling 288
cytokines 309, 311
cytokinesis 121, 126
  animal cells 126–7
  plant cells 127
cytolysis 114
cytoplasm 26
cytosine 69
cytoskeleton 29–30

deforestation 278–9
diaphragm 156
diastole 187
differentiation 138, 139
  replacement of red and
    white blood cells
    139
diffusion 108
  concentration
    difference 108

diffusion across
  membranes
  109–10
  facilitated diffusion
    110
  factors affecting
    diffusion rates in
    model cells 110–11
  factors affecting rate of
    diffusion 108
  rate of diffusion and
    surface area 109
  simple diffusion 108
dipeptides 59
disaccharides 47
disease prevention and
  treatment 304, 317
  antibiotic dilemma
    322–4
  artificial active
    immunity 318
  artificial immunity 317
  artificial passive
    immunity 318
  drug design for the
    future 321–2
  medicines and the
    management of
    disease 320–1
  natural immunity 317
  vaccines and the
    prevention of
    epidemics 318–20
disulfide bonds 61–2
divergent evolution 238
DNA 35, 68–9
  antiparallel strands 69
  antisense strands 76
  base pairing rules 70
  DNA extraction 71
  double helix 69–70
  lagging strands 73
  leading strands 73
  Okazaki fragments 73
  protein synthesis 76–9
  sense strands 76
  template strands 76
DNA helicase 72
DNA polymerase 72
DNA replication 72
  continuous and
    discontinuous
    replication 73
  genetic code 74–5
  replication errors 74
  role of enzymes in
    replication 72–3
  semi-conservative
    replication 72–4
domains 222, 229–31
double helix 69–70
droplet infection 303

drug development 140,
  321–2

ecology 284–5
economic arguments for
  biodiversity 283–4
ectopic heartbeat 190
elastin 65, 66, 155, 175
electrocardiograms (ECG)
  189–90
electrolytes 42
electron microscopy
  19–21
elements 40
emulsion test 57
endocytosis 113
endoplasmic reticulum
  (ER) 30
endosymbiosis 36
environmental variation
  239, 241
enzymes 72–3
  active site 85
  blood-clotting
    mechanism 98
  control of metabolic
    activity within cells
    94–6
  denaturation from
    temperature 88
  digestion of proteins 87
  digestion of starch 86–7
  enzyme inhibitors 94–6
  enzyme–product
    complex 85
  enzyme–substrate
    complex 85
  extracellular enzymes
    86–7
  inactive precursor
    enzymes 97
  induced-fit hypothesis
    85
  intracellular enzymes
    85–6
  investigations into the
    effects of different
    factors on enzyme
    activity 91–3
  lock and key
    hypothesis 85
  mechanism of enzyme
    action 84–5
  optimum temperature
    88–9
  pH 89–91
  precursor activation
    97–8
  renaturation 90
  role of enzymes in
    reactions 84
  temperature 88–91

why are enzymes important? 84
epidemics 318–20
epidermis tissue 136
epithelial tissue 135
erythrocytes 133
ester bonds 54
esterification 54
eubacteria 230
eukaryotes 227
eukaryotic cells 26
  comparison with prokaryotic cells 36–7
evolution 220, 234
  Anolis lizards 258–9
  antibiotic-resistant bacteria 257
  comparative anatomy 237
  comparative biochemistry 237–8
  convergent evolution 253–4
  developing the theory of evolution 234–5
  divergent evolution 238
  evidence for evolution 235
  evolutionary embryology 237–8
  Flavobacterium 258
  natural selection 256
  palaeontology 236
  peppered moths 257
  sheep blowflies 257–8
exchange surfaces 148
  histology of exchange surfaces 166–7
  need for specialised exchange surfaces 150
  specialised exchange surface 152
  surface area: volume ratio (SA:V) 150–1
exocytosis 113
expiratory reserve volume 159
expulsive reflexes 308

families 222
fatty acids 54–5
fevers 310
fibres 50
fibrous proteins 65–6
fish 163
flagella 30, 36
fomites 303
food security 298

forced expiratory volume in 1 second 159
fossil record 236
founder effect 275
frame quadrats 267–8
fungi 227, 228, 295–6
fungicides 298

gametes 128
gaseous exchange 153
  alveoli 155–6
  bronchioles 155
  bronchus 155
  dissecting, examining and drawing 166
  measuring the process 159–60
  nasal cavity 154
  trachea 154
  ventilating the lungs 156–7
gaseous exchange in fish 163
  effective gaseous exchange in water 165–6
  gills 163–4
  water flow over the gills 164–5
gaseous exchange in insects 161
  collapsible enlarged tracheae or air sacs 162
  discontinuous gas exchange in insects 163
  how does gas exchange take place in insects? 161–2
  mechanical ventilation of the tracheal system 162
gene flow 274
genes 74
genetic biodiversity 263
  artificial cloning 275
  captive breeding 275
  founder effect 275
  gene flow 274
  genetic bottlenecks 275
  importance of genetic biodiversity 274
  interbreeding 274
  measuring genetic biodiversity 275–7
  mutations 274
  natural selection 274
  rare breeds 275
  selective breeding 275
genetic bottlenecks 275
genetic code 74

  degenerate code 74–5
  non-overlapping code 74
  triplet code 74
  universal code 74
genetic races 290
genetic variation 239, 241
genus (genera) 222
globular proteins 64
glucanases 307
glucose 46
  alpha and beta glucose 46
  condensation reactions 46–7
  other sugars 47
glycerol 54
glycogen 48–9
glycolipids 104
glycoproteins 104
glycosidic bonds 47
goblet cells 154
Golgi apparatus 31
grana 34
graticules 16–17
grazing 288
guanine (G) 69
guard cells 133, 201

habitat biodiversity 263, 290
haem groups 64
haemocoel 171
haemoglobin 64, 181
  fetal haemoglobin 183
  haemoglobinic acid 184
haemolymph 171
heart 185
  arrhythmia 190
  atria 186
  atrial fibrillation 190
  atrio-ventricular node (AVN) 189
  basic rhythm of the heart 189
  bicuspid valve 186–7
  blood pressure 190–1
  bradycardia 190
  bundle of His 189
  cardiac cycle and the heartbeat 187–8
  dissecting a heart 186
  ectopic heartbeat 190
  electrocardiograms (ECG) 189–90
  heart sounds 188
  hole in the heart 187
  Purkyne fibres 189
  semilunar valves 186
  sino-atrial node (SAN) 189

  structure and function of the heart 186–7
  tachycardia 190
  tricuspid valve 186
  ventricles 186, 187
heterotrophic feeders 228
hexose monosaccharides 46
histamines 309
histones 28, 36
HIV (human immunodeficiency virus) 300
holoenzymes 98
human gaseous exchange system 153–8
humoral immunity 314
hydrogen bonds 44
hydrogen carbonate ions 183
hydrolysis reactions 49, 62, 68, 81
hydrophilic interactions 55–6, 62
hydrophobic interactions 55–6, 62
hydrophytes 213
hydrostatic pressure 114, 179
hydroxyl (OH) groups 44

immune system 312–16
immunoglobulins 312
infections 303
inflammatory response 309
influenza (flu) 300–1, 319–20
ingestion 303
inhibitors 94
innate behaviour 252
inoculation 303
insect repellants 307
insecticides 307
insects 161
inspiratory reserve volume 159
instinctive behaviour 252
insulin 64
interbreeding 274
intercostal muscles 156
interleukins 313
intermediate fibres 29
International Union for the Conservation of Nature (IUCN) 290–1
interspecific variation 239
intraspecific variation 239
invasive species 288

keratin 65
keystone species 285–6
kingdoms 222, 227
    animalia 228–9
    archaebacteria 230
    are there now six
        kingdoms? 229–30
    eubacteria 230
    fungi 228
    plantae 228
    prokaryotae 227
    protocista 227–8

lactose 47
lamellae 34
laser scanning confocal
    microscopy 21–2
learned behaviour 252
leaves 194
    cuticle 211, 213
    ways of conserving
        water 211–12
    ways of living in
        water 213–14
light microscopes 8, 10,
    16–17, 20
line transects 265
Linnaean classification
    222, 225
lipids 42, 54
    health advice 57–8
    identification of
        lipids 57
    phospholids 55–6
    role of lipids 56
    sterols 56
    triglycerides 54–5
lungs 156–7
    components of the lung
        volume
        159–60
    elastic recoil 155
    lung surfactant 156, 158
    measuring the capacity
        of the lungs 159
lymph 179–80
    lymph capillaries 179
    lymph nodes 180
lymphocytes 313
    B lymphocytes 313
    T lymphocytes 313
lysosomes 29

macrofibrils 50
macrophages 310
magnification 15–16
major histocompatibility
    complex (MHC) 311
malaria 301–2
maltose 47

mammalian gaseous
    exchange system
    153–8
marine conservation
    zones 288–9
mast cells 309
medicines 142, 320
    drug design for the
        future 321–2
    sources of medicine
        320–1
meiosis 128, 240
membranes 102, 106
    cell membrane
        components 103–5
    cell membrane theory
        102–3
    cholesterol 105
    diffusion across
        membranes 109–10
    fluid-mosaic model 103
    glycolipids 104
    investigating membrane
        permeability 107
    membrane proteins
        103–4
    membrane
        structure 102
    phospholipid
        bilayer 102
    sites of chemical
        reactions 105
    solvents 106–7
    temperature 106
meningitis 300
meristematic tissue
    (meristems) 140
messenger (m) RNA 76–7
metabolism 27
microfibrils 50
microfilaments 29
microscopy 8–9
    atomic force
        microscopy 23–4
    calibration 16–17
    electron microscopy 19–
        21
    graticules 16–17
    laser scanning confocal
        microscopy 21–2
    magnification 15–16
    resolution 15
    sample preparation
        11, 19
    scientific drawings
        13–14
    super resolved
        fluorescence
        microscopy 24
    using staining 11–13
microtubules 29
mimicry 251

mitochondria 28
mitosis 121
    anaphase 126
    chromosomes 124
    importance of mitosis
        124
    metaphase 125
    prophase 125
    stages of mitosis 124–5
    telophase 126
molecules 26
    biological molecules 42
    bonding 40–1
    building blocks of life 40
    elements 40
    ions 41–2
    polar molecules 44, 55
    polymers 42
monoculture 278, 280
monomers 42
monosaccharides 46
monounsaturated fats 55
MRSA (methicillin-resistant
    Staphylococcus
    aureus) 323
muscle 135
    cardiac muscle 185, 189
    intercostal muscles 156
    muscle tissue 135
    smooth muscle 155
mutations 74, 240, 274

nasal cavity 154
natural selection 256, 174
    advantageous
        characteristics 256
    selection pressures 256
nervous tissue 135
neutrophils 133, 310
nitrogen (N) 40
normal distribution curves
    243–4
nucleic acids 42, 68
    deoxyribonucleic acid
        (DNA) 68–70
    nucleotides and nucleic
        acids 68
    ribonucleic acid
        (RNA) 70–1
nucleolus 28
nucleotides 68
nucleus 28
    nuclear envelope and
        nuclear pores 28

opportunistic sampling
    264
opsonins 311
orders 222
organelles 26, 27

plant cell organelles 33
    protein synthesis 30–1
organs 136
osmosis 114
    effects of osmosis on
        animal cells 114–15
    effects of osmosis on
        plant cells 115–16
    oncotic pressure 179
    osmosis investigations
        116–17
    water potential 114
oxygen 40, 174
    carrying oxygen 181–3
    oxygen dissociation
        curve 182
    oxyhaemoglobin 181
    transporting oxygen 181

palaeontology 236
palisade cells 133
pandemics 319
parasites 295
passive transport 108, 208
pathogens 294
    damaging host tissues
        directly 290
    getting rid of pathogens
        309–11
    identifying pathogens
        302
    keeping pathogens
        out 308–9
    producing toxins
        which damage host
        tissues 290
    transmission between
        animals 303–4
    transmission between
        plants 304–5
    types of pathogens
        294–6
peak flow meters 159
penicillin 320
pentose monosaccharides
    47
peptides 59
    breakdown of
        peptides 62
    peptide bonds 59
    synthesis of peptides
        59–60
peptidyl transferase 78
perforin 313
permeability 107, 109, 110
pH 89–91
    optimum pH 90
phagocytosis 113, 309, 310
    counting blood cells
        311
    helpful chemicals 311

phagolysosomes 310
phagosomes 310
pharmacogenetics 321
phloem 136
  apoplast route 208–9
  companion cells 197
  phloem loading 208–9
  phloem unloading 209
  plasmodesmata 197
  sieve plates 197
  sieve tube elements 197
  structure and functions
    of the phloem 197
  symplast route 208
phosphodiester bonds 68
phospholids 55–6
phosphorus (P) 40
phosphorylation 81
phylogeny 232–3
phylum (phyla) 222
physiological adaptations
  253
pinocytosis 113
pitfall traps 266
plant cells 33
  cellulose cell wall 33
  chloroplasts 34
  organelles 33
  plant stem cells and
    medicines 142
  sources of plant stem
    cells 140
  vacuoles 33
plant defences 305
  chemical defences 307
  physical defences
    306–7
  recognising an attack
    306
plant diseases 297–8
  banana diseases and
    food security 298
  fungi 295–6
plants 136, 227, 228
  adaptations to water
    availability 210–14
  dicotyledonous
    plants 136, 195
  herbaceous plants 195
  metabolic demands 194
  perennials 194
  surface area:volume
    ratio (SA:V) 194
  translocation 207–9
  transpiration 201–6
plasma 178
  plasma cells 313
  plasma membrane
    102–5
plasmolysis 115

poaching 288
polar molecules 44, 55
polymers 42
polymorphic genes 275–6
polynucleotides 68
polypeptides 59, 312
polysaccharides 46
polyunsaturated fats 55
pooters 266
potato blight (tomato
  blight, late blight) 298
proenzymes 98
prokaryotes 227
prokaryotic cells 26, 35
  cell wall 35
  comparison with
    eukaryotic cells
    36–7
  DNA 35
  flagella 36
  prokaryotic cell
    study 37
  ribosomes 35
prosthetic groups 64, 97–8
protein synthesis 30–1, 76
  protein production 31
  transcription 76–7
  translation 77–9
proteins 42, 59
  amino acids 59–61
  breakdown of peptides
    62–3
  carrier proteins 104
  channel proteins 104
  conjugated proteins
    64–5
  digestion of proteins 87
  extrinsic proteins 104
  fibrous proteins 65–6
  globular proteins 64
  glycoproteins 104
  identification of
    proteins 62–3
  intrinsic proteins 104
  levels of protein
    structure 61–2
protocista 227–8, 295
purines 69
Purkyne fibres 189
pyrimidines 69

quadrats 266–8

R-groups 59
random sampling 264
rare breeds 275
receptors 104
reduction division 128
reintroduction of species
  288

reproductive success 256
residual volume 159
ribose 47
ribosomal (r) RNA 77
ribosomes 30–1, 35
ring rot 297
ring worm 302
Rio Convention 291
RNA 70–1, 77
RNA polymerase 76
roots 194
  apoplast pathway 199
  evidence for the role of
    active transport in
    root pressure 200
  root hair cells 133,
    198–9
  symplast pathway 199
  ways of conserving
    water 212
  ways of living in water
    214
rough endoplasmic
  reticulum 30

sample preparation 11, 19
sampling 264
  measuring abiotic
    factors 269
  non-random sampling
    264–5
  quadrats 266–8
  random sampling 264
  reliability 265
  transects 265
sampling animals 266
  capture-mark-release-
    recapture 268
  estimating animal
    population size
    268–9
  kick sampling 266
sampling plants 266–7
saprophytic feeders 228
saturated fats 55
scanning electron
  microscopy (SEM) 19
scientific drawings
  13–14, 20
seasonal behaviours 252
sebum 308
secretion 30
seed banks 289
selective breeding 275
selective toxicity 322
semi-conservative
  replication 72–4
semilunar valves 186
sexual reproduction 240
sinks 207
sino-atrial node (SAN) 189

skin 308
smooth endoplasmic
  reticulum 30
smooth muscle 155
soil contamination 305
solutes 114
solvents 106–7, 114
sources 207
species biodiversity 263
species evenness 263
  measuring 267–8
species richness 263
  measuring 267
specific immune system
  312
  antibodies 312
  autoimmune diseases
    315–16
  autoimmune response
    313
  cell-mediated immunity
    313–14
  humoral immunity
    314–15
  immunological
    memory 313
  lymphocytes and the
    immune response
    313
  primary immune
    response 315
  secondary immune
    response 315
sperm cells 133
spirometers 159
squamous epithelium 135
staining 11–13
standard deviation 244–6
starch 47–9
  digestion of starch 86–7
  hydrolysis reactions 49
statistical tests 246–8
stem cells 138
  developmental biology
    141
  differentiation 139
  drug trials 140
  ethics 141
  gene therapy 141–2
  plant stem cells and
    medicines 142
  sources of animal stem
    cells 139–40
  sources of plant stem
    cells 140
  stem cell potency 138
  treatment of burns 140
  uses of stem cells
    140–1
stems 194, 196
sterols 56

stomata 201, 205
  hydrophytes 213
  sunken stomata 211
stratified sampling 265
stroma 34
succession 288
succulents 211
sucrose 47, 207
sulfur (CS 40
super resolved fluorescence
    microscopy 24
surfactants 56
survival behaviours 252
sustainable development
    287
sweep nets 266
symplastic pathways
    199, 208
synthetic biology 322
systematic sampling 265
systole 187

T lymphocytes 313
  T helper cells 313
  T killer cells 313
  T memory cells 313
  T regulator cells 313
tachycardia 190
taxonomy 222–6
temperature 88–91
  diffusion 108
  membranes 106
  pH 89–91
  temperature coefficient
    Q10 88
  temperature extremes
    89
  transpiration 205
tension 203
thylakoids 34
thymine (T) 69
tidal volume 159
tissue fluid 179
tissues 134–6
  plants 136
tobacco mosaic virus
    (TMV) 298
total lung capacity 160
toxins 290, 307, 312, 322
trachea 154
transcription 76–7
transfer (t) RNA 77
translation 77–9
translocation 207
transmission electron
    microscopy (TEM) 19
transpiration 201, 252
  air movement 205–6
  evidence for the
    cohesion-tension
    theory 203–4

factors affecting
    transpiration 205–6
  light 205
  measuring transpiration
    204
  process of transpiration
    201
  relative humidity 205
  soil-water availability
    206
  stomata 205
  temperature 205
  transpiration pull 203
  transpiration stream
    201–3
transport in animals 148
  mass transport systems
    171
  need for specialised
    transport systems
    170
  types of circulatory
    systems 170–2
transport in plants 148,
    194, 198
  need for plant transport
    systems 194
  structure and functions
    of the phloem 197
  structure and functions
    of the xylem 196–7
  transport systems in
    dicotyledonous
    plants 195
  water transport in
    plants 198–200
transporting carbon
    dioxide 183–4
transporting oxygen 181–3
tricuspid valve 186
triglycerides 54–5
tuberculosis (TB) 299
tubulin 29
turgidity 115
turgor 115

unsaturated fats 55
uracil 71

vaccination 318–20
vacuoles 33
variation 239
  continuous variation
    243–4
  discontinuous variation
    243
  environmental and
    genetic causes of
    variation 241
  environmental causes of
    variation 241

genetic causes of
    variation 240
  variation in identical
    twins 242
vascular tissue (plants) 136
  vascular bundles 195
  vascular system 195
vasoconstriction 175
vasodilation 175
vectors 304, 305
veins 176
  inferior vena cava 176
  pulmonary vein 186
  superior vena cava 176
ventilation 156
  expiration 157
  inspiration 156–7
  ram ventilation 164
  ventilation rate 160
  ventilation to
    maintain diffusion
    gradient 152
ventricles 186, 187
venules 176–7
vesicles 28
viruses 295
vital capacity 159
vitalographs 159

water 44–5
water potential 114
water transport in
    plants 198
  evidence for the role of
    active transport in
    root pressure 200
  movement of water
    across the root 199
  movement of water into
    the root 198–9
  movement of water into
    the xylem 199–200
wildlife reserves 288

xerophytes 210
  marram grass
    (Ammophila spp.)
    252
  ways of conserving
    water 211–12
xylem 136
  bordered pits 197
  movement of water into
    the xylem 199–200
  observing xylem vessels
    in living plant
    stems 196
  structure and functions
    of the xylem 196–7

zoonotic influenza 301
zygotes 128
zymogens 98

Appendix (Statistics data tables)

▼ *Table of values of* t

Degree of freedom (df)	p values			
	0.10	0.05	0.01	0.001
1	6.31	12.71	63.66	636.60
2	2.92	4.30	9.92	31.60
3	2.35	3.18	5.84	12.92
4	2.13	2.78	4.60	8.61
5	2.02	2.57	4.03	6.87
6	1.94	2.45	3.71	5.96
7	1.89	2.36	3.50	5.41
8	1.86	2.31	3.36	5.04
9	1.83	2.26	3.25	4.78
10	1.81	2.23	3.17	4.59
12	1.78	2.18	3.05	4.32
14	1.76	2.15	2.98	4.14
16	1.75	2.12	2.92	4.02
18	1.73	2.10	2.88	3.92
20	1.72	2.09	2.85	3.85
α	1.64	1.96	2.58	3.29

▼ *Critical values for Spearman's rank correlation coefficient,* r_s

	$p = 0.1$	$p = 0.05$	$p = 0.02$	$p = 0.01$	
	5%	$2\frac{1}{2}\%$	1%	$\frac{1}{2}\%$	1-Tail Test
	10%	5%	2%	1%	2-Tail Test
n					
1	–	–	–	–	
2	–	–	–	–	
3	–	–	–	–	
4	1.0000	–	–	–	
5	0.9000	1.0000	1.0000	–	
6	0.8286	0.8857	0.9429	1.0000	
7	0.7143	0.7857	0.8929	0.9286	
8	0.6429	0.7381	0.8333	0.8810	
9	0.6000	0.7000	0.7833	0.8333	
10	0.5636	0.6485	0.7455	0.7939	
11	0.5364	0.6182	0.7091	0.7545	
12	0.5035	0.5874	0.6783	0.7273	
13	0.4835	0.5604	0.6484	0.7033	
14	0.4637	0.5385	0.6264	0.6791	
15	0.4464	0.5214	0.6036	0.6536	
16	0.4294	0.5029	0.5824	0.6353	
17	0.4142	0.4877	0.5662	0.6176	
18	0.4014	0.4716	0.5501	0.5996	
19	0.3912	0.4596	0.5351	0.5842	
20	0.3805	0.4466	0.5218	0.5699	

	$p = 0.1$	$p = 0.05$	$p = 0.02$	$p = 0.01$
21	0.3701	0.4364	0.5091	0.5558
22	0.3608	0.4252	0.4975	0.5438
23	0.3528	0.4160	0.4862	0.5316
24	0.3443	0.4070	0.4757	0.5209
25	0.3369	0.3977	0.4662	0.5108
26	0.3306	0.3901	0.4571	0.5009
27	0.3242	0.3828	0.4487	0.4915
28	0.3180	0.3755	0.4401	0.4828
29	0.3118	0.3685	0.4325	0.4749
30	0.3063	0.3624	0.4251	0.4670

Acknowledgements

p2-3 & **4-5**: Science Photo/Shutterstock; **p6-7**: Sashkin/Shutterstock; **p10**(L): Dr. Jeremy Burgess/Science Photo Library; **p10**(R): Dr. Jeremy Burgess/Science Photo Library; **p12**(T): Biophoto Associates/Science Photo Library; **p12**(B): CNRI/Science Photo Library; **p18**: Biophoto Associates/Science Photo Library; **p12**(C): Dr. Frederick Skvara, Visuals Unlimited/Science Photo Library; **p14**: John Burbidge/Science Photo Library; **p16**: Victor Shahin, Prof. Dr. H. Oberleithner, University Hospital of Muenster/Science Photo Library; **p19**(T): Eye of Science/Science Photo Library; **p19**(B): David Scharf/Science Photo Library; **p20**: Marilyn Schaller/Science Photo Library; **p22**: Heiti Paves/Science Photo Library; **p23**: Victor Shahin, Prof. Dr. H.Oberleithner, University Hospital Of Muenster/Science Photo Library; **p24**(L): IBM and Nature Chemistry; **p24**(R): Peter Eaton-Requimte/University of Porto, Visuals Unlimited /Science Photo Library; **p25**(L): Alfred Pasieka/Science Photo Library; **p25**(R): National Cancer Institute/Science Photo Library; **p27**: Biophoto Associates/Science Photo Library; **p28**(T): Alfred Pasieka/Science Photo Library; **p28**(B): Dr. Gopal Murti/Science Photo Library; **p30**(B): Biomedical Imaging Unit, Southampton General Hospital/Science Photo Library; **p33**: Marilyn Schaller/Science Photo Library; **p34**(L): Dr. Kari Lounatmaa/Science Photo Library; **p30**(T): Dr. Gopal Murti/Science Photo Library; **p34**(R): Dr. David Furness, Keele University/Science Photo Library; **p31**(T): Steve Gschmeissner/Science Photo Library; **p31**(B): Biophoto Associates/Science Photo Library; **p37**: Biology Media/Science Photo Library; **p43**: Science Photo Library; **p45**: Hermann Eisenbess/Science Photo Library; **p51**: Martyn F. Chillmaid/Science Photo Library; **p52**: Martyn F. Chillmaid/Science Photo Library; **p64**(T): lculig/Shutterstock; **p64**(B): Indigo Molecular Images/Science Photo Library; **p65**: Steve Gschmeissner/Science Photo Library; **p71**: Philippe Psaila/Science Photo Library; **p89**: Vasiliy Koval/Shutterstock; **p97**: Laguna Design/Science Photo Library; **p115**: Dr. Stanley Flegler/Science Photo Library; **p124**: Power and Syred/Science Photo Library; **p125**(T): Steve Gschmeissner/Science Photo Library; **p125**(B): Steve Gschmeissner/Science Photo Library; **p126**(T): Steve Gschmeissner/Science Photo Library; **p126**(B): Steve Gschmeissner/Science Photo Library; **p127**(T): Dr. Gopal Murti/Science Photo Library; **p127**(B): Steve Gschmeissner/Science Photo Library; **p129**: Ed Reschke/Photolibrary/Getty Images; **p130**(T): Ed Reschke/Photolibrary/Getty Images; **p130**(C): Ed Reschke/Photolibrary/Getty Images; **p130**(B): Ed Reschke/Photolibrary/Getty Images; **p131**(T): Ed Reschke/Photolibrary/Getty Images; **p131**(CT): Ed Reschke/Photolibrary/Getty Images; **p131**(CB): Ed Reschke/Photolibrary/Getty Images; **p131**(B): Ed Reschke/Photolibrary/Getty Images; **p135**(T): Herve Conge, ISM/Science Photo Library; **p135**(B): Eric Grave/Science Photo Library; **p136**(T): Marek Mis/Science Photo Library; **p136**(C): Dr. Keith Wheeler/Science Photo Library; **p136**(B): Dr. Keith Wheeler/Science Photo Library; **p137**: Eye of Science/Science Photo Library; **p138**: Science Photo Library; **p139**(T): Biophoto Associates/Science Photo Library; **p139**(B): Steve Gschmeissner/Science Photo Library; **p140**: Dr. John Runions/Science Photo Library; **p150**: Four Oaks/Shutterstock; **p152**: Dr. Fred Hossler, Visuals Unlimited/Science Photo Library; **p154**: Dr. Fred Hossler, Visuals Unlimited/Corbis; **p155**: Biophoto Associates/Science Photo Library; **p158**: Vivid Pixels/Shutterstock; **p159**: Image Point Fr/Shutterstock; **p161**: Martin Dohrn/Science Photo Library; **p170**: Anthony Short; **p174**: Steve Gschmeissner/Science Photo Library; **p175**: Ed Reschke/Getty Images; **p176**: Biophoto Associates/Science Photo Library; **p178**: Biophoto Associates/Science Photo Library; **p191**: Photographee.eu/Shutterstock; p194: Anthony Short; **p195**(T): Dr. Keith Wheeler/Science Photo Library; **p195**(C): Stan Elems/Visuals Unlimited/Corbis; **p195**(B): Jubal Harshaw/Shutterstock; **p197**: Anthony Short; **p200**: David Cavagnaro/Visuals Unlimited/Science Photo Library; **p209**(L): D. Fischer; **p209**(R): D. Fischer; **p209**(L): D. Fischer; **p210**: Anthony Short; **p211**(T): Power and Syred/Science Photo Library; **p211**(C): Anthony Short; **p211**(B): Anthony Short; **p212**(T): Anthony Short; **p212**(B): Anthony Short; **p213**(TL): Martin Fowler/Shutterstock; **p213**(TR): Dr. Keith Wheeler/Science Photo Library; **p213**(B): Digital Vision; **p214**: Dr. Keith Wheeler/Science Photo Library; **p224**(L): Chantelle Bosch/Shutterstock; **p224**(C): Lenkadan/Shutterstock; **p224**(R): DragoNika/Shutterstock; **p225**(L): Howard Marsh/Shutterstock; **p225**(C): Stephen Rees/Shutterstock; **p225**(R): Olha Insight/Shutterstock; **p226**(T): Beata Aldridge/Shutterstock; **p226**(B): Kamonrat/Shutterstock; **p227**: Scimat/Science Photo Library; **p228**(T): Lebendkulturen.de/Shutterstock; **p228**(C): Power and Syred/Science Photo Library; **p228**(B): Liza1979/Shutterstock; **p230**: Jim David/Shutterstock; **p233**: Gary Hincks/Science Photo Library; **p234**(T): Nicku/Shutterstock; **p234**(B): Dr. Jeremy Burgess/Science Photo Library; **p235**: Vladimir Sazonov/Shutterstock; **p236**: Sinclair Stammers/Science Photo Library; **p237**: Paul D. Stewart/Science Photo Library; **p238**: Eric Isselee/Shutterstock; **p239**(CL): Bernadette Heath/Shutterstock; **p239**(C): Grigoriy Pil/Shutterstock; **p239**(CR): Bildagentur Zoonar GmbH/Shutterstock; **p239**(TR): Jeanne White/Science Photo Library; **p239**(BR): Dr. Keith Wheeler/Science Photo Library; **p240**: Jeff Foott/ Getty Images; **p241**(L): Debu55y/Shutterstock; **p241**(R): Diana Taliun/Shutterstock; **p244**: TAGSTOCK1/Shutterstock; **p245**: Madlen/Shutterstock; **p246**: Dr. Linda Stannard, UCT/Science Photo Library; **p251**(T): Krzysztof Wiktor/Shutterstock; **p251**(C): Matt Jeppson/Shutterstock; **p251**(B): Matt Jeppson/Shutterstock; **p252**(T): Dr. Keith Wheeler/Science Photo Library; **p252**(B): Frans Lanting/Mint Images/Science Photo Library; **p253**(T): Diccon Alexander/Science Photo Library; **p253**(C): Eric Isselee/Shutterstock; **p253**(B): BMJ/Shutterstock; **p254**(T): Michael & Patricia Fogden/Corbis; **p254**(BL): Grimplet/Shutterstock; **p254**(BR): Adrian Thomas/Science Photo Library; **p255**(T): Nelik/Shutterstock; **p255**(C): Karel Gallas/Shutterstock; **p255**(B): Adam Van Spronsen/Shutterstock; **p257**(T): Michael W. Tweedie/Science Photo Library; **p257**(B): Michael W. Tweedie/Science Photo Library; **p258**: Louise Murray/Science Photo Library; **p259**: GJones Creative/Shutterstock; **p262**(T): Vlad61/Shutterstock; **p262**(B): Pchais/Shutterstock; **p263**(T): Anneka/Shutterstock; **p263**(B): Marietjie/Shutterstock; **p264**: Public Health England/Science Photo Library; **p265**: Martyn F. Chillmaid/Science Photo Library; **p266**(T): Martyn F. Chillmaid/Science Photo Library; **p266**(B): Nigel Cattlin/Science Photo Library; **p268**: Chris Johnson/Alamy; **p272**: Anthony Short; **p273**(T): Sinclair Stammers/Science Photo Library; **p273**(B): Mary Terriberry/Shutterstock; **p278**: Frans Lanting/Mint Images/Science Photo Library; **p279**(T): Aseph/Shutterstock; **p279**(B): Alan Bryant/Shutterstock; **p280**: Art Wolfe/Mint Images/Science Photo Library; **p281**(T): Agap/Shutterstock; **p281**(B): PHOTO FUN/Shutterstock; **p282**(T): Helen Hotson/Shutterstock; **p282**(B): Piotr Krzeslak/Shutterstock; **p283**(T): Calin Tatu/Shutterstock; **p283**(B): Nickolay Khoroshkov/Shutterstock; **p285**: David Wrobel/Visuals Unlimited/Science Photo Library; **p286**(T): Nsemprevivo/Shutterstock; **p286**(B): Henk Bentlage/Shutterstock; **p287**: Nelik/Shutterstock; **p288**(T): Helen Hotson/Shutterstock; **p288**(C): Louise Murray/Science Photo Library; **p288**(B): A40757/Shutterstock; **p289**: Frans Lanting/Mint Images/Science Photo Library; **p290**: Ronald van der Beek/Shutterstock; **p291**: 360b/Shutterstock; **p297**(B): Nigel Cattlin/Visuals Unlimited/Getty Images; **p298**(TL): Niko Grigorieff/Visuals Unlimited, Inc./Science Photo Library; **p298**(TR): Norm Thomas/Science Photo Library; **p298**(CT): Vadym Zaitsev/Shutterstock; **p302**(T): Amir Ridhwan/Shutterstock; **p302**(B): Eye of Science/Science Photo Library; **p303**(T): Denis Kuvaev/Shutterstock; **p303**(CT): Bork/Shutterstock; **p303**(C): Slawomir Fajer/Shutterstock; **p303**(CB): Reddogs/Shutterstock; **p303**(B): Rio Patuca/Shutterstock; **p304**: Amir Ridhwan/Shutterstock; **p305**: Schubbel/Shutterstock; **p307**: James King-Holmes/Science Photo Library; **p309**: Steve Gschmeissner/Science Photo Library; **p310**(T): Don W.Fawcett/Science Photo Library; **p310**(B): Steve Gschmeissner/Science Photo Library; **p318**: Monkey Business Images/Shutterstock; **p321**(T): Brian Maudsley/Shutterstock; **p321**(C): Jassada Watt/Shutterstock; **p321**(B): AN NGUYEN/Shutterstock; **p167**(L): Astrid & Hanns-Frieder Michler/Science Photo Library; **p167**(B): Dr. Keith Wheeler/Science Photo Library; **p167**(R): Herve Conge, ISM/Science Photo Library; **p186**(T): Neil Carveth, Allison Daley and Nathalie Feiner from Department of Zoology, University of Oxford; p186(B): Neil Carveth, Allison Daley and Nathalie Feiner from Department of Zoology, University of Oxford; **p220-221**: Pablo Hidalgo - Fotos593/Shutterstock; **p275**: Steve McWilliam/Shutterstock; **p297**(T): 1000 Words/Shutterstock; **p298**(B): Chantal de Bruijne/Shutterstock; **p298**(CB): Nigel Cattlin/Visuals Unlimited/Corbis; **p299**: Eduard Kyslynskyy/Shutterstock; **p301**: Raj Creationzs/Shutterstock; **p162**: Neil Carveth, Allison Daley and Nathalie Feiner from Department of Zoology, University of Oxford; **p165**(T): Neil Carveth, Allison Daley and Nathalie Feiner from Department of Zoology, University of Oxford; **p165**(C): Neil Carveth, Allison Daley and Nathalie Feiner from Department of Zoology, University of Oxford; **p165**(B): Neil Carveth, Allison Daley and Nathalie Feiner from Department of Zoology, University of Oxford; **p172**: Neil Carveth, Allison Daley and Nathalie Feiner from Department of Zoology, University of Oxford; **p148-149**: Anthony Short; **p223**(L): MidoSemsem/Shutterstock; **p223**(R): Nicku/Shutterstock; **p219**: PhotographybyMK/Shutterstock; **p327**: Nagui Antoun; **p38**: Claire Ting/Science Photo Library; **p261**: Becky Stares/Shutterstock; **p260**: Eye of Science/Science Photo Library;

Artwork by Q2A Media

Although we have made every effort to trace and contact all copyright holders before publication this has not been possible in all cases. If notified, the publisher will rectify any errors or omissions at the earliest opportunity.

Links to third party websites are provided by Oxford in good faith and for information only. Oxford disclaims any responsibility for the materials contained in any third party website referenced in this work.